Advances in
Nuclear Dynamics

Advances in Nuclear Dynamics

Edited by

Wolfgang Bauer

Michigan State University
East Lansing, Michigan

and

Alice Mignerey

University of Maryland
College Park, Maryland

Plenum Press • New York and London

Library of Congress Cataloging-in-Publication Data

Advances in nuclear dynamics / edited by Wolfgang Bauer and Alice
Mignerey.
      p.   cm.
   "Proceedings of the 11th Winter Workshop on Nuclear Dynamics, held
February 1-3, 1995, in Key West, Florida"--T.p..verso.
   Includes bibliographical references and index.

   1. Heavy ion collisions--Congresses. 2. Nuclear reactions-
-Congresses. 3. Nuclear fragmentation--Congresses. 4. Nuclear
excitation--Congresses. I. Bauer, W. (Wolfgang), 1959-   .
II. Mignerey, A. (Alice) III. Winter Workshop on Nuclear Dynamics
(11th : 1995 : Key West, Fla.)
QC794.8.H4A38 1996
539.7'234--dc20   96-18856
                                                         CIP

Proceedings of the 11th Winter Workshop on Nuclear Dynamics,
held February 1–3, 1995, in Key West, Florida

ISBN-13: 978-1-4613-8019-1   e-ISBN-13: 978-1-4613-0367-1
DOI: 10.1007/978-1-4613-0367-1

© 1996 Plenum Press, New York
Softcover reprint of the hardcover 1st edition 1996

A Division of Plenum Publishing Corporation
233 Spring Street, New York, N. Y. 10013

PREFACE

The study of nuclear dynamics is now in one of its most interesting phases. The theory is in the process of establishing an increasingly reliable transport description of heavy ion reactions from the initial violent phase dominated by first collisions to the more thermalized later stages of the reaction. This is true for the low-to-medium energy reactions, where the dynamics is formulated in terms of nucleonic, or in general hadronic, degrees of freedom. And it is also becoming a reality in ultrarelativistic heavy-ion reactions, where partonic elementary degrees of freedom have to be used.

Experiments are now able to utilize the existing accelerators and multiparticle detection systems to conduct unprecedented studies of heavy-ion collisions on an event-by-event basis. In addition, the field anticipates the completion of the construction of the Relativistic Heavy Ion Collider and the proposed upgrade of the National Superconducting Cyclotron Laboratory, promising qualitatively new data for the near future.

All of these efforts are basically directed to the exploration of the change the nuclear medium provides for the properties and interactions of individual nucleons and, ultimately, the exploration of the nuclear matter phase diagram. The investigation of this phase diagram, including all of the interesting phase transitions predicted from theoretical grounds, is the focus of most of the theoretical and experimental investigations of nuclear dynamics conducted today.

A central feature of this series of workshops has been the attempt to define unifying concepts in the description of the underlying dynamics from the lowest to the highest energies. The present one was no exception, and these proceedings of the 11th Winter Workshop on Nuclear Dynamics try to give the reader a relevant cross section through the present state of the art in the field of nuclear dynamics, both in theory and in experiment.

Wolfgang Bauer
Michigan State University

Alice Mignerey
University of Maryland

PREVIOUS WORKSHOPS

The following table contains a list of the dates and locations of the previous Winter Workshops on Nuclear Dynamics as well as the members of the organizing committees. The chairpersons of the conferences are in italics.

1. Granlibakken, California, 17–21 March 1980
   W. D. Myers, J. Randrup, G. D. Westfall

2. Granlibakken, California, 22–26 April 1982
   W. D. Myers, J. J. Griffin, J. R. Huizenga, J. R. Nix, F. Plasil, V. E. Viola

3. Copper Mountain, Colorado, 5–9 March 1984
   W. D. Myers, C. K. Gelbke, J. J. Griffin, J. R. Huizenga, J. R. Nix, F. Plasil, *V. E. Viola*

4. Copper Mountain, Colorado, 24–28 February 1986
   J. J. Griffin, J. R. Huizenga, J. R. Nix, *F. Plasil*, J. Randrup, *V. E. Viola*

5. Sun Valley, Idaho, 22–26 February 1988
   J. R. Huizenga, *J. I. Kapusta*, J. R. Nix, J. Randrup, V. E. Viola, *G. D. Westfall*

6. Jackson Hole, Wyoming, 17–24 February 1990
   B. B. Back, J. R. Huizenga, *J. I. Kapusta*, J. R. Nix, J. Randrup, V. E. Viola, *G. D. Westfall*

7. Key West, Florida, 26 January–2 February 1991
   B. B. Back, W. Bauer, J. R. Huizenga, *J. I. Kapusta*, J. R. Nix, J. Randrup

8. Jackson Hole, Wyoming, 18–25 January 1992
   B. B. Back, *W. Bauer*, J. R. Huizenga, J. I. Kapusta, J. R. Nix, J. Randrup

9. Key West, Florida, 30 January–6 February 1993
   B. B. Back, W. Bauer, J. Harris, J. I. Kapusta, A. Mignerey, J. R. Nix, G. D. Westfall

10. Snowbird, Utah, 16–22 January 1994
    B. B. Back, W. Bauer, *J. Harris*, A. Mignerey, J. R. Nix, G. D. Westfall

CONTENTS

THE TOPOLOGY OF INTERMEDIATE MASS FRAGMENT EMISSION

W. J. Llope*

T. W. Bonner Nuclear Laboratory
Rice University
Houston, TX

ABSTRACT

The study of the patterns in momentum and/or coordinate space, i.e., the topology, of fragment emission in central heavy-ion collisions allows one to distinguish between the sequential binary (SB) and multifragmentation (MF) disassembly modes. We present results on fragment azimuthal correlations using two different methods, and each implies beam energy dependent transitions between these two disassembly modes. An experimental data set consisting of 40 reactions in the entrance channels $^{12}C + ^{12}C$, $^{20}Ne + ^{27}Al$, $^{40}Ar + ^{45}Sc$, $^{84}Kr + ^{93}Nb$, and $^{129}Xe + ^{139}La$ was used for these analyses, as well as our previous studies of central event topology using the sphericity variable. The SB to MF transitional behavior extracted from all of these analyses will be compared to independently measured disassembly time scales. It will be shown that, for increasing energies, the SB to MF transitions that we observe occur at energies consistent with those leading to strong decreases in the disassembly time scales. This lends credence to the identification that the present SB to MF transitions are the physical manifestation of a nuclear liquid–gas phase transition.

1. INTRODUCTION

The identification of the liquid and gaseous phases of some excited thermodynamic system of A particles via the characteristics of its disassembly is straight-forward. Excited liquids evaporate, or undergo fission, while gases rapidly expand to fill their container. The application of such classical concepts for the understanding of the excited systems formed

---

*For the MSU 4π Group: National Superconducting Cyclotron Laboratory, Michigan State University; T. W. Bonner Nuclear Laboratory, Rice University; Department of Chemistry, State University of New York — Stony Brook; Department of Physics, U. of Michigan — Dearborn; Cyclotron Institute, Texas A&M University; Department of Physics and Astronomy, U. of Iowa.

in heavy-ion collisions is made difficult by the quantum-mechanical nature and charge, Z, of these systems, as well as the fact that A is some twenty-three orders of magnitude less than the thermodynamic limit. Detailed model calculations[1] nonetheless imply that the equivalent of a proper liquid–gas phase transition occurs in finite nuclear systems at excitation energies between ~ 4 to ~ 10 MeV/nucleon, depending on A and Z. Experimental searches for such a phase transition are complicated by pre-equilibrium emission, non-zero angular momenta, and the inability to strictly control Z, A, and the excitation energy via the impact parameter. Systematic experimental data sets collected using a nearly hermetic apparatus allow one to overcome such effects, and to search for the nuclear liquid–gas phase transition. One compelling signal of the nuclear gaseous phase following the central collision of two nuclear liquids at intermediate beam energies is the observation of extremely short disassembly time scales,[2] i.e., $\lesssim 100$ fm/c. Indeed, a number of experiments[3,4] have observed *decreasing* time scales with increases in the beam energy over the range from some tens to ~ 100 MeV/nucleon.

Beam energy dependent transitions from Sequential Binary (SB) disassembly to multifragmentation (MF) also appear in this beam energy range for central collisions.[5] Sequential binary disassembly proceeds via a cascade of two-body decay steps, which in many ways resembles the dissociation of an excited liquid. For excitation energies above ~ 1–2 MeV/nucleon and the typical angular momenta involved in central heavy-ion collisions, the initial stage of this disassembly mode involves an (a)symmetric fission, which results in two excited pre-fragments moving back-to-back in the rest frame of the decaying system. These pre-fragments may, depending on the excitation energy, subsequently undergo further fissions or particle evaporation. Overall, the largest fragments in the final state reflect the back-to-back emission of the two primordial pre-fragments. On the other hand, multifragmentation proceeds via a more violent "explosion" of the excited nucleus. This mode produces more than two excited pre-fragments, which may evaporate particles as they rapidly move apart under strong inter-particle Coulomb repulsions. The largest fragments in the final states of this mode of disassembly are more isotropically distributed in the rest frame of the decaying system.

The study of the topology, i.e., the patterns in momentum and/or coordinate space, of intermediate mass fragment (IMFs, for which $3 \leq Z \lesssim 20$) emission can therefore distinguish between the SB and MF scenarios.[6] In previous work, we have obtained evidence for SB to MF transitions using many different experimental observables in a systematic set of experimental data.[7,8,9,10] The majority of this evidence was obtained from the study of the fragment topology in a momentum space coordinate system that spatially coincides with the center of momentum (CM) frame as quantified by the sphericity variable.[8,10] In this Contribution, additional evidence for SB to MF transitions is presented, which was obtained from the study of the fragment coordinate space topology via the quantification of the IMF azimuthal distributions. Two different methods will be described. The SB to MF transitional beam energies that are extracted will be compared with the transitional energies that were implied by the previous analyses.

Of principal interest is the evaluation of the extent to which the observed SB to MF transitions are indeed a reflection of a nuclear liquid–gas phase transition. To this end, we will finally describe the comparison of the SB to MF transitional energies that we observe with the independently measured time scales for the disassembly of the excited systems formed in the same or similar central reactions.

The experimental data were collected with the Michigan State University 4π Array[11] at the National Superconducting Cyclotron Laboratory (NSCL) using beams extracted from the K1200 cyclotron. The reactions that were studied include ^{12}C + ^{12}C at 55, 75, 95, 105, 115, 125, 135, 145, and 155 MeV/nucleon, ^{20}Ne + ^{27}Al at 55, 75, 95, 105, 115, 125, 135, and

140 MeV/nucleon, ^{40}Ar + ^{45}Sc at 15, 25, 35, 45, 65, 75, 85, 95, 105, and 115 MeV/nucleon, ^{84}Kr + ^{93}Nb at 35, 45, 55, 65, and 75 MeV/nucleon, and ^{129}Xe + ^{139}La at 25, 30, 35, 40, 45, 50, 55, and 60 MeV/nucleon. Detailed information on the data collection may be found elsewhere.[3,7,8,9,10,12,13]

2. TRANSITIONS AS VIEWED BY AZIMUTHAL CORRELATIONS

The study of the opening angles between the largest fragments in the final states of central heavy-ion reactions should provide a clear separation between the SB and MF disassembly modes.[6] By assumption, SB disassembly results in larger average opening angles between the two heaviest fragments as compared to those following MF. The calculation of fragment opening angles in the rest frame of the decaying system, however, involves the assumption of the laboratory velocity of this system. For the most central collisions, this source velocity could be assumed to be the velocity of the CM frame, or experimentally measured as the weighted average of the components of the final state particle velocities along the beam direction. The study of the relative *azimuthal* angles of the emitted fragments in such reactions has the advantage that the specification of a source velocity is not necessary. Furthermore, the effects of the experimental acceptance on a study of relative azimuthal angles are of little concern if an azimuthally symmetric apparatus, e.g., the MSU 4π Array, was used to collect the data. If the selection of central collisions is sufficiently strict, the contributions from directed transverse flow[12,13] and collective rotational motion[13] are strongly suppressed. This allows the one to assume that central events with azimuthally back-to-back heavy fragments follow SB disassembly, while those with azimuthally isotropic fragments follow MF.

The central events were selected by a two-dimensional cut on the total transverse kinetic energy and the total mid-rapidity charge, giving $b_{max} \sim 0.2[R_P + R_T]$ geometrically. This cut is significantly more central than that used in Refs. 12 and 13, where the contributions of transverse directed flow and rotational motion were observed in the same data set studied herein. This centrality cut does not autocorrelate[7] with the analyses described in the following two Sections.[9]

2.1. Azimuthal Correlations of the Three Heaviest Fragments

The relative azimuthal angles of the three largest fragments in each central event, $\Delta\phi_i$, are measured in such a way that $\sum \Delta\phi_i = 1$. The events are plotted *a lá* Dalitz[14] in a triangle so that the perpendicular distances to each of the the three axes are $\Delta\phi_i$. Such plots of relative azimuthal angles are not affected by the "Z_{sum}" dependences intrinsic to the relative charge Dalitz plots.[9] The relative azimuthal angle Dalitz plots for the central ^{40}Ar + ^{45}Sc and ^{129}Xe + ^{139}La events are shown in Figure 1.

Events near the corners of the triangles are those in which all three of the largest fragments have nearly the same laboratory azimuthal angle. Events populating the sides of these triangles include two nearly back-to-back fragments, with the third fragment relatively close in azimuthal angle to one of the other two. A population of events near the center have three fragments that are azimuthally isotropic.

At the lowest beam energies shown in Figure 1, a predominance of events near the sides of these triangles is apparent. As the beam energy is increased, both entrance channels shown an accompanying increase in the number of events near the centers of these triangles, and a depletion of the events near the sides. These plots are thus consistent with a transition

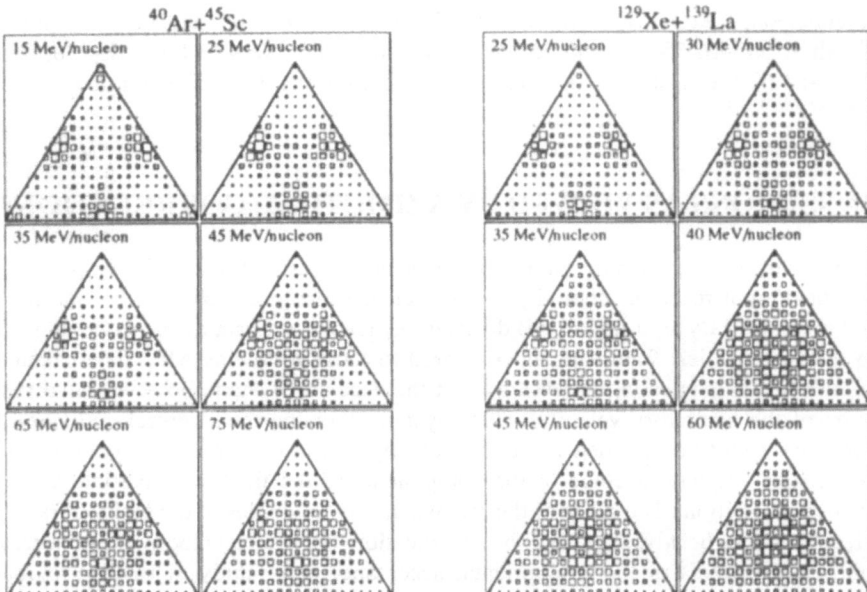

Figure 1. The relative azimuthal angle Dalitz plots for the three largest fragments in the central ^{40}Ar + ^{45}Sc and ^{129}Xe + ^{139}La reactions at six different beam energies each.

from sequential binary disassembly to multifragmentation, as the data indicates a change from predominantly back-to-back heavy fragment emission to fragment azimuthal isotropy for increasing beam energies.

However, it is difficult to locate a transitional beam energy from the plots shown in Figure 1. To quantify these plots, we measure the values of the distances D_{cent}^{ϕ} and D_{edge}^{ϕ} for each central event. These variables were first used as a means of quantifying the relative charge Dalitz plots that were discussed in Ref. 9. The quantity $D_{cent}^{\phi}(D_{edge}^{\phi})$ is the distance from the position of the event in the triangle to the center(nearest edge). According to the assumptions above, a sample of predominantly SB events has $\langle D_{cent}^{\phi} \rangle > \langle D_{edge}^{\phi} \rangle$, while a sample of MF events has $\langle D_{cent}^{\phi} \rangle < \langle D_{edge}^{\phi} \rangle$. A minimum in the beam energy dependence of the quantity $\langle D_{cent}^{\phi} \rangle$, with a maximum in the same dependence of $\langle D_{edge}^{\phi} \rangle$, indicates a beam energy at which the central events are maximally isotropic azimuthally.

The beam energy dependence of $\langle D_{cent}^{\phi} \rangle$ and $\langle D_{edge}^{\phi} \rangle$ is shown in Figure 2. The results for the central events for all of the available entrance channels and beam energies are shown with the different point styles as labelled in the Figure.

Relatively large(small) values of $\langle D_{cent}^{\phi} \rangle(\langle D_{edge}^{\phi} \rangle)$ are apparent at beam energies near and below ~ 35 MeV/nucleon. Minima(Maxima) in $\langle D_{cent}^{\phi} \rangle(\langle D_{edge}^{\phi} \rangle)$ are observed at beam energies near ~ 50 MeV/nucleon in the central ^{40}Ar + ^{45}Sc reactions, and near ~ 45 MeV/nucleon in the central ^{129}Xe + ^{139}La reactions. Consistent trends are observed for the other reactions. Above the beam energies leading to the minimal(maximal) values of $\langle D_{cent}^{\phi} \rangle(\langle D_{edge}^{\phi} \rangle)$, relatively more gradual increases(decreases) are observed for increasing beam energies. A SB to MF transitional beam energy near ~ 45 MeV/nucleon, depending slightly on the entrance channel mass, is therefore implied. An alternative means of quantifying the azimuthal distributions of the emitted fragments, which has two advantages as compared to the Dalitz method described above, is described in the next Section.

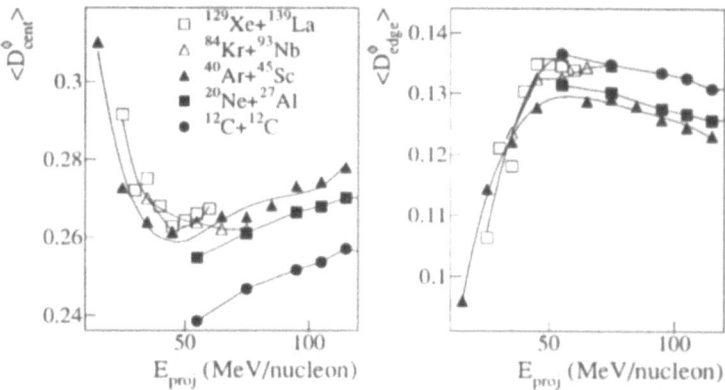

Figure 2. The average values of D_{cent}^{ϕ} (left frame) and D_{edge}^{ϕ} (right frame) obtained from the relative azimuthal angle Dalitz plots versus the beam energy in the central events for the reactions listed in the left frame.

2.2. Azimuthal Correlations of N_{IMF} Fragments

The EMU01 Collaboration proposed[15] a pair of complementary observables for the study of the azimuthal patterns of particles emitted in 200 GeV/nucleon collisions at the CERN/SPS. We will apply these two observables to quantify azimuthal distributions in a way that differs somewhat from that used in Ref. 15. This method has two important advantages over the Dalitz azimuthal correlations analysis described in the previous Section. First, the azimuthal patterns of the particle emission can be quantified for an arbitrary number of particles, and second, the mean values of each variable can be derived in the limit of purely stochastic emission of independent particles in a simple way for each multiplicity under consideration.

These variables are applied for the study of IMF emission in the present data by first measuring the azimuthal angles of the N_{IMF} fragments in a central event in the same way as done in the previous Section, i.e., so that $\sum_{i=1}^{N_{IMF}} \Delta\phi_i = 1$. The variables S_1 and S_2 are then defined as:

$$S_1 = -\sum_{i=1}^{N_{IMF}} \log(\Delta\phi_i), \quad \text{and} \quad S_2 = \sum_{i=1}^{N_{IMF}} \Delta\phi_i^2. \tag{2.1}$$

The variable $S_1(S_2)$ is large if there are small(large) gaps in the particle azimuthal distributions, implying "jet-like" azimuthal distributions according to Ref. 15. Thus, under the same assumptions used in the previous Sections, large values of these variables imply SB disassembly, while small values imply "ring-like" distributions,[15] indicating multifragmentation. The mean values of these variables in the purely stochastic limit are given by:

$$S_1^{stoch} = N_{IMF} \sum_{k=1}^{N_{IMF}-1} \frac{1}{k} \quad \text{and} \quad S_2^{stoch} = \frac{2}{N_{IMF}+1}. \tag{2.2}$$

The experimental distributions of S_1/S_1^{stoch} and S_2/S_2^{stoch} have been extracted from all of the present reactions and for all IMF multiplicities in the central events.

The truncated icosahedron, or "soccer ball," geometry of the MSU 4π Array has several groups of detectors with centers at different polar angles but the same azimuthal angle. Particle hits in two such detectors that are included in the calculation of S_1 and S_2 result in a divergence in the calculation of S_1. However, the variables S_1 and S_2 are complementary,

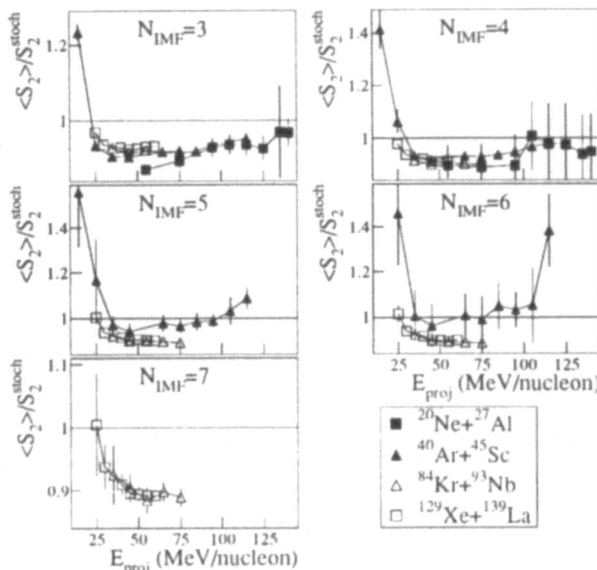

Figure 3. The experimental values of $\langle S_2 \rangle / S_2^{stoch}$ versus the beam energy for the reactions listed on the lower right. The quantity S_2 is calculated only for the IMFs in the central events. Each frame corresponds to a specific IMF multiplicity as labelled.

like the variables D_{cent}^{ϕ} and D_{edge}^{ϕ}. We will therefore simply concentrate on the variable S_2 in the discussion below.

The experimental values of $\langle S_2 \rangle / S_2^{stoch}$ versus the beam energy are shown in Figure 3. Mean values of this ratio that are below(above) unity imply azimuthal distributions that are more isotropic(planar) than the stochastic limit. In similarity to the results extracted in the previous Section, the ^{40}Ar + ^{45}Sc results imply SB disassembly at beam energies below ~ 35 MeV/nucleon, as $\langle S_2 \rangle / S_2^{stoch}$ exceeds unity. The results for the other reactions imply a universal trend with the beam energy. Minimal values of $\langle S_2 \rangle / S_2^{stoch}$, i.e., maximal IMF azimuthal isotropy, occurs at beam energies near ~ 35–45 MeV/nucleon for $N_{IMF} = 3$, and ~ 45–55 MeV/nucleon for $N_{IMF} \geq 5$. These minima are below unity, i.e., the IMFs are more azimuthally isotropic than the stochastic average, which implies that the MF mechanism governed the disassembly at the beam energies near and above these minima.

3. COMPARISON TO TIME SCALE MEASUREMENTS

The azimuthal-correlation analyses of the previous Sections, and our previous studies of the topology of IMF emission,[8,10] imply SB to MF transitions for increasing beam energies in the central ^{40}Ar + ^{45}Sc, ^{84}Kr + ^{93}Nb, and ^{129}Xe + ^{139}La reactions. The beam energies at which these transitions are observed are summarized in Figure 4. Also included in this Figure is the SB to MF transitional energies observed via a Dalitz charge correlations analysis of the present data,[9] and an event shape analysis of central ^{40}Ar + ^{51}V reactions.[16] The transitional beam energies implied by all of the analyses for each reaction are quite similar, despite the fact that many different observables were studied. The transitional beam energies decrease from ~ 50 MeV/nucleon to ~ 40 MeV/nucleon, with increases in the entrance channel mass

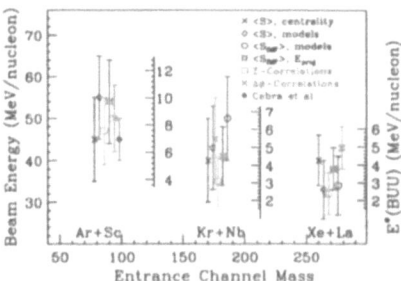

Figure 4. The SB to MF transitional beam energies extracted from all of the analyses of the present data for the central ^{40}Ar + ^{45}Sc, ^{84}Kr + ^{93}Nb, and and ^{129}Xe + ^{139}La reactions. The various analyses are described in Refs. 8, 9, 10, and in the previous Sections. Predictions for the excitation energies reached in these central reactions obtained from BUU calculations[7] are shown on the right-side axes.

from $\sim$ 80 to $\sim$ 280, for reasons that were discussed in Ref. 10.

To evaluate the extent to which the observed SB to MF transitions are the physical manifestation of a nuclear liquid–gas phase transition, we compare the observed SB to MF transitional energies with independent measurements for the disassembly time scales.[3,4] These time scale measurements involved generally different symmetric and asymmetric entrance channels than those used for the results of Figure 4. Thus, to allow comparisons with the present results, the time-scales are plotted not as a function of the beam energy, but instead versus the excitation energy for each reaction obtained in the limit of perfectly inelastic collisions. This limiting excitation energy is defined as $\frac{E_{proj}}{A_{proj}} \frac{V_{CM}}{V_{proj}} (1 - \frac{V_{CM}}{V_{proj}})$, where E_{proj} is the beam energy, A_{proj} is the projectile mass number, and $V_{CM}(V_{proj})$ is the laboratory velocity of the CM frame(projectile).

The disassembly time scales from these independent measurements are shown in Figure 5. The present SB to MF transitional energies, i.e., Fig. 4, are enclosed in the shaded region, while the time scales in fm/c are shown as the large numbers, labelled by the various point styles. The time scale measurements shown in this Figure are representative of a larger number of such measurements that have been published. The measurements of Lisa *et al.* and Bauge *et al.* involved symmetric reactions measured with the MSU 4π Array, while the remainder of the results were obtained in asymmetric collisions measured with different detector systems. The results of Bauge *et al.* were in fact obtained from exactly the same ^{84}Kr + ^{93}Nb reactions discussed in the previous Sections and in Refs. 7, 8, 9, 10. For the other time scale measurements, the differences in the entrance channel nuclei and the experimental devices, as well as the different methods used to extract the time scales in each of these analyses, demand that the comparisons in Figure 5 be taken as qualitative only.

Central reactions populating the region below the shaded band in Figure 5 form excited systems that predominantly disassemble via the SB mechanism. The time scales for the reactions in this region are in the range from $\sim$ 150 fm/c to $\sim$ 500 fm/c. Above the shaded band, i.e., in the MF domain, time scales from $\sim$ 30 fm/c to $\sim$ 125 fm/c have been observed. Despite the caveats noted above, these trends imply that the SB to MF transitions observed in our analyses are indeed accompanied by significant decreases in the independently measured disassembly time scales. This lends credence to the identification that the present SB to MF transitions are indeed indicative of a nuclear phase transition from an excited quantum liquid phase to a phase of liquid–gas coexistence with disassembly time scales consistent with that of a quantum gas.

The present systematic results clearly indicate an entrance channel mass dependence

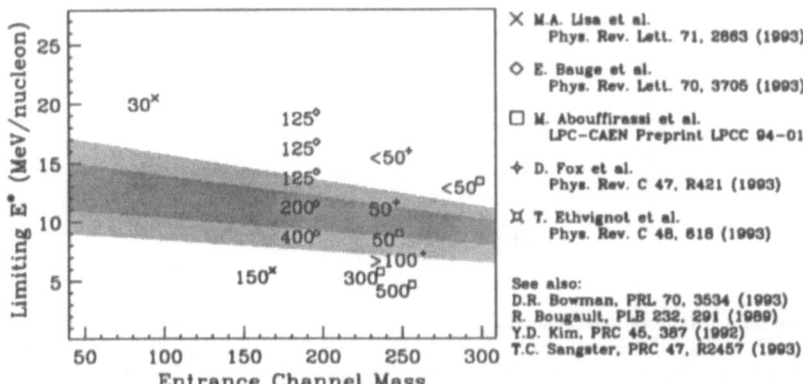

Figure 5. The comparison of the SB to MF transitional energies, shown as the shaded region (from Fig. 4), with the independently measured time scales (in fm/c) for the same or similar reactions.[3,4] To allow this comparison, the present results and the time scales are plotted on the vertical axis using the limiting excitation energy (see text) for each projectile/target combination and beam energy.

of the transitional energies in central symmetric heavy-ion collisions.[10] The excitation energies for these transitions can be obtained from, e.g., BUU calculations,[7] and are $\sim$ 8–10 MeV/nucleon for systems with $A \sim 70$ and $\frac{Z}{A} \sim 0.5$, and $\sim$ 3–5 MeV/nucleon for systems with $A \sim 250$ and $\frac{Z}{A} \sim 0.4$. We note that the systems formed in the heavier symmetric collisions studied herein are generally more proton-rich than ground state nuclei of the same total mass. Further work should therefore concentrate on the dependence of the transition energy on the total charge, Z, of systems of fixed total mass, A. This will allow the disentangling of the presumably different dependences of the transition energies on the total mass and on the Coulomb energy in the excited system. Additional work should extend the systematics of the time scale measurements in such a way that the energy dependence of the disassembly time scales can be more quantitatively compared to the present results on SB to MF transitions.

ACKNOWLEDGMENTS

We thank Thomas Glasmacher and M. Betty Tsang for helpful comments concerning time scale measurements. This work was supported by the U. S. Department of Energy under the Grant No. DE-FG03-93ER40772, and the U. S. National Science Foundation under Grants No. PHY 89-13815 and No. PHY 92-14992.

REFERENCES

1. J. P. Bondorf *et al.*, *Nucl. Phys.* **A448**, 753 (1986), and references therein; D. H. E. Gross, *Prog. Part. Nucl. Phys.* **30**, 155 (1993), and references therein; A. S. Botvina *et al.*, *Nucl. Phys.* **A475**, 663 (1987); W. Bauer, *Phys. Rev. C* **38**, 1297 (1988).
2. S. Pratt and M. B. Tsang, *Phys. Rev. C* **36**, 2390 (1987).
3. E. Bauge *et al.*, *Phys. Rev. Lett.* **70**, 3705 (1993).
4. M. A. Lisa *et al.*, *Phys. Rev. Lett.* **71**, 2863 (1993); T. Ethvignot *et al.*, *Phys. Rev. C* **48**, 618 (1993); D. Fox *et al.*, *Phys. Rev. C* **47**, R421 (1993); M. Abouffirasi *et al.*, LPC-CAEN Preprint LPCC 94-01 (1994); D. Durand *et al.*, LPC-CAEN Preprint LPCC 94-02 (1994).

5. L. G. Moretto and G. J. Wozniak, *Ann. Rev. Nucl. Part. Sci.* **43**, 379 (1993), and references therein; H. Fuchs and K. Möhring, *Rep. Prog. Phys.* **57**, 231 (1994), and references therein.
6. J. A. López and J. Randrup, *Nucl. Phys.* **A491**, 477 (1989).
7. W. J. Llope *et al.*, *Phys. Rev. C*, in press.
8. W. J. Llope *et al.*, MSU Preprint MSUCL-959, *Phys. Rev. C*, submitted (1995).
9. N. T. B. Stone, W. J. Llope, and G. D. Westfall, MSU Preprint MSUCL-916, *Physical Review C*, submitted (1994).
10. W. J. Llope *et al.*, "Advances in Nuclear Dynamics, Proc. of the 10th Winter Workshop on Nuclear Dynamics", eds. J. Harris, A. Mignerey, and W. Bauer, Snowbird, Utah (1994).
11. G. D. Westfall *et al.*, *Nucl. Inst. and Methods* **A238**, 347 (1985).
12. G. D. Westfall *et al.*, *Phys. Rev. Lett.* **71**, 1986 (1993).
13. R. A. Lacey *et al.*, *Phys. Rev. Lett.* **70**, 1224 (1993); J. Lauret *et al.*, *Phys. Lett.* B **339**, 22 (1994).
14. R. H. Dalitz, *Phil. Mag.* **44**, 1068 (1953); E. Fabri, *Nuovo Cim.* **11**, 479 (1954).
15. M. I. Adamovich *et al.*, (The EMU01 Collaboration), *J. Phys. G*, **19**, 2035 (1993); E. Stenlund *et al.*, (The EMU01 Collaboration), *Nucl. Phys.*, **A498**, 541c (1989).
16. D. A. Cebra *et al.*, *Phys. Rev. Lett.* **64**, 2246 (1990).

2

REDUCIBLE EMISSION PROBABILITIES AND THERMAL SCALING IN MULTIFRAGMENTATION

L. Phair,[1] K. Tso,[1] R. Ghetti,[1] N. Colonna,[1*] K. Hanold,[1†] M. A. McMahan,[1]
G. J. Wozniak,[1] L. G. Moretto,[1] D. R. Bowman,[2‡] N. Carlin,[2§] R. T. de
Souza,[2¶] C. K. Gelbke,[2] W. G. Gong,[2||] Y. D. Kim,[2**] M. A. Lisa,[26]
W. G. Lynch,[2] G. F. Peaslee,[2††] M. B. Tsang,[2] C. Williams,[2] and F. Zhu[2‡‡]

[1]Nuclear Science Division
Lawrence Berkeley Laboratory, Berkeley, CA
[2]National Superconducting Cyclotron Laboratory
and Department of Physics and Astronomy
Michigan State University, East Lansing, MI

ABSTRACT

Intermediate-mass-fragment multiplicity distributions for ^{36}Ar + ^{197}Au reactions at intermediate energies are shown to be binomial and thus reducible at all measured transverse energies. From these distributions a single binary event probability can be extracted that has a thermal dependence. A strong thermal signature is also found in the charge distributions. The n-fold charge distributions are reducible to the 1-fold charge distributions through a simple scaling that is dictated by fold number and charge conservation.

1. INTRODUCTION

At low excitation energies, complex fragments are emitted with low probability by a compound nucleus mechanism 1, 2. At increasingly larger energies, the probability of complex

*Present address: Instituto Nazionale Fisica Nucleare, V. Amendola, Bari, Italy.
†Present address: Department of Chemistry, University of California, San Diego, La Jolla, CA 92093-0314.
‡Present address: Chalk River Laboratories, Chalk River, Ontario K0J 1J0, Canada.
§Present address: Instituto de Fisica, Universidade de Sao Paulo, C. P. 20516, CEP 01498, Sao Paulo, Brazil.
¶Present address: Department of Chemistry, Indiana University, Bloomington, IN 47405.
||Present address: Lawrence Berkeley Laboratory, Berkeley, CA 94720.
**Present address: National Laboratory for High Energy Physics, 1-1 Oho, Tsukuba, Ibaraki 305, Japan.
††Present address: Physics Department, Hope College, Holland, MI 49423.
‡‡Present address: Brookhaven National Laboratory, Upton, NY 19973.

Advances in Nuclear Dynamics
Edited by Wolfgang Bauer and Alice Mignerey, Plenum Press, New York, 1996

fragment emission increases dramatically, until several fragments are observed within a single event 3, 4, 5. The nature of this multifragmentation process is at the center of much current attention. For example, the time-scale of fragment emission and the associated issue of sequentiality versus simultaneity are the objects of intense theoretical 3, 4, 5, 6, 7, 8 and experimental 9, 10, 11, 12, 13, 14, 15, 16, 17 study.

Recent experimental work 18, 19 has shown that the excitation functions for the production of two, three, four, etc. fragments give a characteristically linear Arrhenius plot, suggesting a statistical energy dependence.

A fundamental issue, connected in part to those mentioned above, is that of reducibility. Can multifragmentation be reduced to a combination of (nearly) independent emissions of fragments? If the answer is yes, what are the implications for the resulting charge distributions? In what follows we show evidence that the n-fragment emission probabilities are indeed reducible to an elementary binary emission probability. We show that the energy dependence of the extracted elementary probabilities give a linear Arrhenius plot implying that these probabilities are probably thermal. We also show that the experimental charge distributions have similar reducible and thermal scaling properties.

2. REDUCIBLE EMISSION PROBABILITIES

The partial decay width Γ associated with a given binary channel can be approximated by

$$\Gamma = \hbar\omega_0 e^{-B/T}, \tag{2.1}$$

where ω_0 is a frequency characteristic of the channel under consideration, B is the barrier associated with the channel and T is the temperature. The elementary probability p for a binary decay to occur at any given "try" (defined by the channel period $\tau_0 = 1/\omega_0$) is

$$p = \frac{\Gamma}{\hbar\omega_0} = e^{-B/T}. \tag{2.2}$$

We note that the elementary binary probability p can be directly related to the experimental branching ratios for binary, ternary, quaternary, etc., decay (shown in the bottom panel of Fig. 1 for the reaction ^{36}Ar + ^{197}Au at $E/A = 80$ MeV).

For simplicity, let us assume that the system has the opportunity to try m times to emit an "inert" fragment with constant probability p. The probability P_n^m of emitting exactly n fragments is given by the binomial distribution

$$P_n^m = \frac{m!}{n!(m-n)!}p^n(1-p)^{m-n}. \tag{2.3}$$

The average multiplicity and variance of fragment distribution are then

$$\langle n \rangle = mp \qquad \sigma^2 = \langle n \rangle(1-p). \tag{2.4}$$

From the experimental values of $\langle n \rangle$ and σ^2 (top panel of Fig. 1) one can extract values for p and m, at any transverse energy E_t (assumed to be proportional to the excitation energy). In Fig. 2 we plot on a log scale $1/p$ vs. $1/\sqrt{E_t}$ for the fragment distributions (Arrhenius plot). If the probability is thermal, as given in Eq.2.2, this plot ought to be linear since $T \propto \sqrt{E^*} \propto \sqrt{E_t}$. The linearity of this plot over nearly two orders of magnitude is stunning and strongly suggests the thermal nature of p.

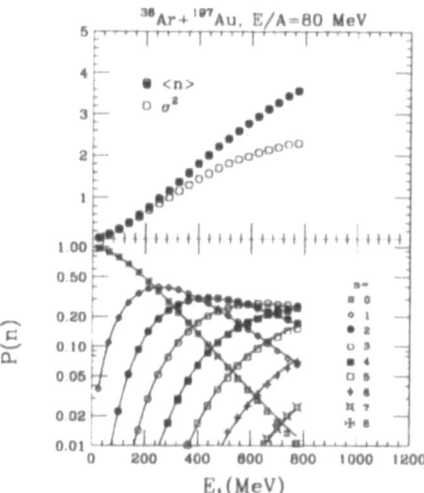

Figure 1. The probability of emitting n fragments as a function of the transverse energy E_t is plotted in the bottom panel ($E_t = \sum E_i \sin^2 \theta_i$ where E_i and θ_i are the kinetic energy and polar angle of particle i in an event). The solid lines represent the binomial distributions described in the text. The excitation functions are reduced to their mean value (solid circles) and variance (open circles) as a function of E_t in the top panel.

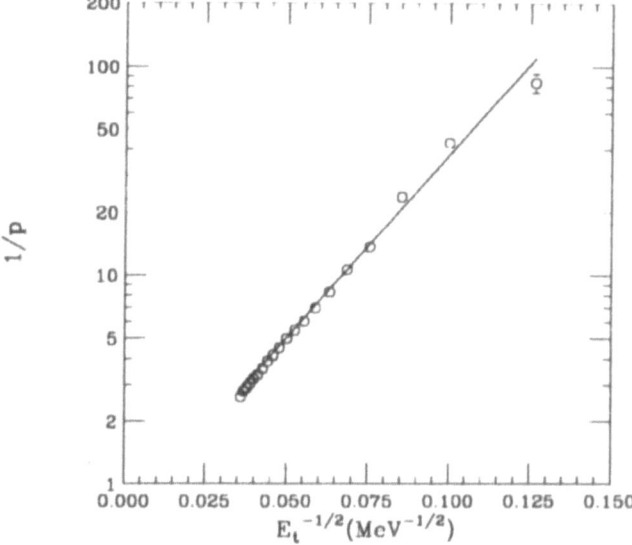

Figure 2. The reciprocal of the binary decay probability $1/p$ (calculated from the mean and variance of the IMF distributions) as a function of $1/\sqrt{E_t}$ for the reaction ^{36}Ar + ^{197}Au at $E/A = 80$ MeV. The solid line is a linear fit to $\ln(1/p)$.

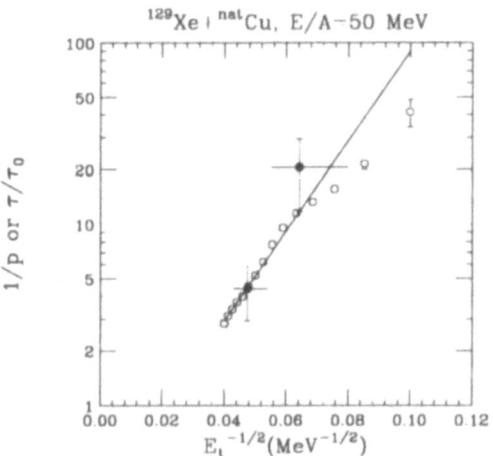

Figure 3. A comparison between the measured fragment emission time scales 12 (solid circles) and $1/p$ (open circles) for the reaction ^{129}Xe + ^{nat}Cu at $E/A = 50$ MeV. The time scale measurements have been normalized to the $1/p$ data at the highest E_t point.

We show a comparison (bottom panel of Fig. 1) between the experimental excitation functions and those calculated for a binomial distribution using the values of p obtained from the linear fits of Fig. 2 and the associated values of m from Eq. (2.4). The extraordinary quantitative agreement between the calculation and the experimental data confirms the binomiality of the multifragmentation process. Details of this analysis can be found in reference 20.

The more directly interpretable physical parameter contained in this analysis is the binary barrier B (proportional to the slope of the data in Fig. 2). One may wonder why a single binary barrier suffices, since mass asymmetries with many different barriers may be present. This is an old problem. Let us consider a barrier distribution as a function of mass asymmetry x of the form $B = B_0 + ax^s$, where B_0 is the lowest barrier in the range considered. Then,

$$p = \frac{\Gamma}{\hbar\omega_0} = \int e^{-B_0/T} e^{-ax^s/T} dx \approx \left(\frac{T}{a}\right)^{1/s} e^{-B_0/T} \tag{2.5}$$

Thus the simple form of Eq. (2.2) is retained with a small and renormalizable pre-exponential modification.

One possible interpretation of the reducibility discussed above is sequential decay with constant probability p. Assuming that the (small) fragments, once produced, do not generate additional fragments or disappear, the binomial distribution follows directly. In this framework, it is possible to translate the probability p into the mean time separation between fragments. The corresponding time associated with Eq. (2.1) is (from the uncertainty principle)

$$\tau = \tau_0 e^{B/T} = \frac{\tau_0}{p}. \tag{2.6}$$

In other words, we can relate the n-fragment emission probabilities to the mean time separations between fragments. The validity of this interpretation can be tested by experiment.

Lifetime estimates were made for the reaction ^{129}Xe + ^{nat}Cu at $E/A = 50$ MeV 12 using intensity interferometry techniques. A comparison of these measured time scales can be made with the $1/p$ data for this system by normalizing one of the time-scale points to the $1/p$ data and then comparing the results at a lower excitation energy. This comparison is made in Fig.

3. The solid circles represent the time scales measured in reference 12, normalized to $1/p$ data at the highest transverse energy. The error bars in the y direction represent the uncertainty in the lifetime measurements. The error bars in the x direction represent the FWHM of the distribution in E_t for the two total charged particle multiplicity cuts used in reference 12 to construct the correlation functions. We observe reasonable agreement within the uncertainty of these measurements.

The final proof for or against sequentiality must rest on independent time measurements. The establishment of an agreement between the times inferred from the emission probabilities and from the fragment-fragment correlations would go a long way toward resolving this issue.

3. REDUCIBLE CHARGE DISTRIBUTIONS

The aspects of reducibility and thermal scaling in the integrated fragment emission probabilities lead naturally to the question: Is the charge distribution itself reducible and scalable? In particular, what is the charge distribution form that satisfies the condition of reducibility and of thermal dependence?

Let us first consider the aspect of reducibility as it applies to the charge distributions. In its broadest form, reducibility demands that the probability $p(Z)$, from which an event of n fragments is generated by m trials, is the same at every step of extraction. The consequence of this extreme reducibility is straightforward: the charge distribution for the one-fold events is the same as that for the n-fold events and equal to the singles distributions, i.e.:

$$P_{(1)}(Z) = P_{(n)}(Z) = P_{singles}(Z) = p(Z). \tag{3.1}$$

We now consider the consequences of the thermal dependence of p on the charge distributions. If the one-fold = n-fold = singles distribution is thermal, then

$$P(Z) \propto e^{-\frac{B(Z)}{T}} \tag{3.2}$$

or $T \ln P(Z) \propto -B(Z)$. This suggests that, under the usual assumption $E_t \propto E^*$, the function

$$\sqrt{E_t} \ln P(Z) = D(Z) \tag{3.3}$$

should be independent of E_t.

In the ^{36}Ar + ^{197}Au reaction considered here, the charge distributions are empirically found to be nearly exponential functions of Z

$$P_n(Z) \propto e^{-\alpha_n Z} \tag{3.4}$$

as shown in Fig. 4. In light of the above considerations, we would expect for α_n the following simple dependence

$$\alpha_n \propto \frac{1}{T} \propto \frac{1}{\sqrt{E_t}} \tag{3.5}$$

for all folds n. Thus a plot of α_n vs. $1/\sqrt{E_t}$ should give nearly straight lines. This is shown in Fig. 5 for ^{36}Ar + ^{197}Au at $E/A = 110$ MeV.

The expectation of thermal scaling appears to be met quite satisfactorily. For each value of n the exponent α_n shows the linear dependence on $1/\sqrt{E_t}$ anticipated in Eq. (3.5). On the other hand, the extreme reducibility condition demanded by Eq. (3.1), namely that

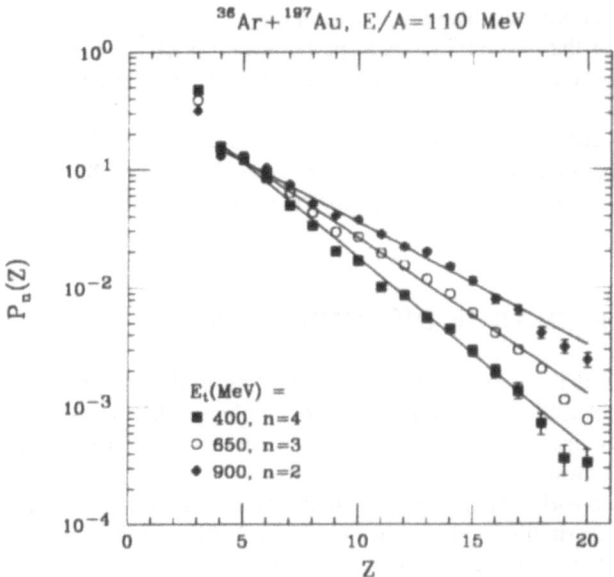

Figure 4. The n-fold charge distributions $P_n(Z)$ for intermediate mass fragments (IMF: $3 \leq Z \leq 20$) are plotted for the indicated cuts on transverse energy E_t and IMF multiplicity n. The width of the cuts ΔE_t is 37.5 MeV. The solid lines are exponential fits over the range $Z = 4$–20.

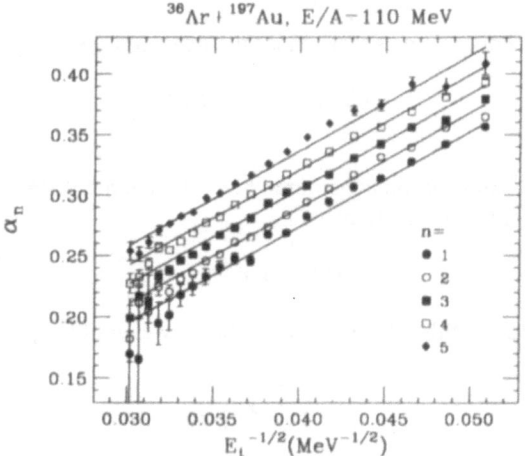

Figure 5. The exponential fit parameter α_n (from fits to the charge distributions, see Eq. (3.4)) is plotted as a function of $1/\sqrt{E_t}$. The solid lines are a fit to the values of α_n using Eq. (3.6).

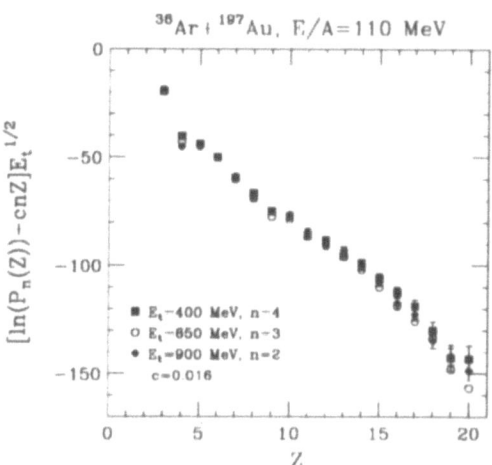

Figure 6. The "reduced" charge distributions (see Eq. (3.9)), are plotted for the same cuts on E_t and n as Fig. 4. The different data sets are normalized at $Z = 6$. The value of $c = 0.016$ is the spacing between the curves shown in Fig. 5.

$\alpha_1 = \alpha_2 = \ldots = \alpha_n = \alpha$, is not met. Rather than collapsing on a single straight line, the values of α_n for the different fragment multiplicities are offset one with respect to another by what appears to be a constant quantity.

In fact, one can fit all of the data remarkably well, assuming for α_n the form:

$$\alpha_n = \frac{K'}{\sqrt{E_t}} + nc \qquad (3.6)$$

which implies:

$$\alpha_n = \frac{K}{T} + nc \qquad (3.7)$$

or more generally, for the Z distribution:

$$P_n(Z) \propto e^{-\frac{B(Z)}{T} - ncZ}. \qquad (3.8)$$

Thus, we expect a more general reducibility expression for the charge distribution of any form to be:

$$[\ln P_n(Z) + ncZ]\sqrt{E_t} = F(Z) \qquad (3.9)$$

for all values of n and E_t. This equation indicates that it should be possible to reduce the charge distributions associated with any intermediate mass fragment multiplicity to the charge distribution of the singles. As a demonstration of this reducibility, we have compared $P_n(Z)$ and $F(Z)$ in Figures 4 and 6. Fig. 4 compares three charge distributions for different cuts on E_t and n; their slopes are clearly different. The reduced quantity $F(Z)$, on the other hand, collapses to a single line in Fig. 6.

What is the origin of the regular offset that separates the curves in Fig. 2? In our specific case the spacing may be related to an asymptotic combinatorial structure of the multifragmentation process in the high temperature limit. As an example, we consider the

Euler problem of an integer Z_0 to be written as the sum of smaller integers Z. It can be shown 21 that the resulting integer distribution has the form

$$n_Z = \frac{n^2}{Z_0} e^{-\frac{nZ}{Z_0}}. \tag{3.10}$$

This expression has the correct asymptotic structure for $T \to \infty$ required by Eq. (3.8). The significance of this form is transparent. First, the overall scale for the fragment size is set by the total charge Z_0. Second, for a specific multiplicity n, the scale is reduced by a factor n to the value Z_0/n.

Thus the offset introduced in Eq. (3.8) with increasing the multiplicity n may just be due to this scale reduction. If this is so, the quantity c in Eq. (3.8) takes the meaning $c = 1/Z_0$. The empirical value from Fig. 5 is $c \approx 0.016$ which corresponds to a value of $Z_0 \approx 60$ which is quite reasonable for the source size.

4. CONCLUSIONS

The multifragment emission probability has been found to be binomial and thus reducible to an elementary binary probability. This binary elementary probability p is observed to have a "thermal" energy dependence. The reducible and thermal features observed in the n-fragment emission probabilities extend consistently to the charge distributions. We find strong evidence for a thermal scaling of the Z-distributions and reducibility of the n-fold charge distributions to the 1-fold distributions through Eq. (3.8).

The implications of the experimental evidence presented above are far reaching. On the one hand, the observed thermal features in the n-fragment emission probabilities and the charge distributions strengthen the hypothesis of the dominant role of phase space in multifragmentation. On the other hand, the reducibility (of the n-fold-event charge distributions to that of the singles distribution and the n-fragment emission probabilities to single fragment emission) highlights the near independence of individual fragment emission, limited only by the constraint of charge conservation.

ACKNOWLEDGMENTS

This work was supported by the Director, Office of Energy Research, Office of High Energy and Nuclear Physics, Nuclear Physics Division of the US Department of Energy, under contract DE-AC03-76SF00098 and by the National Science Foundation under Grant Nos. PHY-8913815, PHY-90117077, and PHY-9214992.

REFERENCES

1. L. G. Sobotka, et. al., Phys. Rev. Lett. **51**, 2187 (1983).
2. L. G. Moretto and G. J. Wozniak, Prog. Part. & Nucl. Phys. **21**, 401 (1988).
3. D. Guerreau, *Formation and Decay of Hot Nuclei: The Experimental Situation* ed. (Plenum Publishing Corp., 1989).
4. D. H. E. Gross, Rep. Prog. Phys. **53**, 605 (1990).
5. L. G. Moretto and G. J. Wozniak, Ann. Rev. Part. Nucl. Sci. **43**, 379 (1993).
6. J. Aichelin, Phys. Rep. **202**, 233 (1991).
7. B. Borderie, Ann. Phys. Fr. **17**, 349 (1992).

8. O. Schapiro and D. H. E. Gross, Nucl. Phys. A **573**, 143 (1994).
9. T. Ethvignot, et al., Phys. Rev. C **48**, 618 (1993).
10. D. Fox, et al., Phys. Rev. C **47**, R421 (1993).
11. E. Bauge, et al., Phys. Rev. Lett. **70**, 3705 (1993).
12. D. R. Bowman, et al., Phys. Rev. Lett. **70**, 3534 (1993).
13. T. C. Sangster, et al., Phys. Rev. C **47**, R2457 (1993).
14. M. Louvel, et. al., Phys. Lett. B **320**, 221 (1994).
15. M. Aboufirassi, et al., LPC Caen preprint LPCC 94-02 (1994).
16. A. Lleres, et al., ISN Grenoble ISN 94-33 (1994).
17. T. Glasmacher, et al., Phys. Rev. C **50**, 952 (1994).
18. L. G. Moretto, D. N. Delis, and G. J. Wozniak, Phys. Rev. Lett. **71**, 3935 (1993).
19. J. Pouliot, et al., Phys. Rev. C **48**, 2514 (1993).
20. L. G. Moretto, et al., Phys. Rev. Lett. **74**, 1530 (1995).
21. L. Phair, et al., LBL preprint LBL-36730 UC-413 (1995).

3

HEAVY RESIDUE PRODUCTION IN DISSIPATIVE ^{197}Au + ^{86}Kr COLLISIONS AT E/A = 35 MeV

W. Skulski,[1]* B. Djerroud,[1] D. K. Agnihotri,[1] S. P. Baldwin,[1]
W. U. Schröder,[1] J. Tõke,[1] X. Zhao,[1] L. G. Sobotka,[2] R. J. Charity,[2]
J. Dempsey,[2] D. G. Sarantites,[2] B. Lott,[3] W. Loveland,[4] and K. Aleklett[5]

[1]Dept. of Chemistry and NSRL
University of Rochester
Rochester, NY
[2]Dept. of Chemistry
Washington University
St. Louis, MO
[3]Laboratoire National GANIL, BP 5027
Caen 14021, France
[4]Oregon State University
Corvallis, OR
[5]Uppsala University
S-611 82 Nyköping, Sweden

ABSTRACT

Massive residues of projectile — and target-like fragments from the ^{197}Au + ^{86}Kr reaction at E/A = 35 MeV have been measured in coincidence with neutrons, as well as light — and intermediate-mass charged products, using a highly efficient detector setup including two 4π devices — the Rochester SuperBall neutron detector and the St. Louis Microball. The observed joint distribution of neutron and charged-particle multiplicities, the emission patterns of charged-particles and projectile-like fragments, and the yield of slow, massive residues, are all indicative of binary dissipative collisions, followed by statistical decay of the primary massive fragments. To a large extent, the slow massive residues are found to be remnants of target-like fragments, produced even in the most dissipative collisions identified in the present experiment.

---

*On leave of absence from Heavy Ion Laboratory, Warsaw University, Poland.

Advances in Nuclear Dynamics
Edited by Wolfgang Bauer and Alice Mignerey, Plenum Press, New York, 1996

1. INTRODUCTION

Heavy-ion reactions at intermediate incident energies per nucleon of several tens of MeV have been of considerable experimental and theoretical interest in the last decade. This energy domain offers the opportunity to observe the evolution of the reaction mechanism from that governed by the mean-field dynamics, in the low-energy domain,[1] to the one defined by two-body nucleon–nucleon collisions, in the high-energy domain.[2] It also offers the opportunity of producing and observing the decay modes of nuclear systems under extreme conditions of excitation energy and angular momentum.

In a previous experiment,[3,4] exclusive measurements of massive projectile-like fragments (PLF), intermediate-mass fragments (IMF), light charged particles (LCP), and neutrons have been performed for the reaction ^{209}Bi + ^{136}Xe at E/A = 28 MeV. Characteristic correlations between the emission angle of the PLF and the degree of energy damping, as well as the bimodal character of the emission patterns of LCP's found in this previous study, pointed to a reaction scenario substantially similar to that of damped collisions known from heavy-ion reaction studies at lower bombarding energies.[1] In this scenario, kinetic energy of relative motion is converted in a binary process into excitation energies of the primary projectile– and target-like fragments. Subsequently, these primary fragments undergo deexcitation via particle and γ-decay, with a clear signature of the dissipative reaction mechanism reflected in the distributions of the secondary products.

In order to extend the study of the collision dynamics to the region of higher relative energies per nucleon and larger entrance-channel mass asymmetries, an experiment was performed for the system ^{197}Au + ^{86}Kr at E/A = 35 MeV, in which massive reaction products were measured, along with neutrons and lighter charged products. The choice of this system was also motivated by the recent radiochemical study[5] of the system ^{197}Au + ^{84}Kr, in which a large cross section was reported for the production of very slow ($E/A \approx$ 0.1–0.5 MeV) reaction products with masses comparable to the target mass. It has been argued[5,6,7] that the production of these heavy residues may constitute an important reaction mode, that could inhibit or replace fission at sufficiently high excitation energies. Therefore, in this work, an exclusive study has been performed of these residues measured in coincidence with PLF's, LCP's, and neutrons. Preliminary results of this study are presented below.

2. EXPERIMENTAL PROCEDURE

The experiment was performed at the K1200 Cyclotron of the National Superconducting Cyclotron Laboratory at Michigan State University. A 35-MeV/nucleon ^{86}Kr beam bombarded a 0.3-mg/cm^2 thick ^{197}Au target. The experimental setup consisted of the Rochester SuperBall, the St. Louis Microball, and a number of silicon detectors.

Massive projectile-like fragments were detected with two position-sensitive Si-detector telescopes covering an angular range between 2° and 8°. Both telescopes provided atomic-number resolution in the range of Z = 2–38. Slow, heavy reaction products were detected with three 4 cm × 7 cm, 300-μm thick multi-strip silicon detectors, positioned between θ = $-7°$ and $-46°$, i.e., on the side of the beam opposite to that of the PLF telescopes. The time-of–flight measurement was performed, using a timing signal derived from either the PLF telescopes, or from a thin 17-mg/cm^2 plastic scintillator detector placed between the target and the telescopes.

Neutrons were detected using the SuperBall, a 16-m^3 gadolinium-loaded liquid-scintillator detector enclosing the scattering chamber in a 4π geometry. The SuperBall

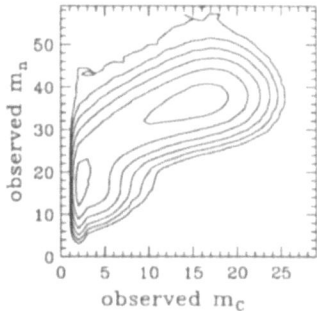

Figure 1. Logarithmic contour plots of the observed joint distribution of neutron multiplicity m_n and charged particle multiplicity m_c.

measured the multiplicity of neutrons and provided information on the summed neutron kinetic energy.

Light- and intermediate-mass charged products were measured with the Microball detector, which covered angles from $14°$ to $171°$. The Microball, configured with 86 CsI(Tl) crystal detectors, was able to resolve elements of hydrogen, helium, lithium, and beryllium, as well as three isotopes of hydrogen. Fragments with $Z > 4$ were counted, but not resolved according to their atomic number.

3. RESULTS

In the following sections, correlations between various experimental observables are discussed, which reflect underlying reaction and deexcitation scenarios. The experimental observables include neutron and charged-particle multiplicities, m_n and m_c, charged-particle atomic numbers and velocity vectors, PLF energy, atomic number and deflection angle, and massive-residue energy. Unless stated otherwise, the raw neutron multiplicities presented in this work are corrected neither for the detection efficiency, nor for background (typically ≈ 7). A rather large background correction is due to the long time gate (128 μs), used online to count neutron capture pulses. In the further offline analysis, a shorter time gate will be applied, and the background multiplicity will be properly unfolded from the measured value.

3.1. Joint Distribution of Neutron and Charged-Particle Multiplicities

The observed joint distribution of neutron and charged-particle multiplicities, is plotted in Fig. 1 in the form of a logarithmic contour plot. As seen in this figure, charged particle emission from the Au + Kr system contributes substantially only, when at least also ≈ 13 neutrons are emitted. This is in agreement with the results of statistical-model calculations (code EVAP[8]), showing that in the present system charged particle emission sets in for m_n larger than $m_n \approx 10$. Only from this point on charged particles compete significantly with neutron emission. While at higher degrees of energy damping both, m_n and m_c, are good and independent measures of dissipated energy, at lower dissipation only m_n is correlated with the excitation energy deposited in the system.

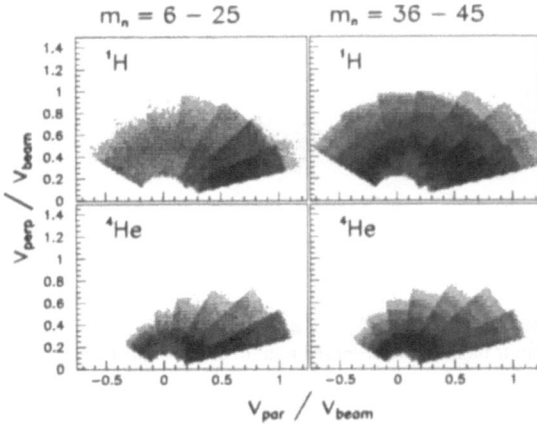

Figure 2. Emission patterns of protons (upper row) and α–particles (lower row) in a Galilei-invariant representation of v_{par} versus v_{perp} velocity components. Events of different degree of dissipation are selected based on the observed associated neutron multiplicity. The left column corresponds to low and medium energy dissipation, while the right column corresponds to highly dissipative collisions, sampling the high-multiplicity peak of the joint multiplicity distribution. The velocity of the center of mass of the system is equal to $0.3 \cdot v_{beam}$.

3.2. Velocity Distributions of Light Charged Particles

Four representative Galilei-invariant LCP velocity distributions are shown in Fig. 2 in the form of logarithmic contour diagrams, plotted versus velocity components v_{perp} and v_{par}, perpendicular and parallel to the beam, respectively. For a scenario, where LCP's are emitted from the excited PLF and TLF formed in a binary dissipative collision, one expects the LCP yield to be distributed along two semicircular "Coulomb ridges," centered at the average PLF and TLF velocities. The ridge due to emission from the TLF source is clearly observed in the data, whereas the one corresponding to emission from the PLF is not clearly visible. Due to the angular resolution of the Microball, and to the thickness of the Microball detectors at forward angles, the circular patterns of LCP's emitted from the PLF (centered close to the beam velocity v_{beam}), could not have been observed clearly in this experiment. However, a hint to their possible presence is offered by an increased LCP yield at v_{par} close to v_{beam}.

In addition, for low and intermediate degrees of energy dissipation (left column of Fig. 2), an enhancement of the proton yield is observed for $v_{par} \approx 0.6 \cdot v_{beam}$. For α–particles, enhanced yield is seen at $v_{par} \approx 0.3 \cdot v_{beam}$, consistent with the laboratory velocity of the center of mass of the projectile-target system. The enhancement may partially be due to yield pileup at the intersection of the Coulomb rings associated with the PLF or TLF. On the other hand, such a feature is consistent with emission of α-particles from a participant or "neck" region, a process previously reported[9] for the Bi + Xe reaction at $E/A = 28$ MeV. At the present stage of analysis, these two possible contributions have not yet been separated.

The right column of Fig. 2 shows the emission patterns associated with the peak $(m_n, m_c) \approx (35, 16)$ of the joint multiplicity distribution (see Fig. 1), and, hence, corresponding to the most dissipative collisions observed in the present experiment. These patterns appear to be inconsistent with a scenario involving only one emitting source moving with the velocity of the center of mass of the system. Most of the observed LCP's seem to be emitted from the target-like source, as evidenced by a Coulomb ridge discernible in the velocity distributions of both protons and α–particles. As stated previously, the emission pattern from a fast-moving PLF source would have been largely missed by the present experimental setup.

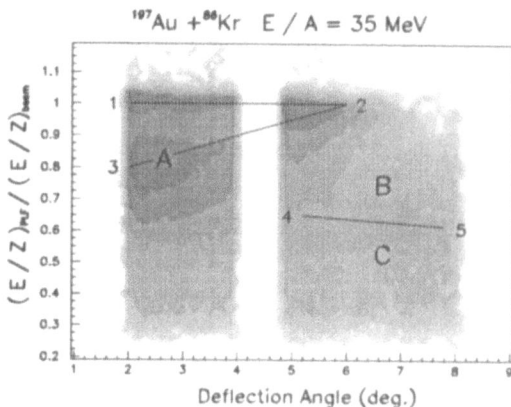

Figure 3. Yield of massive fragments with $Z \geq 10$ from the reaction ^{197}Au + ^{86}Kr at E/A = 35 MeV, plotted as a function of the fragment E/Z value, and the deflection angle in the lab.

3.3. The Deflection Function

The yield of massive charged fragments with $Z \geq 10$, observed with two silicon-detector telescopes at forward angles, is plotted in Fig. 3 versus laboratory deflection angle and fragment E/Z value (E/Z is the measure of a fragment velocity). Most of the yield in this figure is distributed along a characteristic Z-shaped ridge. From the small-angle limit of the telescope aperture to the grazing angle of $\approx 6°$ (line segment "1 $\rightarrow$ 2" in Fig. 3), a ridge of mainly elastic yield runs horizontally. Then the ridge bends back towards smaller angles and turns to smaller E/Z values (line segment "2 $\rightarrow$ 3"), as energy is dissipated in the reaction. The third, returning part of the ridge associated with high degrees of dissipation, is only discernible at angles larger than the grazing angle (line segment "4 $\rightarrow$ 5"). This observed yield pattern is obviously very similar to the Wilczyński diagram known from heavy-ion reaction studies at lower energies to represent the deflection function.[1] Such diagrams have also been reported[4,10,11] for several systems at intermediate bombarding energies. The correlation of fragment energy with angle, shown in Fig. 3, is characteristic of a dissipative orbiting process, where the two reaction partners rotate about their common center of gravity for a fraction of a revolution, while kinetic energy of relative motion is dissipated. Subsequently, the reaction partners disengage as primary TLF and PLF and deexcite, while proceeding along Coulomb trajectories.

The dissipative character of the reaction is clearly demonstrated by the increase of the associated particle multiplicities along the ridge "1 $\rightarrow$ ... $\rightarrow$ 5" in Fig. 3. One-dimensional multiplicity distributions associated with three regions "A," "B," and "C," defined in Fig. 3, are shown in Fig. 4. The horizontal segment of the yield ridge in Fig. 3 (line segment "1 $\rightarrow$ 2") represents elastic and quasielastic scattering, with both associated m_n and m_c consistent with background. Passing point "2" on the ridge, both multiplicities rise quickly above the background to $m_n = 9$ and $m_c = 2.5$, respectively. They continue increasing along the middle portion of the ridge, reaching (background-corrected) values of $m_n = 18$ and $m_c = 5$, respectively, at the small-angle limit of the telescope aperture (region "A" in Fig. 3). The third segment of the yield ridge discernible backward of the grazing angle could correspond to either positive or negative deflection angles (line segment "4 $\rightarrow$ 5"). Along this portion of the ridge, with increasing deflection angle ("4 $\rightarrow$ 5") one observes a further increase

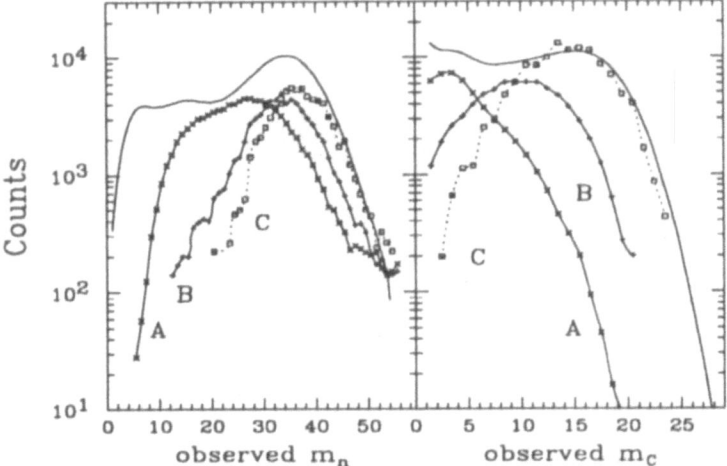

Figure 4. Neutron and charged particle multiplicities associated with different parts of the deflection plot of Fig. 3. The distributions were arbitrarily normalized to facilitate comparison with the minimum-bias distributions (thin lines). Small values of m_n are strongly suppressed as a result of an on-line (trigger) requirement that at least two charged particles are detected in Microball.

of the associated (background-corrected) multiplicities to $m_n = 25$ and $m_c = 9$ (in region "B"). Hence, the observed trends in associated particle multiplicities suggest, that the last segment "4 → 5" of the ridge in Fig. 3 is a continuation of the ridge "1 → 2 → 3," which is seen at high values of E/Z. Both, the E/Z value and the associated multiplicities, are correlated with the dissipated energy and, consequently, with the impact parameter. Therefore, the yield pattern depicted in Fig. 3, with the above trends in associated particle multiplicities, can be viewed as an experimental representation of the deflection function for collisions governed by dissipative nuclear reaction dynamics.

The highest associated particle multiplicities, $m_n = 30$ and $m_c = 14$ (after background correction), are observed for the region "C" in Fig. 3, which is consistent with the high-dissipation peak of the joint multiplicity distribution of Fig. 1. It appears, however, that most of the yield in region "C" may not exclusively represent projectile-like fragments.

3.4. Production of Slow Heavy Residues

Heavy residues were measured using three multistrip silicon detectors placed on the side of the beam opposite with respect to the PLF telescopes, in order to be able to detect heavy residues in coincidence with PLF's. The detectors covered the angular range between 7° and 46° in the laboratory. Heavy fragments were selected based on the observed correlations between their energy and time of flight. The energy spectrum of these fragments, measured in all strip detectors, is displayed as the upper curve in the left panel of Fig. 5 (diamonds). This spectrum features a narrow low-energy peak "A," attributed to TLF evaporation residues, and a broader peak "B" located between 50 and 150 MeV. The latter peak is consistent with sequential TLF fission fragments. The bimodal structure of the fragment energy spectrum is enhanced in the spectrum measured in coincidence with PLF's with $Z_{PLF} > 25$, as selected by the fast plastic scintillator detector. This latter spectrum is displayed as the lower curve (squares) in the left panel of Fig. 5.

E/Z spectra of PLF's, measured in coincidence with either of the two groups of heavy

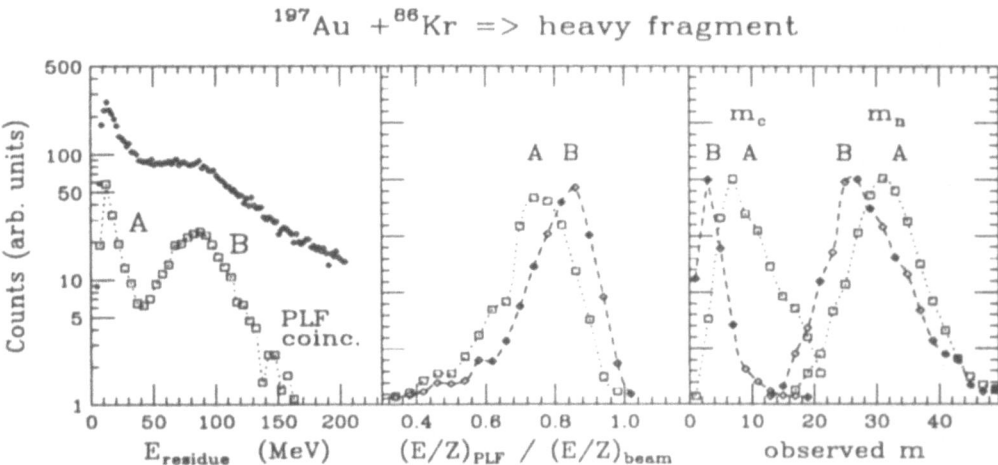

Figure 5. *Left panel:* Energy spectra of heavy fragments observed in the angular range between 7° and 46° in "singles" mode (diamonds) and in coincidence with PLF's with $Z_{PLF} > 25$ (squares). *Middle panel:* E/Z-spectra of PLF's in coincidence with TLF residues (A) and TLF fission fragments (B), as defined by two peaks in the left panel. *Right panel:* Multiplicity distributions of neutrons and charged-particles associated with production of heavy TLF residues (A) and TLF fission fragments (B).

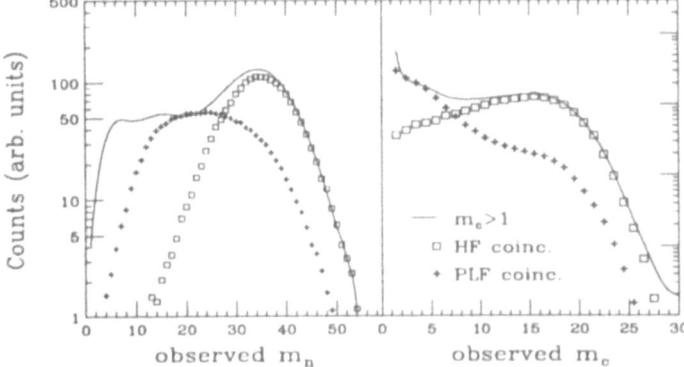

Figure 6. A comparison of the m_n and m_c distributions coincident with heavy fragments (squares) and with the $Z_{PLF} > 25$ (crosses). The ungated distributions (thin lines) are plotted for reference. The ungated distributions were obtained under the minimum-bias condition $m_c > 1$, imposed by the Microball on-line trigger condition. The coincidence spectra were multiplied by arbitrary factors to facilitate comparison with the ungated distributions.

fragments, are shown in the middle panel of Fig. 5. As seen in this panel, PLF's coincident with low-energy TLF residues have, on the average, lower E/Z-values than those coincident with TLF fission fragments. An approximate kinematic reconstruction, based on the observed average E/Z-values, leads to estimated average dissipated energies of 810 MeV and 660 MeV, for events associated with TLF residues and fission fragments, respectively. These estimates are consistent with the results of previous radiochemical studies.[5] One should note, that the values of the dissipated energy, reported in the present section for the PLF–TLF coincidence events, are representative for rather peripheral reactions. This is due to the requirement, that a heavy PLF ($Z_{PLF} > 25$) was detected in coincidence, favoring peripheral collisions.

The right panel of Fig. 5 shows m_n and m_c distributions associated with the production of TLF residues (A) or TLF fission fragments (B), both measured in coincidence with PLF's of $Z_{PLF} > 25$. Similar to the middle panel of Fig. 5, the distributions shown in the right panel demonstrate again that at least some of the slowest massive fragments are remnants of highly dissipative binary collisions.

The presence of either a heavy TLF residue or a TLF fission fragment in the exit channel signals that a primary hot TLF has survived the primary reaction and, hence, that the system did not undergo an instantaneous disintegration into many small fragments. Fig. 6 presents another piece of evidence for the fact, that such hot TLF's can survive even the most dissipative collisions identified in this experiment. This figure shows m_n and m_c distributions (right and left panel, respectively), observed in single or coincidence modes. The distributions measured in coincidence with heavy residues are plotted as squares, while those measured in coincidence with $Z_{PLF} > 25$ are plotted as crosses. In both panels, the high-multiplicity portions of the "singles" distribution (solid lines "$m_c > 1$"), coincide surprisingly well with the distribution measured in coincidence with heavy fragments, clearly pointing to the survival of massive TLF's in the most dissipative collisions. On the other hand, the coincidence requirement with a massive PLF depletes the high-multiplicity portions of the associated m_n and m_c distributions, due to experimental detection efficiences discussed above.

4. CONCLUSIONS

The present analysis of various correlations between experimental observables has shown that ^{197}Au + ^{86}Kr collisions at $E/A = 35$ MeV follow a binary dissipative scenario in the first step, followed by the sequential decay of primary fragments. This process is reminiscent of that known from heavy-ion reaction studies at lower energies, and similar to the one reported for the Bi + Xe reaction at $E/A = 28$ MeV. The analysis of the exclusive yield of massive, slow-moving residues in coincidence with PLF's observed at forward angles and/or as a function of the associated particle multiplicities, has shown that these residues are also produced mostly in binary dissipative collisions, including the most dissipative collisions observed in this experiment.

This work was supported by the U. S. Dept. of Energy under Grant No. DE-FG02-88ER40414 and DE-FG02-87ER40316.

REFERENCES

1. W. U. Schröder and J. R. Huizenga, in *Treatise in Heavy-Ion Science*, ed. D. A. Bromley, Plenum Press, New York and London, 1984, Vol. 2, p. 113, and references therein.
2. C. K. Gelbke and D. H. Boal, Prog. Part. Nucl. Phys. **33**, (1987), and references therein.
3. B. Lott, et. al., Phys. Rev. Lett. **68**, 3141 (1992).

4. S. P. Baldwin, et. al., in Proc. 9th Winter Workshop on Nuclear Dynamics, Key West, World Scientific, Singapore (1993), p. 36.
    S. P. Baldwin, et. al., Phys. Rev. Lett. **74**, 1299 (1995).
5. K. Aleklett, et. al., Phys. Lett. B **236**, 404 (1990).
    W. Loveland, et. al., Phys. Rev. C **41**, 973 (1990).
6. E. Schwinn et. al., Nucl. Phys. A **568**, 169 (1994).
    E. C. Polacco, et. al., Nucl. Phys. A **583**, 441 (1995).
7. D. Utley, et. al., Phys. Rev. C **49**, R1737 (1994).
8. N. G. Nicolis, D. G. Sarantites, and J. R. Beene, computer code EVAP (unpublished); derived from the computer code PACE by A. Gavron, Phys. Rev. C **21**, 230 (1980).
9.' J. Tõke, et. al., Nucl. Phys. A **583**, 519 (1995).
10. J. F. LeColley, et. al., Phys. Lett. B **325**, 317 (1994).
11. R. J. Charity, et. al., Z. Phys. A **341**, 53 (1991).

4

EVOLUTION OF FRAGMENT DISTRIBUTIONS AND REACTION MECHANISMS FOR THE ^{36}Ar + ^{58}Ni SYSTEM FROM 32 TO 95 A.MeV

L. Nalpas,[1] J-L. Charvet,[1] R. Dayras,[1] E. De Filippo,[1] G. Auger,[3]
Ch. O. Bacri,[2] A. Benkirane,[3] J. Benlliure,[3] B. Berthier,[1] B. Borderie,[2]
R. Bougault,[4] P. Box,[2] R. Brou,[4] Y. Cassagnou,[1] A. Chbihi,[3] J. Colin,[4]
D. Cussol,[4] A. Demeyer,[5] D. Durand,[4] P. Ecomard,[3] P. Eudes,[6]
A. Genoux-Lubain,[4] D. Gourio,[6] D. Guinet,[5] L. Lakehal-Ayat,[2] P. Lautesse,[5]
P. Lautridou,[6] J. L. Laville,[6] L. Lebreton,[5] C. Le Brun,[4] J. F. Lecolley,[4]
A. Le Fèvre,[3] R. Legrain,[1] O. Lopez,[4] M. Louvel,[4] N. Marie,[3] V. Métivier,[4]
T. Nakagawa,[4] A. Ouatizerga,[2] M. Parlog,[2] J. Péter,[4] E. Plagnol,[2]
E. C. Pollacco,[1] J. Pouthas,[3] A. Rahmani,[6] R. Regimbart,[4] M. F. Rivet,[2]
T. Reposeur,[6] E. Rosato,[4] F. Saint-Laurent,[3] M. Squalli,[2] J. C. Steckmeyer,[4]
B. Tamain,[4] L. Tassan-Got,[2] E. Vient,[4] C. Volant,[1] J. P. Wieleczko,[3]
A. Wieloch,[4] and K. Yuasa-Nakagawa[4]

[1]DAPNIA/SPhN, CEA Saclay, 91191 Gif/Yvette, France
[2]IPN Orsay, CNRS-Université, 91406 Orsay, France
[3]GANIL, B. P.5027, 14021 Caen, France
[4]LPC Caen, CNRS-ISMRA, 14050 Caen, France
[5]IPN Lyon, CNRS-Université, 69622 Villeurbanne, France
[6]SUBATECH, CNRS-Université, 44072 Nantes, France

ABSTRACT

Within the framework of flow and multifragmentation study, the ^{36}Ar + ^{58}Ni experiment has been performed at seven incident energies from 32 to 95 A.MeV with the INDRA detector at GANIL. After a brief description of the experimental set-up, the main trends as well as the evolution of fragment distributions will be presented. Some results about reaction mechanisms for particular classes of events will conclude this report.

1. INTRODUCTION

The study of the limiting temperature of hot nuclei has been a major theme of research at intermediate energies for the past decade 1. At low excitation energy, nuclei usually

Advances in Nuclear Dynamics
Edited by Wolfgang Bauer and Alice Mignerey, Plenum Press, New York, 1996

disintegrate by emitting neutrons and light charged particles ($Z = 1$ and $Z = 2$). When increasing excitation energy, an emission of intermediate mass fragments (IMF namely $Z \geq 3$) has been experimentally observed 2. However up to now the fragment emission mechanism remains poorly understood. Indeed this emission could be described either in continuity of the statistical evaporation process at high excitation energy, or as a multifragmentation process for which fragments are emitted almost simultaneously. Some theoretical calculations support this new interpretation and predict an initial compression phase of the nuclear matter followed by an expansion phase during which the fragments are created 3, 4. Moreover a liquid–gas phase transition could be associated with this multifragmentation process 5. The new generation of 4π detector such as INDRA should give some reliable answers to these open questions in the next years.

2. EXPERIMENTAL SET-UP

The INDRA detector is dedicated to the measurement of the charge and of the kinematical properties of light particles and fragments. The overall characteristics of the detector and of its electronics as well as the identification and calibration methods are presented in details in Ref. 6. In few words, the detector is composed of 336 detection modules over 17 rings covering 90% of 4π. The first ring ($2°–3°$) is made of 12 plastic scintillator phoswiches (NE102/NE115) able to endure high counting rates. The forward rings ($3°–45°$) are made of three-stage telescopes composed of a low pressure ionization chamber followed by a wafer of three or four 300 μm thick silicon detectors, each followed by a CsI(Tl) scintillator stopping all fast products. For the backward rings ($45°–176°$) the silicon detectors are removed due to the smaller velocity range of the fragments.

Such a set-up provides high granularity and geometrical efficiency as well as a large energy range with low detection thresholds for light particles and fragments in order to study high multiplicity events coming from the de-excitation of hot nuclei. The charge identification is performed by the well known $\Delta E/E$ technique up to $Z \sim 25$ for the slow products and up to $Z \sim 50$ for the fastest ones. A shape discrimination method for CsI signals allows an isotopic separation for light charged particles up to $Z = 4$.

For the Ar + Ni experiment performed in 1993, a large range of bombarding energies from 95 to 32 A.MeV was obtained by slowing down the beam through a thick ^{12}C target placed before a magnetic analysis. The multiple interaction rate was kept below 10^{-4} by means of a low ^{36}Ar beam intensity ($\sim 2–3 \times 10^7$ particles/sec) as well as a thin ^{58}Ni target (193 μg/cm^2). The data acquisition was activated only when at least 3 modules were hit for low incident energies (4 hits for the highest). In these conditions, the acquisition dead time was around 20%. For this experiment, the rings beyond 92° were not equipped with ionization chambers. Therefore only identification of light charged particles is possible at backward angles. Moreover the energy calibration for fragments are not available beyond 45° at the present time.

3. GENERAL TRENDS

The total detected charge (Z_{TOT}) versus the charged particle multiplicity, plotted in Fig. 1, shows an overview of the INDRA efficiency. Indeed the total charge of the system ($Z_0 = 46$) is well reconstructed for a large multiplicity range. However, at low multiplicity, two classes of badly detected events corresponding to peripheral collisions can be found. Firstly, at low

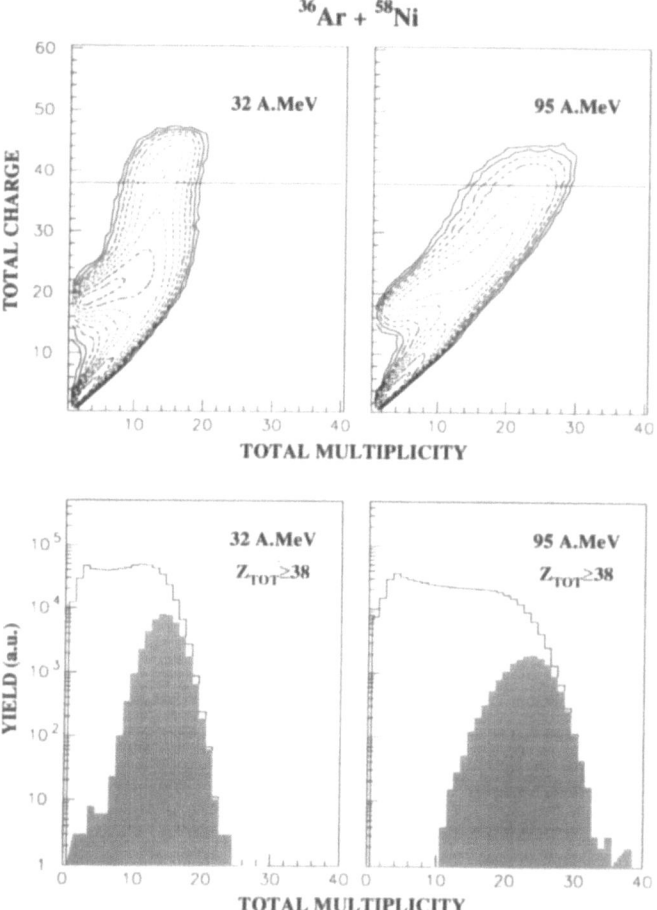

Figure 1. a) Top: Total detected charge as a function of charged particle multiplicity at 32 and 95 A.MeV. The solid line (Z_{TOT} = 38) indicates the selection for complete events (80% of the total charge of the system). b) Bottom: Multiplicity distributions for all events (black curve) and complete events (dark area).

Z_{TOT}, the projectile-like fragment (PLF) emitted at small angles as well as the low energy target-like fragment (TLF) are not detected owing to the geometrical inefficiency at forward angles on the one hand and the detection thresholds on the other hand. For $Z_{TOT} \sim 18$, the PLF is detected but the TLF is still missing due to the thresholds. In the following only complete events for which Z_{TOT} is greater than 80% of the total charge of the system ($Z_{TOT} \geq 38$) are taken into account. This cut-off eliminates about 80% of the events at 32 A.MeV and close to 90% at 95 A.MeV. However a large range of impact parameter is selected as some results about reaction mechanisms will confirm.

The IMF distributions shown in Fig. 2a exhibit the same bell shape with a maximum obtained for 2 and 3-IMF events at any bombarding energy. Fig. 2b displays the evolution of the relative proportion of events with different IMF multiplicities as a function of the incident energy. The partitions are very similar for all incident energies. The lack of a large variation of the IMF number is an intriguing result which means that the IMF multiplicity is independent of the available energy in the center of mass system. Otherwise minor fluctuations show the

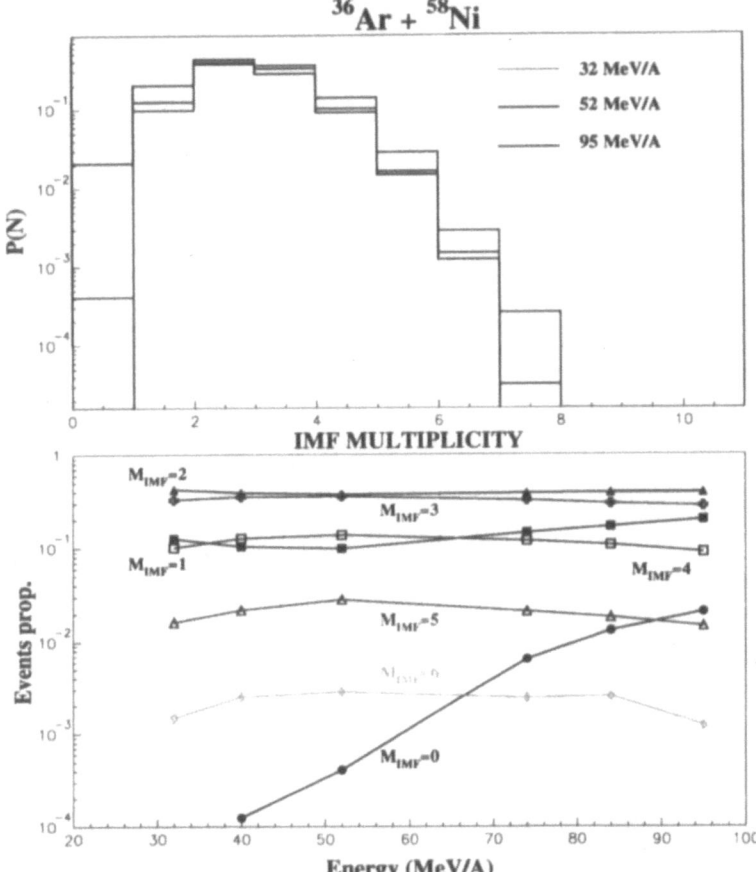

Figure 2. a) Top: IMF normalized distributions at 32, 52 and 95 A.MeV. b) Bottom: Evolution of the IMF partitions with the bombarding energy.

existence of a transition around 52 A.MeV in the IMF yield. Indeed from 32 to 52 A.MeV the proportions of 1 and 2-IMF events decrease slightly whereas the other partitions increase. This rise of the IMF yield with energy is predicted by statistical models 5, 7. However owing to the total charge conservation, the IMF charge falls as the IMF multiplicity goes up. Therefore the high IMF multiplicities reach a maximum around 52 A.MeV. Beyond this energy, a fraction of fragments disappears in light particles (generally $Z = 2$) leading to a steady fall of the high IMF partitions to the benefit of the low IMF multiplicity events. The most striking trend is obtained for the no IMF ratio which increases significantly from 52 to 95 A.MeV expressing the trend towards the vaporization of the nuclear system 8.

4. EVOLUTION OF REACTION MECHANISMS

An efficient detector such as INDRA allows to reconstruct event by event the kinematics of the collision and to obtain some information about the reaction mechanisms. In central collisions, at low energies, the fusion process is generally involved in the formation

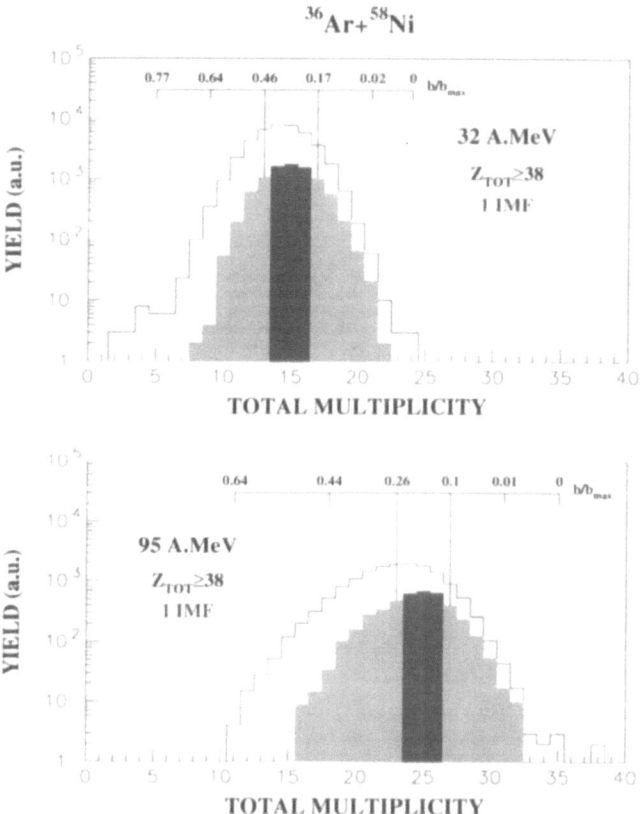

Figure 3. Multiplicity distributions for complete events with 1-IMF at 32 and 95 A.MeV. Each distribution is divided in three multiplicity slices (dark areas). The upper scale gives a reduced impact parameter range ($b_{max} \sim 10$ fm).

of hot nuclei (at least for the light systems). This process is characterized by the presence of a source close to the center of mass rapidity in Lorentz invariant cross section $\frac{\partial^2\sigma}{\partial y\partial p_\perp}$ plot. However some results have shown that binary processes become dominant with increasing energy 9. They are characterized by two sources at low rapidity on the one hand and near the projectile rapidity on the other hand. The partition of the total cross section between these reaction mechanisms remains an open question which is linked to the problem of source characterization. Some qualitative results for the particular classes of events including 1 and 2-IMF are now presented.

A relationship between the total multiplicity and an estimated impact parameter has been used to select the most dissipative collisions from the less dissipative ones by means of three multiplicity slices (Fig. 3). For the following analysis, only the two extreme mutiplicity slices are considered. The Lorentz invariant cross section plot for alpha particles associated to the 1-IMF events is displayed at 32 A.MeV in Fig. 4a. A source can be distinguished around the center of mass rapidity. The IMF velocity and charge distributions confirm the assumption of a fusion process (Fig. 4b). Indeed the remaining IMF is relatively heavy and slow ($\langle < V \rangle \sim 2$ cm/ns close to the detection threshold). Moreover the IMF angular distribution is peaked at forward angle. These overall characteristics are compatible with an incomplete fusion process

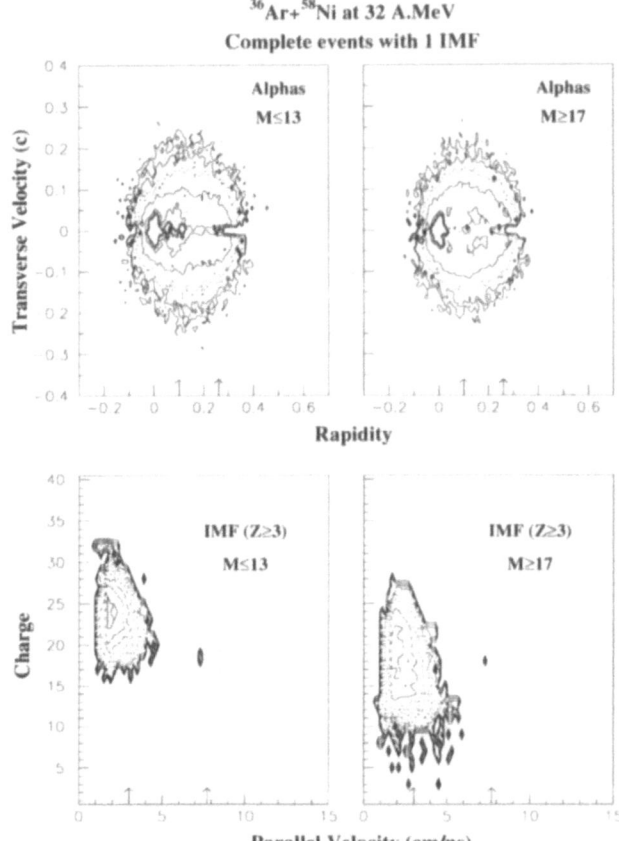

Figure 4. a) Top: Lorentz invariant cross section in the transverse velocity versus rapidity plane for alpha particles associated to the 1-IMF events at 32 A.MeV for the lowest multiplicity slice and the highest one. b) Bottom: Charge versus parallel velocity distributions of the IMF (extreme multiplicity slices). Arrows on the horizontal axis indicate respectively the center of mass and the projectile rapidity (or velocity).

leading to an evaporation residue with 60% of transferred momentum 10. The cross section for fusion is estimated at 70 mbarn (2% of the reaction cross section).

When increasing energy, the invariant cross section plot for alpha particles at 95 A.MeV clearly displays two separated sources for the lowest multiplicity slice (Fig. 5a). Notice the rapidity range of the fast source for the most dissipative collisions. Otherwise the IMF velocity distribution presents a main peak below the center of mass velocity associated with a TLF residue on the one hand and a secondary bump around the projectile velocity on the other hand (Fig. 5b). Therefore the remaining fragment comes from a binary reaction in which one of the partners has been completely disintegrated in light particles. For light systems in central collisions, Landau–Vlasov calculations predict a final binary behavior at high incident energy owing to a transparency effect 11.

The 2-IMF events which belong to the most probable partition allow to precise the contribution of binary mechanisms to the total cross section. As previously, multiplicity slices have been used to estimate the centrality of the collisions. Fig. 6a displays at 32 A.MeV the relative velocity (V_R) of the two detected fragments versus their parallel velocities. At high V_R,

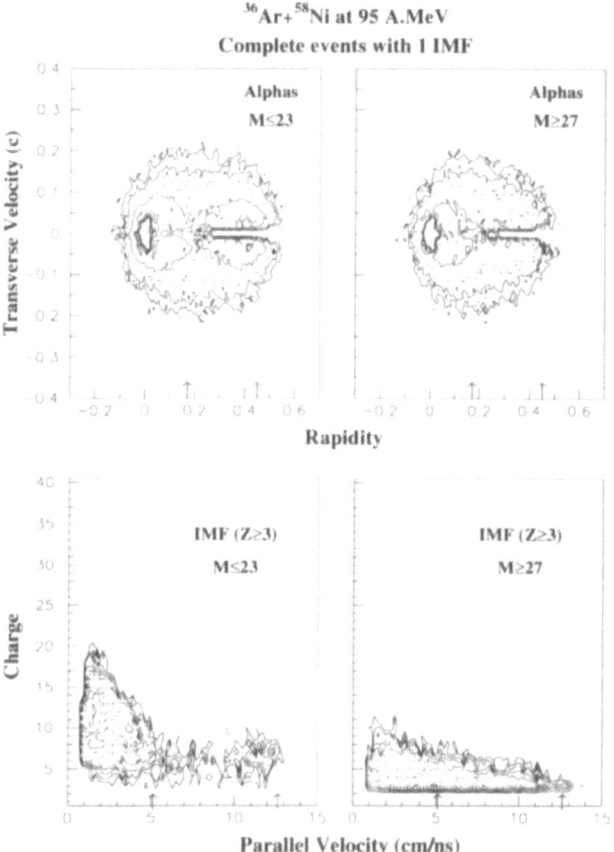

^{36}Ar+^{58}Ni at 95 A.MeV
Complete events with 1 IMF

Figure 5. a) Top: Lorentz invariant cross section for alpha particles associated to the 1-IMF events at 95 A.MeV (extreme multiplicity slices). b) Bottom: Charge versus parallel velocity distributions of the IMF (extreme multiplicity slices).

both partners of a binary reaction are clearly seen: respectively the TLF residue peaked at low parallel velocity and the PLF residue with a range of high velocity. The relative angle (Θ_R) is defined as the angle between the relative velocity and the beam axis. As V_R decreases, Θ_R increases which is interpreted in terms of increasing dissipation (Fig. 6b). The arrow on the vertical axis indicates the Coulomb velocity corresponding to completely damped reactions which are observed beyond 50°. The similar behavior is obtained for the high multiplicity slice with larger dissipation. At last, the strong correlation between the deflection angle and the relative velocity is in good agreement with a deep inelastic process 10.

5. CONCLUSION

In the Ar + Ni system, the evolution of IMF distributions from 32 to 95 A.MeV has shown that the probability to have a given IMF multiplicity is almost independent of the available energy. On the other hand, the analysis of reaction mechanisms for complete events with 1 and 2-IMF in central collision underlines the dominance of binary dissipative reactions.

Complete events with 2 IMFs
for ^{36}Ar+^{58}Ni at 32 A.MeV

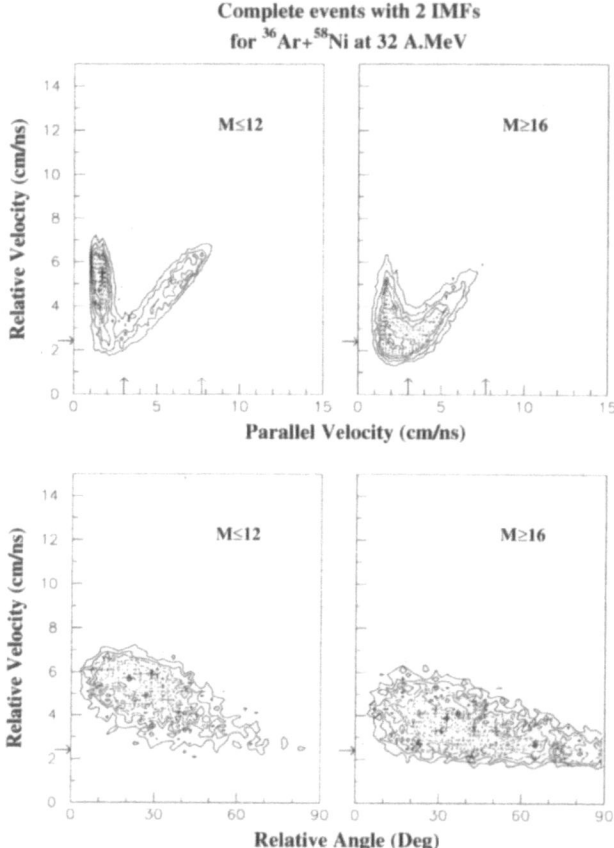

Figure 6. a) Top: Relative velocity between the two IMF as a function of their parallel velocity for the 2-IMF events at 32 A.MeV (extreme multiplicity slices). b) Bottom: IMF relative velocity as a function of the deflection angle (extreme multiplicity slices). Arrow on the vertical axis indicates the Coulomb velocity (damped reactions).

In this way, a weak cross section for fusion has been measured at 32 A.MeV. Future analysis will be dedicated to a sensitive determination of the primary sources in order to obtain an event classification for the study of hot nuclei de-excitation.

REFERENCES

1. B. Borderie, Ann. de Phys. 17 (1992) 349.
2. L. G. Moretto, G. J. Wozniak, Ann. Rev. Nucl. Part. Sci. 43 (1993) 379.
3. J. Cugnon, Phys. Lett. B135 (1984) 374.
4. E. Suraud et al., Nucl. Phys. A495 (1989) 73c;
    E. Suraud et al., Phys. Lett. B229 (1989) 359.
5. J. Bondorf et al., Nucl. Phys. A443 (1985) 321;
    J. Bondorf et al., Nucl. Phys. A444 (1985) 460.
6. J. Pouthas et al., to be published in Nucl. Inst. and Meth.
7. D. H. E. Gross et al., Rep. Prog. Phys. 53 (1990) 605.
8. Ch.O. Bacri et al., submitted to Phys. Lett. B.
9. M. F. Rivet et al., Phys. Lett. B215 (1988) 55.

10. M. Lefort, C. Ngô, Ann. de Phys. 3 (1978) 5.
11. V. de la Mota et al., Phys. Rev. C46 (1992) 677.

5

CRITICAL EXPONENTS FROM THE MULTIFRAGMENTATION OF 1A GeV Au NUCLEI

N. T. Porile,[1] S. Albergo,[2] F. Bieser,[3] F. P. Brady,[4] Z. Caccia,[2] D. A. Cebra,[4] A. D. Chacon,[5] J. L. Chance,[4] Y. Choi,[1]* S. Costa,[2] J. B. Elliott,[1] M. L. Gilkes,[1] J. A. Hauger,[1] A. S. Hirsch,[1] E. L. Hjort,[1] A. Insolia,[2] M. Justice,[6] D. Keane,[6] J. C. Kintner,[4] V. Lindenstruth,[7] M. A. Lisa,[3] U. Lynen,[7] H. S. Matis,[3] M. McMahan,[3] C. McParland,[3] W. F. J. Müller,[7] D. L. Olson,[3] M. D. Partlan,[4] R. Potenza,[2] G. Rai,[3] J. Rasmussen,[3] H. G. Ritter,[3] J. Romanski,[2] J. L. Romero,[4] G. V. Russo,[2] H. Sann,[7] R. Scharenberg,[1] A. Scott,[6] Y. Shao,[6] B. K. Srivastava,[1] T. J. M. Symons,[3] M. Tincknell,[1] C. Tuvé,[2] S. Wang,[6] P. Warren,[1] H. H. Wieman,[3] and K. Wolf[5]

[1]Purdue University West Lafayette, IN
[2]Universitá di Catania
and Istituto Nazionale di Fisica Nucleare-Sezione di Catania
95129 Catania, Italy
[3]Nuclear Science Division
Lawrence Berkeley Laboratory
Berkeley, CA
[4]University of California
Davis, CA
[5]Texas A&M University
College Station, TX
[6]Kent State University
Kent, OH
[7]GSI
D-64220 Darmstadt, Germany

ABSTRACT

The breakup of 1A GeV gold nuclei incident on carbon has been studied with a detector system that permitted exclusive event reconstruction of nearly all charged reaction products. The moments of the resulting charged fragment distribution

*The original file had an asterisk here which would normally indicate a footnote stating the present address of this author. That footnote is missing.

and fluctuations therein provide strong evidence that nuclear matter possesses a critical point observable in finite nuclei. Values of the critical exponents γ, β, and τ have been determined. These values are close to those for liquid–gas systems but differ significantly from those for other 3-dimensional systems. Assumptions in our analysis have been examined and confirmed by dynamical information.

1. INTRODUCTION

The suggestion that nuclear multifragmentation might be a critical phenomenon, analogous to the vaporization of a liquid at its critical point, was made over a decade ago on the basis of the observed power law distribution in the yield of intermediate mass fragments produced in high-energy proton-nucleus collisions 1. Considerable progress has been made since then in understanding how a system with such few constituents can exhibit critical behavior. Thus, Campi showed that the moments of the fragment charge or mass distribution should exhibit features characteristic of critical phenomena if the above suggestion were to be valid 2. The determination of these moments requires a measurement of all the charged fragments, $Z = 1$ to Z_{max}, from a given interation. In a recent experiment, the EOS Collaboration performed such a measurement for the interation of $1A$ GeV ^{197}Au nuclei incident on carbon, and reported the first determination of three critical exponents associated with the multifragment breakup 3.

We report here the results of a redetermination of these exponents based on a larger data set. We also examine the validity of the assumptions in the analysis on the basis of dynamical information. The latter is preliminary in nature and based on a subset of the data.

2. EXPERIMENTAL RESULTS

The experiment, which was performed at the Lawrence Berkeley Laboratory Bevalac, has been described previously 3. Charged particles were identified using a time projection chamber for $1 \leq Z \leq 6$ 4, a time-of-flight wall for $7 \leq Z \leq 10$, and a multiple sampling ionization chamber for $11 \leq Z \leq Z_{beam}$ 5. Monte Carlo simulation indicated that the overall detection system had an acceptance $\gtrsim 90\%$ for projectile fragments, whereas that for target fragments was low. The total reconstructed charge, Z_{sum}, shown in Fig. 1, consequently peaked at $Z = 79$. The analysis presented here required $Z_{sum} = 79 \pm 3$. Approximately 50% of the available 8.0×10^4 events met this criterion.

For each event, we determine the multiplicity of charged fragments, m, and the number of charged fragments of nuclear charge Z, n_Z. Figure 2 shows the fragment charge yield distribution, $n_Z(Z)$, for different multiplicity intervals. At low multiplicities, the charge yield distribution shows the presence of a heavy residue and a number of light particles; some fission fragments may also be formed. In terms of a phase transition model, these interactions correspond to the formation of the "liquid" phase. At high multiplicities, there is no heavy remnant and the charge yield distribution is an exponentially decreasing function of Z, corresponding to the formation of the "vapor" phase. The broadest yield distribution is obtained at intermediate multiplicities, as expected if critical fluctuations are present.

A clearer picture of these fluctuations can be seen from an examination of the dependence on m of the mean charge of the largest fragment, Z_{max}, and in the standard deviation of the Z_{max} distribution. Figure 3 shows the results, where the number of events at each m is approximately independent of m. The standard deviation of Z_{max}, ΔZ_{max}, peaks sharply at

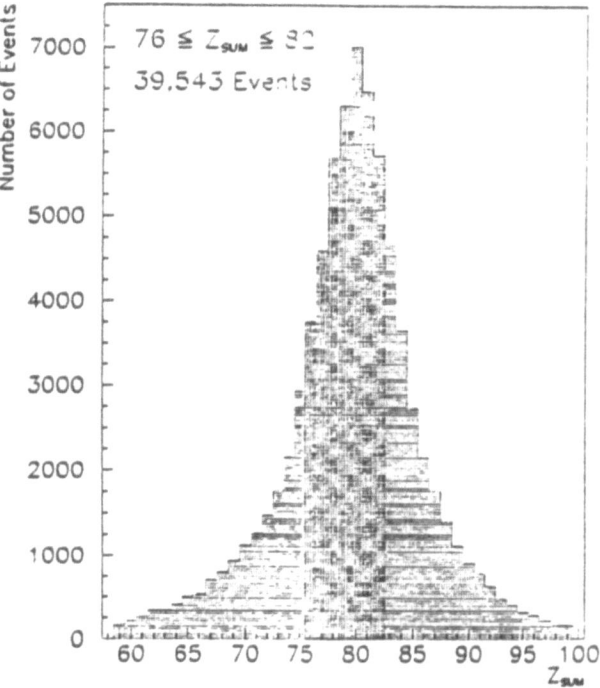

Figure 1. Distribution of total reconstructed charge. Events chosen for subsequent analysis are highlighted.

intermediate multiplicities, suggesting that critical fluctuations may be present in this multiplicity regime. Similar results are obtained for the standard deviation of the second moment of the charge yield distribution.

3. DETERMINATION OF CRITICAL EXPONENTS

The presence of fluctuations is one of the characteristic features of a phase transition in the vicinity of a critical point. Another such feature is the power-law behavior of various quantities as the critical point is approached. This behavior is governed by the values of the critical exponents, which are the exponents in the power-law relations. We have already described 3, following the approach of Campi 2, the procedure for the determination of the critical exponents γ, β, and τ in multifragmentation. The small size of the nuclear system requires that the procedure used to determine critical exponents in ordinary macroscopic systems 6 be modified. We have previously shown 7 that percolation theory 8 provides a useful model for the determination of critical exponents for systems of different size and have applied this procedure to multifragmentation. Our procedure assumes that (1) m is a linear measure of the distance from the critical point, which is characterized by a critical multiplicity m_c, i.e. that multiplicity is proportional to temperature T for T close to T_c; and (2) that the values of the critical exponents are not sensitive to the inclusion of preequilibrium particles. The validity of these assumptions is discussed below.

Briefly, our procedure involves, as a first step, the evaluation of the moments of the

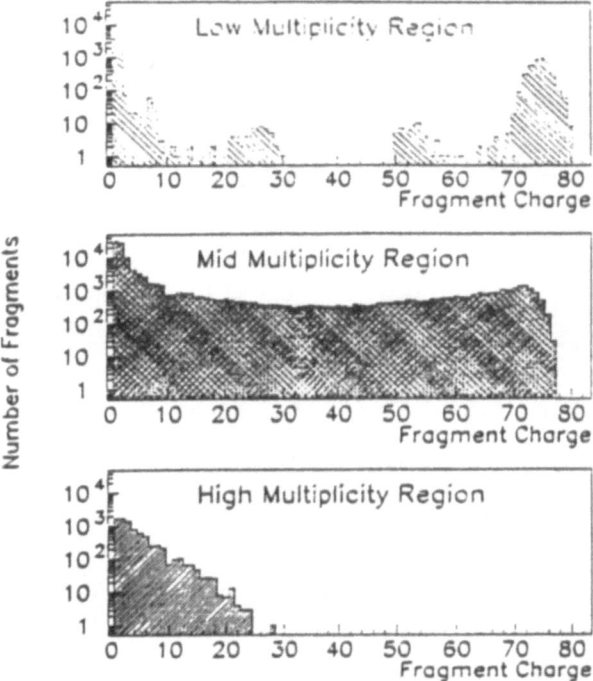

Figure 2. Fragment charge yield distribution for various multiplicity intervals. Top: $1 \leq m \leq 5$; middle: $6 \leq m \leq 50$; bottom: $m > 50$.

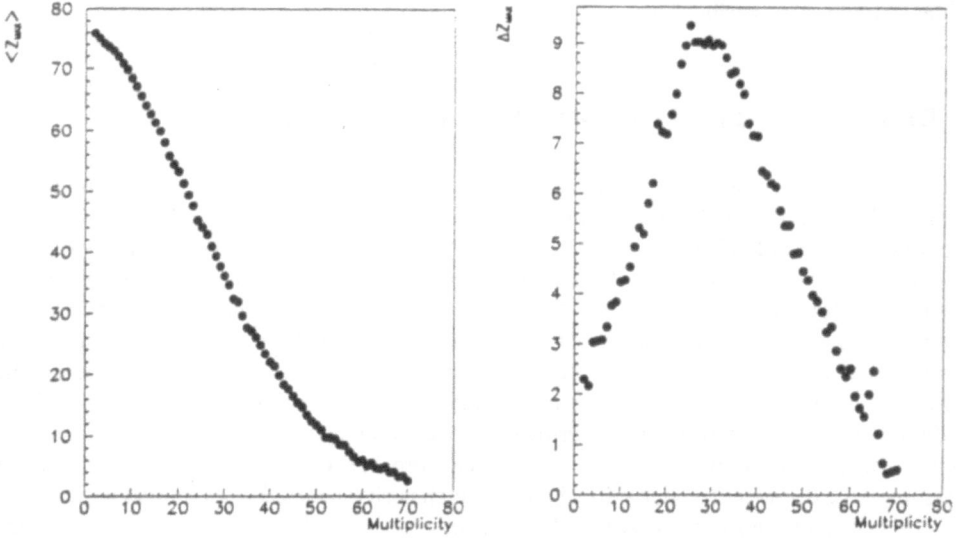

Figure 3. Dependence on multiplicity of (left) the mean charge of the largest fragment, Z_{max}, and (right) the standard deviation of the Z_{max} distribution.

Table 1. Critical Multiplicity and Exponents of Gold Projectile Fragments and Other Three-Dimensional Systems

Quantity	Experiment	Liquid–gas	Percolation	Liquid–gas mean field
m_c	25 ± 3			
γ	1.3 ± 0.1	1.23	1.8	1
β	0.27 ± 0.04	0.33	0.41	$\frac{1}{2}$
τ	2.03 ± 0.04	2.21	2.18	$2\frac{1}{3}$

fragment charge distribution:

$$M_k(m) = \sum_Z Z^k n_Z(m) \tag{3.1}$$

where Z_{max} is not included on the liquid side of the phase transition 8. Next, the values of γ and m_c are determined by the γ-matching technique 3, 7. The exponent γ determines the power-law behavior of the second moment, M_2:

$$M_2(m) \sim |m - m_c|^{-\gamma} \tag{3.2}$$

Values of γ and m_c were obtained by fitting the liquid and gas branches of M_2 according to Eq. (2). Some 3×10^4 fits were performed in which the widths and locations of the fitted multiplicity intervals, as well as m_c, were varied. Those fits for which γ_{liquid} and γ_{gas} differed by less than 10% and which had χ^2 values smaller than at least 50% of the events meeting the γ-matching criterion were accepted. Figure 4 shows the distribution of $\langle \gamma \rangle = (\gamma_{liquid} + \gamma_{gas})/2$ and m_c for the $\sim 10^3$ fits that met these criteria. Note that the distributions are quite narrow. Table 1 lists the average values and their standard deviations. The exponents β and τ were obtained from these same fits in the previously described manner 3 and their average values are included in Table 1. We have varied the selection criteria described above and find that the results are insensitive to reasonable variations.

Table 1 also includes exponent values for several three-dimensional systems possessing a scalar order parameter: liquid–gas 6, percolation 8, and the mean field limit of the liquid–gas system 9. While the values of τ are insensitive to the nature of the system, the multifragmentation values of γ and β lie much closer to those of a real fluid system than they do to either the percolation or mean field fluid values. The comparison thus supports the conclusion that finite nuclei exhibit critical behavior when they undergo multifragmentation.

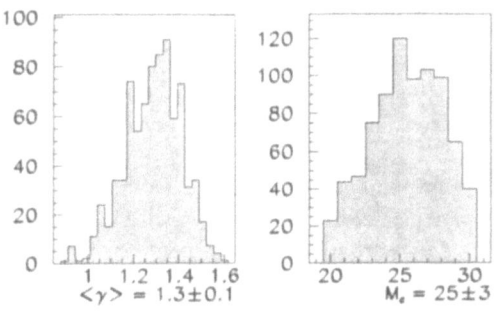

Figure 4. Distribution of $\langle \gamma \rangle$ and m_c values that meet the criteria discussed in the text.

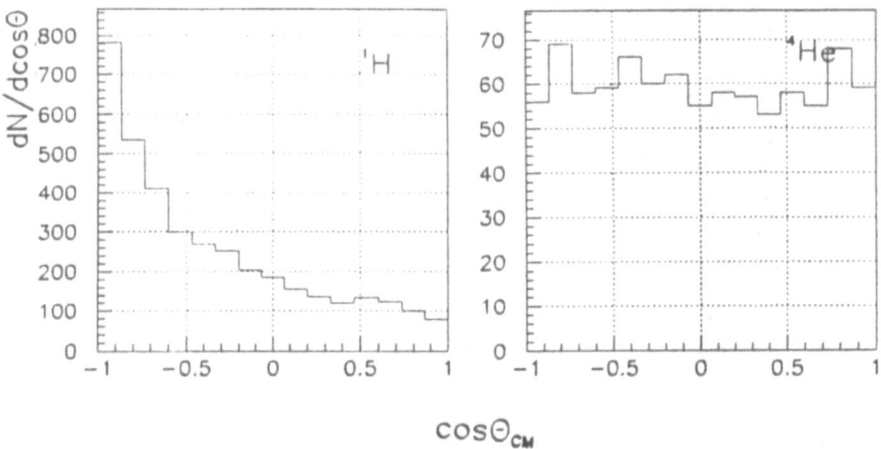

Figure 5. Angular distributions of protons and α-particles in the moving system.

4. EFFECT OF PREEQUILIBRIUM EMISSION

The above analysis is based on the use of the total charge multiplicity. In actuality, m consists of two separable components. One component originates in the initial collision between the gold projectile and the carbon target, producing a preequilibrium source of light particles and subsequently a thermally equilibrated remnant 9, 10. The second component of the multiplicity, m', is associated with the multifragmentation of this remnant. We have separated these first and second stage components following the procedure of Cussol 11, in which a transformation to the moving system of the Au nucleus allows a separation into an isotropic component and an asymmetric preequilibrium component. Figure 5 shows the angular distributions of protons and α-particles in the moving system. While the α-particles display an isotropic angular distribution, the protons peak at backward angles, as expected for preequilibrium emission from a source moving with beam rapidity. The deuteron and triton distributions are qualitatively similar to that for protons. Evidently, preequilibrium emission is primarily important for $Z = 1$ particles. Consequently, these particles were not used in the determination of the moving source velocity.

While the forward half of the proton angular distribution will be enriched in second stage protons, it is apparent from Fig. 5 that this angular interval still contains some preequilibrium protons. This result is corroborated by an examination of the kinetic energy spectrum of the forward protons, which shows a discontinuity in slope at $\sim 50\text{MeV}$. See Fig. 6. We attribute this discontinuity, which is not seen in the α-particle spectrum, to the change in mechanism. Preequilibrium protons can therefore be removed by means of a spherical cut in velocity space made at a velocity corresponding to this discontinuity. The same velocity cut is made for deuterons and tritons, and $Z = 1$ particles with higher velocities are attributed to preequilibrium emission.

The second stage multiplicity m' is obtained from the total charge multiplicity by removal of the prompt $Z = 1$ particles. Figure 7 shows the dependence of m' on m. The dependence is linear, $m = fm'$, particularly close to m_c, where the critical exponents are determined. It follows that the use in the determination of the exponents of the quantity $\ln|m - m_c| = \ln(f|m' - m_c'|) = \ln(f) + \ln|m' - m_c'|$ results in a simple translation along the multiplicity axis. Likewise, the values of M_2 and M_3, which are primarily determined by the

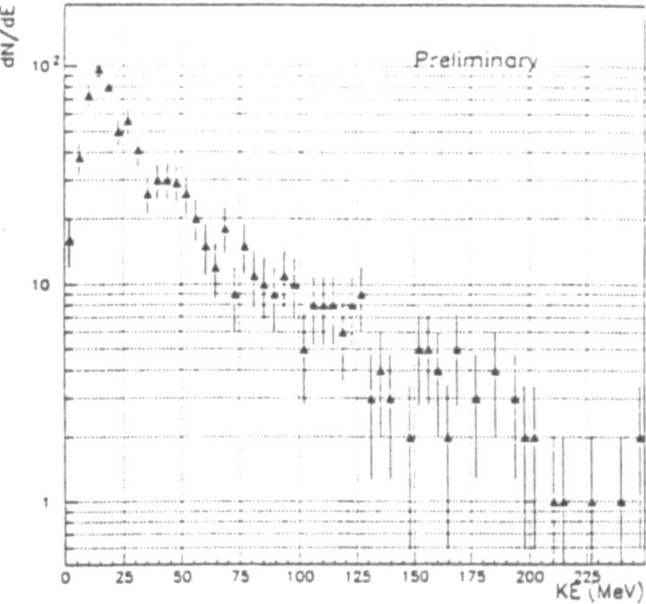

Figure 6. Energy spectrum of protons emitted at forward angles in the moving system.

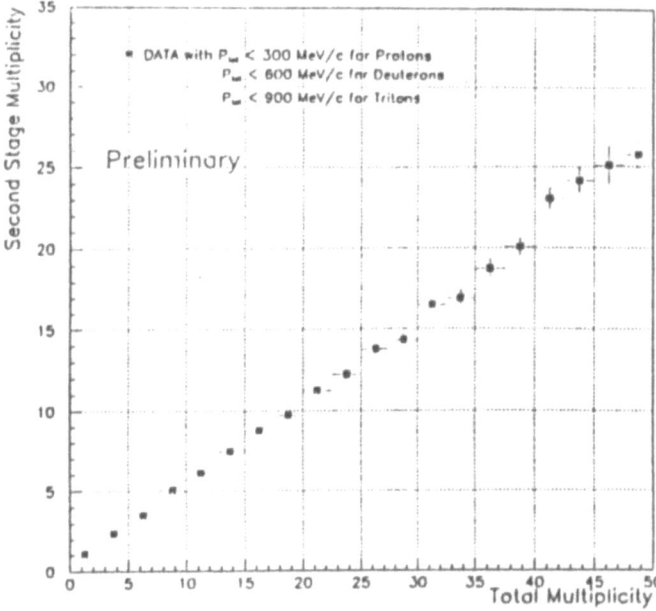

Figure 7. Dependence of second-stage multiplicity on total multiplicity.

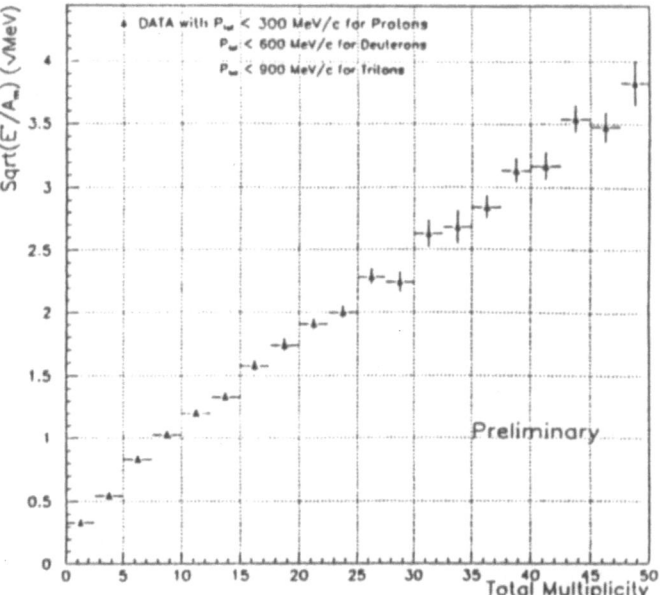

Figure 8. Relation between $[(E^*/A)^{1/2}]$ and charge multiplicity.

contribution of high-Z fragments, are virtually unaffected by the inclusion of prompt $Z = 1$ particles. We conclude that the extraction of critical exponents is unaffected by the inclusion of preequilibrium particles.

5. MULTIPLICITY AS A MEASURE OF TEMPERATURE

In extracting the critical exponents, we have assumed that multiplicity is proportional to temperature in the critical region. We now examine the validity of this assumption. From the measured fragment kinetic energies we can calculate the excitation energy of the remnant on an event-by-event basis 11. We use an energy balance relation:

$$E^*/A = (1/A_{meas}) \left[\sum_i E_i + Q + 2nT \right] \tag{5.1}$$

where E_i is the kinetic of fragment i in the moving system, Q is the mass balance, and A_{meas} is the mass of the remnant, obtained from the reconstructed charge on the assumption that the fragments are formed along the stability line. Only second stage particles are included in A_{meas}. Eq. (3) assumes that n neutrons are emitted in the second stage and that their average kinetic energy is $2T$. We estimate n by assuming that the charge to mass ratio of the remnant is the same as that of gold. According to the Fermi gas model, $T = (E^*/a)^{1/2}$, where the level density parameter a is typically $A/10$. A quadratic equation in $(E^*/a)^{1/2}$ is obtained with this substitution into Eq. (3). One physical solution for E^*/A exists.

The results of this calculation are shown in Fig. 8 as a plot of $(E^*/A)^{1/2}$ vs. m. A linear dependence in the vicinity of the critical multiplicity is observed. Thus, within the context of the Fermi gas model, charge multiplicity provides a linear measure of the distance from criticality. To be sure, considerable evidence has accumulated in recent years indicating that the level density parameter decreases with increasing temperature for hot compound systems 12. Our results would therefore require some modification if this decreases. also occurs for the expanding system of present interest. For example, a decrease of *a* from *A*/8 at 3 MeV to *A*/15 at 6 MeV 13 would change the slope of the *T* vs. *m* line but might not affect the linearity substantially. In any case, we plan to circumvent this problem by obtaining *T*(*m*) directly from the slopes of fragment spectra and, if possible, also from the ratios of isotopic yields.

6. CONCLUSIONS

We have found evidence for critical fluctuations in the multifragment breakup of 1*A* GeV gold nuclei. Three critical exponents, γ, β, and τ, have been obtained from the dependence of the moments of the fragment charge distribution on charge multiplicity. Finite size effects have been taken into account in this procedure. Various assumptions made in our approach are supported by an analysis of the reaction dynamics. Our results provide strong evidence for critical behavior in finite nuclei, with the critical exponents being close to the nominal liquid–gas values.

This work was supported in part by the U. S. Department of Energy under contracts or grants DE-AC03-76SF00098, DE-FG02-89ER40531, DE-FG02-88ER40408, DE-FG02-88ER40412, DE-FG05-88ER40437, and by the U. S. National Science Foundation under grant PHY-91-23301.

REFERENCES

1. J. E. Finn *et al.*, *Phys. Rev. Lett.* **49**, 1321 (1982); R. W. Minich *et al.*, *Phys. Lett. B* **118**, 458 (1982).
2. X. Campi, *J. Phys. A* **19**, L917 (1986); *Phys. Lett. B* **208**, 351 (1988).
3. M. L. Gilkes *et al.*, *Phys. Rev. Lett.* **73**, 1590 (1994).
4. G. Rai *et al.*, *IEEE Trans. Nucl. Sci.* **37**, 56 (1990).
5. W. Christie *et al.*, *Nucl. Instrum. Methods Phys. Res. A* **255**, 46 (1987).
6. H. E. Stanley, *Introduction to Phase Transitions and Critical Phenomena* (Oxford University Press, Oxford, 1971).
7. J. B. Elliott *et al.*, *Phys. Rev. C.* **49**, 3185 (1994).
8. D. Stauffer and A. Aharony, *Introduction to Percolation Theory*, (Taylor and Francis, London, 1992) 2nd ed.
9. A. S. Hirsch *et al.*, *Phys. Rev. C* **29**, 508 (1984).
10. K. Nakai *et al.*, *Phys. Lett B* **121**, 373 (1983).
11. D. Cussol *et al.*, *Nucl. Phys. A* **561**, 298 (1993).
12. G. Nebbia *et al.*, *Phys. Lett B* **176**, 20 (1986); K. Hagel *et al.*, *Nucl. Phys. A* **486**, 429 (1988); M. Gonin *et al.*, *Phys. Lett B* **217**, 406 (1989).
13. O. Civitarese and M. Schvellinger, *J. Phys G* **20**, 1933 (1994).

6

DYNAMICS OF MULTIFRAGMENTATION

M. Belkacem, A. Bonasera, and V. Latora

INFN — Laboratorio Nazionale del Sud
Viale Andrea Doria (ang. Via S. Sofia), 95123 Catania, Italy

ABSTRACT

We investigate the possibility of occurrence of a liquid–gas phase transition in a finite system. Through a study of mass distributions, scaled factorial moments and moments of cluster mass distributions, we find evidence for the presence of a critical behavior of our finite system. Furthermore, by studying scaling invariance of hydrodynamical equations in the framework of classical molecular dynamics, it is shown that hydrodynamical scaling is valid at high beam energies and not at low beam energies. At the beam energy where the violation of the scaling occurs, one observes a mass distribution exhibiting a power law which corresponds to the occurrence of a phase transition.

1. INTRODUCTION

Recent experiments in heavy-ion reactions at energies around the Fermi energy have revealed the creation of many fragments in the final stages of the reaction exhibiting a power law in fragment mass distributions.[1] Such a power law, as described by the droplet model of Fisher,[2] is expected for droplet condensation near the critical temperature, indicating a liquid–gas phase transition. Strictly speaking, sharp phase transitions can only occur in the thermodynamic limit for systems with a very large number of particles. In particular in small systems like two colliding nuclei, the fluctuations can completely wash out the phase transition. Furthermore, assuming that a critical behavior is possible, the problem is how to find evidence for it from the large amount of experimental data. In the first part of this paper, we address both problems and demonstrate that finite systems may in fact exhibit a critical behavior that can be revealed through a study of inclusive mass distributions, scaled factorial moments and through the analysis of conditional moments as developed by Campi.[3] Such a critical behavior is connected, by the use of Fisher's droplet model and Campi analysis, to a liquid–gas phase transition.

Some years ago, Bonasera, Csernai and Schürmann while studying the scaling behavior of observables in heavy-ion collisions, predicted that the inversion of collective flow and hence the violation of hydrodynamical scaling in heavy-ion collisions is due to the predominance of

Advances in Nuclear Dynamics
Edited by Wolfgang Bauer and Alice Mignerey, Plenum Press, New York, 1996

Table 1. Values of the Fitting Parameters Y_0, X, Y and τ Entering the Formula Eq. (2.1)

T (MeV)	2	3	4	5	7	10	15	20
Y_0	442.8	146.1	30.7	39.5	69.7	97.0	290.7	450.5
X	0.042	0.10	1.00	1.00	1.00	1.00	1.00	1.00
Y	2.01	1.83	1.012	0.995	0.87	0.70	0.43	0.31
τ	2.23	2.23	2.23	2.23	2.23	2.23	2.23	2.23

the nuclear attraction resulting in an attractive mean-field and in the occurrence of a liquid–gas phase transition which suddenly decreases the repulsive pressure.[4] In the second part of this paper, we show that within the framework of Classical Molecular Dynamics and for the reaction $Nb^{93} + Nb^{93}$ at an impact parameter $b = 4$ fm, the scaling picture of hydrodynamical equations is valid at high beam energies and not at low beam energies. At the beam energy where the violation of the scaling occurs, one observes a fragment mass distribution exhibiting a power law which corresponds to the occurrence of a critical behavior related to a liquid–gas phase transition in this classical model.

The model used for our study is Classical Molecular Dynamics (CMD) model which is described in details in Ref. 5. Within the framework of this model, we have studied the disassembly of a system with 100 particles starting from an initial density $\rho = 0.125$ fm^{-3} and with different values of the initial temperature. In our calculations, the Coulomb interation is not taken into account.

2. FISHER'S DROPLET MODEL AND FRAGMENT MASS DISTRIBUTIONS

The droplet model of Fisher is a model of liquid to gas transition in which one determines when a liquid–gas system favors the growth of large liquid drops and when it favors the dissociation of those droplets into a vapor.[2] In this model, the probability of formation of a droplet containing A particles is proportional to $exp(-\Delta G/T)$ where $\Delta G = G_{\text{with drop}} - G_{\text{no drop}}$ and G the Gibbs free energy, and results in a mass yield given by:

$$dN/dA = Y_0 A^{-\tau} X^{A^{2/3}} Y^A \tag{2.1}$$

X is proportional to the exponential of the surface tension and it is equal to 1 for temperatures larger or equal to the critical temperature T_c because the surface tension vanishes and Y is proportional to the exponential of the difference between the chemical potentials of the liquid and gas phases and hence is equal to 1 at the critical temperature (because the two chemical potentials are equal).[2,5] The term $A^{-\tau}$ was introduced by Fisher to take into account the fact that the droplet surface closes on itself.

In the upper part of Figure 1 we plot the mass distributions obtained in the expansion of $A = 100$ nucleus starting with three different initial temperatures $T = 4$, 5 and 10 MeV, and with the ground state density $\rho = 0.125$ fm^{-3}. Depending on the initial temperature, the system shows different dynamical evolutions, from evaporation-like process (for small temperatures) to fragmentation and complete vaporization processes (large temperatures). We fitted these mass distributions together with others (not plotted here, see Ref. 5) with initial temperatures ranging from 2 to 20 MeV according to Eq. (2.1) where Y_0, X, Y and τ are fitting parameters. The methodology used for the fits is explained in Ref. 5. The values of the fitting parameters

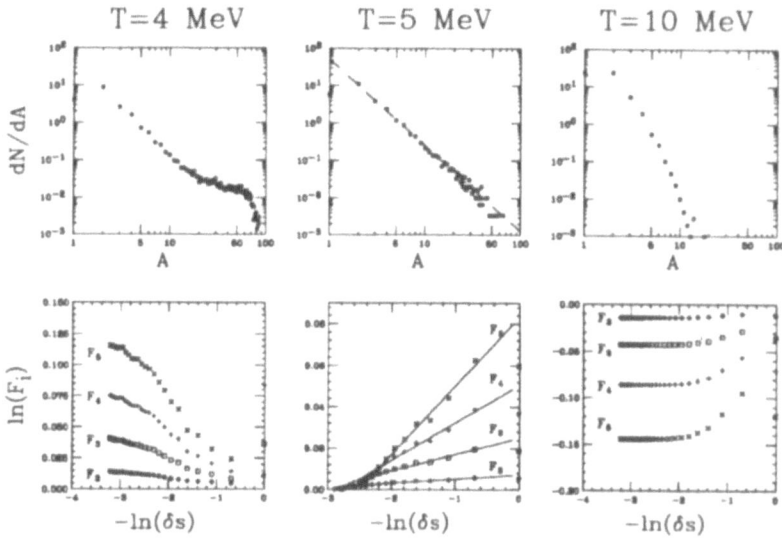

Figure 1. Mass distributions and the corresponding scaled factorial moments $\ln(F_i)$ versus $-\ln(\delta s)$ for events with initial temperatures $T = 4, 5$ and 10MeV. In the central plot of the upper part, the dashed line represents the curve $dN/dA \propto A^{-2.23}$. In the central plot of the lower part, the solid lines are drawn to guide the eye.

are listed in Table I. From this table, we see that the point $X = 1$, $Y = 1$ is obtained in our calculations at $T \approx 4$–5 MeV.

Of course the mass yield shape alone cannot be considered as a conclusive proof for a critical behavior that recalls a phase transition.

3. INTERMITTENCY ANALYSIS

One of the most powerful and promising methods developed to analyze the fluctuations and the correlations for various physical quantities seems to be the analysis of event by event data in terms of intermittency. Intermittency is a statistical concept used to analyze the fluctuations and correlations of a distribution. Bialas and Peschanski introduced this idea to study the dynamical fluctuations in rapidity distributions of particles from high-multiplicity events produced in ultra-relativistic reactions.[6] More recently Ploszajczak and Tucholski suggested looking for intermittency in the fragment distributions in nuclear multifragmentation at intermediate energies. They were able to see evidence for intermittent pattern of fluctuations in the fragment charge distributions both in data and in models.[7] Furthermore, lots of efforts have been devoted to find evidence for the occurrence of a phase transition of nuclear matter in the intermittent behavior of the multiplicity distributions.[5,8,1]

Generally, the occurrence of intermittency corresponds to the existence of large non statistical fluctuations which have self-similarity over a broad range of scales. This signal can be deduced from the scaled factorial moments which measure the properties of dynamical fluctuations without the bias of statistical fluctuations[6]:

$$F_i(\delta s) = \frac{\sum_{k=1}^{X_{max}/\delta s} \langle n_k \cdot (n_k - 1) \cdot \ldots \cdot (n_k - i + 1) \rangle}{\sum_{k=1}^{X_{max}/\delta s} \langle n_k \rangle^i} \tag{3.1}$$

Here X_{max} is an upper characteristic value of the system (i.e., total mass or charge, maximum transverse energy or momentum, etc...) and i is the order of the moment. The total interval 0–X_{max} ($1 - A_{max}$, Z_{max} in the case of mass or charge distributions) is divided in $M = X_{max}/\delta s$ bins of size δs, n_k is the number of particles in the k-th bin for an event, and the brackets $\langle \cdot \rangle$ denote the average over many events. If self-similar fluctuations exist at all scales δs, the scaled factorial moments follow the power law $F_i(\delta s) \propto (\delta s)^{-\lambda_i}$ where λ_i are called intermittency exponents. So the intermittent behavior is defined as a linear rise in a plot of $\ln(F_i)$ versus $-\ln(\delta s)$.

In the lower part of Figure 1 we plot the scaled factorial moments $\ln(F_i)$ versus $-\ln(\delta s)$. At $T = 10$ MeV, the system goes into complete vaporization and the mass distribution has a rather steep slope. The logarithm of the scaled factorial moments $\ln(F_i)$ is always negative and independent on δs (i.e., variances are smaller than poissonian[7]) and almost independent on δs and we have no intermittency signal. The situation is different for the case $T = 5$ MeV. The logarithms of the scaled factorial moments are positive and almost linearly increasing versus $-\ln(\delta s)$ and the intermittency signal is observed. It is possible to relate the initial temperature of the expanding system in this model to the q parameter for bond percolation model. At $T = 4$ MeV and $T = 7$ MeV, the behavior of the scaled factorial moments is the same as, in percolation, for subcritical ($q > q_c$) and overcritical ($q < q_c$) events respectively. The intermittent pattern found at $T = 5$ MeV corresponds to $q \simeq q_c$ case.[7,10]

4. MOMENTS ANALYSIS

The presence of large fluctuations as indicated by the intermittency analysis plus the power law in the mass distribution for initial temperatures between 4 and 5 MeV for the system $A = 100$ indicate a self similar behavior both for fluctuations and for averages. These features might be connected to a second order phase transition in an infinite system. But our system contains few hundreds of constituents only. To better clarify this point we consider the method of conditional moments developed by Campi.[3] Sharp phase transitions are well defined only for an infinite system in that the singularities that are associated with such a transition can only be observed in the thermodynamic limit. If the system has a finite size, the singularities become finite and the transition can be completely washed out by the finite size effects. To overcome this inherent problem of nuclear systems, Campi suggested the method of conditional moments which aims at characterizing the finite system near the critical region in terms of critical exponents and relations between them are derived for the infinite percolation model. One of the tests proposed by Campi is to look at the relative variance γ_2 defined as[3,7]:

$$\gamma_2 = \frac{M_2 M_0}{M_1^2} \tag{4.1}$$

where the moments of the fragment mass distribution M_k are defined as:

$$M_k = \sum_A A^k N(A) \tag{4.2}$$

It was shown by Campi that this quantity presents a peak around the critical point which means that the fluctuations in the fragment size distributions are the largest near the critical point.[3]

In Figure 2, we plot the relative variance γ_2 calculated in two different ways versus the initial temperature T. For the relative variance $\bar{\gamma}_2$ (dashed line), we calculate the average

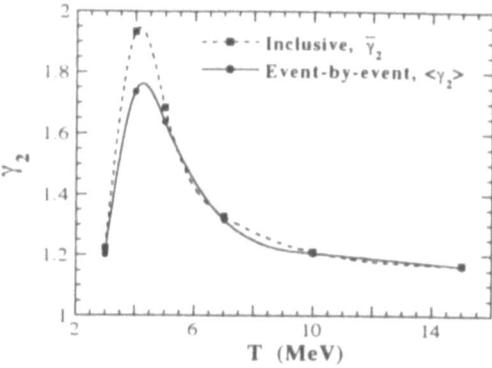

Figure 2. Reduced variance γ_2 versus the initial temperature T of the expansion of the system $A = 100$. Dashed line shows the results for $\bar{\gamma}_2$ and solid line $\langle \gamma_2 \rangle$.

fragment mass distribution $\bar{N}(A, T)$ from all the events starting with the initial temperature T and calculate the moments M_0, M_1 and M_2 as

$$\bar{M}_k = \sum_A A^k \bar{N}(A, T) \tag{4.3}$$

and $\bar{\gamma}_2$ is given by

$$\bar{\gamma}_2 = \frac{\bar{M}_2 \bar{M}_0}{\bar{M}_1^2} \tag{4.4}$$

For the relative variance $\langle \gamma_2 \rangle$ (solid line), we calculate the relative variance $\gamma_2^{(i)}$ for each event i:

$$\gamma_2^{(i)} = \frac{M_2^{(i)} M_0^{(i)}}{M_1^{(i)}}; \quad M_k^{(i)} = \sum_A A^k N^{(i)}(A, T) \tag{4.5}$$

and $\langle \gamma_2 \rangle$ is given by

$$\langle \gamma_2 \rangle (T) = \frac{1}{N_{ev}(T)} \sum_i \gamma_2^{(i)} \tag{4.6}$$

where i runs over all events $N_{ev}(T)$ having the initial temperature T. One notes from the figure that the relative variance γ_2 presents a peak for temperatures between 4 and 5 MeV.

5. SCALING PROPERTIES OF HYDRODYNAMICAL EQUATIONS

In non-relativistic dynamics, the scale invariant equations of motion for a viscous flow are[11]:

$$\frac{\partial \tilde{\rho}}{\partial \tilde{t}} + \tilde{\nabla} \left(\tilde{\rho} \tilde{u} \right) = 0 \tag{5.1}$$

$$\frac{\partial \tilde{u}}{\partial \tilde{t}} + \left(\tilde{u} \cdot \tilde{\nabla} \right) \tilde{u} = -\tilde{c}^2 \frac{\tilde{\nabla} \tilde{\rho}}{\tilde{\rho}} - \frac{1}{R_e} \left[\tilde{\triangle} \tilde{u} + (q + 1/3) \tilde{\nabla} \cdot \left(\tilde{\nabla} \tilde{u} \right) \right] \tag{5.2}$$

where $R_e = l_1 u_1 / \nu$ is the Reynolds number, ν the viscosity and c the sound velocity. The tilde denotes dimensionless quantities. From these equations, we see that if two different systems have the same Reynolds number R_e, they will have the same physical behavior. In order to investigate the validity of hydrodynamical scaling invariance in experimental data, Bonasera *et al* expressed measured quantities in a scale-invariant way.[4] One of these quantities they looked at is the scale-invariant collective flow defined by:

$$\tilde{F} = F/P^{CM}_{proj} \tag{5.3}$$

From the experimental data, they plotted the curves of constant scale-invariant flow in the mass number CM energy (A, E_{CM}) plane and found that these curves show a striking similarity with the curves of constant Reynolds number calculated in the limit of small density and temperature plotted in the same (A, E_{CM}) plane, which confirms the scaling picture of hydrodynamical equations in experimental data. However in the low energy low mass region, they observed a net difference between the two quantities in that the scale-invariant flow drops suddenly and becomes negative while the Reynolds number is always positive. They explained this inversion of the collective flow and hence violation of hydrodynamical scaling by the predominance of the nuclear attraction resulting in an attractive mean-field and in the occurrence of a liquid–gas phase transition in the system which decreases suddenly the repulsive pressure.[4] It is the aim of this work to study, within the framework of Classical Molecular Dynamics, the validity of classical hydrodynamical scaling and to understand, if a violation of this scaling is observed, whether this violation corresponds to the occurrence of a critical behavior in the system.

6. SCALING ANALYSIS WITH THE CMD MODEL

In the classical limit (Boltzmann statistics), the Reynolds number is given by[12]:

$$R_e \propto A^{1/3} \rho \sigma \tag{6.1}$$

where A is the total mass of the system, σ is the particle-particle cross-section and ρ the density. Note that in this limit, the Reynolds number does not depend on the beam energy.

In the following, we consider the reaction $Nb^{93} + Nb^{93}$ at an impact parameter $b = 4$ fm. For this reaction, we calculate the scale-invariant transverse momentum of all fragments defined by:

$$\tilde{P}_\perp = P_\perp/P_{CM}, \quad P_\perp = \frac{1}{N_{frag}} \sum_{j}^{N_{frag}} P_\perp(j) \tag{6.2}$$

If the scaling picture of hydrodynamics applies to our classical model, one should find the scale-invariant transverse momentum $\tilde{P}_\perp$ independent on the beam energy as it is the case for the classical Reynolds number R_e.

In the upper part of Figure 3, we plotted the scale-invariant transverse momentum $\tilde{P}_\perp$ versus the beam energy for the collision $Nb^{93} + Nb^{93}$ at $b = 4$ fm. The dashed line is drawn to guide the eye. One notes from the figure that $\tilde{P}_\perp$ is almost constant from 50 MeV/nucleon to higher energies (Actually we did other calculations at higher energies than those plotted here and $\tilde{P}_\perp$ remains constant). Above 50 MeV/nucleon, the scale-invariant transverse momentum does not depend on the beam energy as the classical Reynolds number and the scaling picture of hydrodynamics applies to our classical model.

At energies lower than 50 MeV/nucleon, one starts to observe a net deviation from the constant value which means that the scaling invariance is violated for these energies. In the lower part of the same figure, we plotted the fragment mass distribution for the same reaction but at the beam energy $E_{Lab}/A = 40$ MeV where the violation of the scaling is observed. we see that this fragment mass distribution exhibits a power law with $\tau = 2.23$ which, as we have seen in the first part of the paper, corresponds to the occurrence of a critical behavior related to a liquid–gas phase transition in this classical model. Other analyses (intermittency, moments analyses,...) are under investigation to understand more precisely this critical behavior.

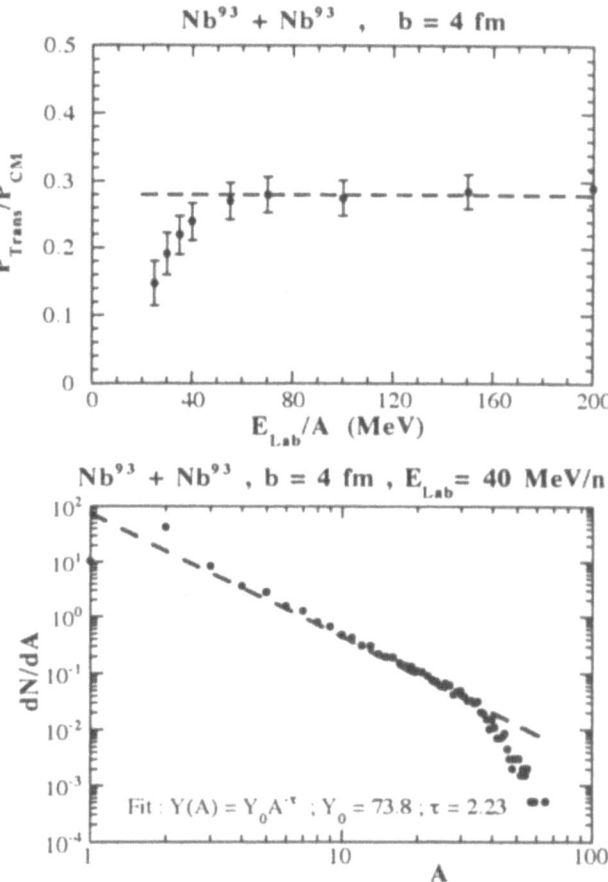

Figure 3. Upper part: Scale-invariant transverse momentum of fragments versus beam energy for the collision $Nb^{93} + Nb^{93}$ at an impact parameter $b = 4$ fm. The dashed line is drawn to guide the eye. Lower part: Fragment mass distribution the same reaction at an incident energy of 40 MeV/nucleon. The data are fitted by a power law with $\tau = 2.23$.

7. CONCLUSIONS

We have shown that it is possible to observe a critical behavior related to a liquid–gas phase transition in a finite classical system. Under some initial conditions, the dynamical evolution creates a self-similar structure both in the average mass distribution (power law) and in the fluctuations (intermittency signal). We have used the moments analysis developed by Campi to better identify the large fluctuations which appear in our system at the critical point. In the second part of the paper and from the preliminary results presented here, we found that the scaling invariance of hydrodynamical equations applies to our classical molecular dynamics model for energies higher than 40–50 MeV (for the reactions considered here) and a violation of this scaling is observed at lower energies. At the beam energy where the violation is observed, we have shown that one gets a fragment mass distribution exhibiting a power law which, as we have seen in the first part of the paper, corresponds to the occurrence of a critical behavior related to a liquid–gas phase transition in this classical model. Other analyses (intermittency, moments analyses,...) are under investigation to better identify this phase transition.

REFERENCES

1. J. E. Finn *et al*, Phys. Rev. Lett. **49**, 1321 (1982); Phys. Lett. **B118**, 458 (1982);
    H. H. Gutbrod, A. I. Warwick and H. Wieman, Nucl. Phys. **A387**, 177c (1982);
    M. Mahi *et al*, Phys. Rev. Lett. **60**, 1936 (1988);
    J. B. Elliot *et al*, Phys. Rev. C **49**, 3185 (1994);
    M. L. Gilkes *et al.*, Phys. Rev. Lett. **73**, 1590 (1994).
2. M. E. Fisher, Rep. Prog. Phys. **30**, 615 (1967); Proceedings of the Interational School of Physics, Enrico Fermi Course LI, Critical Phenomena, edited by M. S. Green (Academic, New York, 1971).
3. X. Campi, J. of Phys. **A19**, L917 (1986);
    X. Campi, Phys. Lett. **B208**, 351 (1988); J. de Phys. **50**, 183 (1989).
4. A. Bonasera and L. P. Csernai, Phys. Rev. Lett. **59**, 630 (1987);
    A. Bonasera, L. P. Csernai and B. Schürmann, Nucl. Phys. **A478**, 159 (1988).
5. V. Latora, M. Belkacem and A. Bonasera, Phys. Rev. Lett. **73**, 1765 (1994);
    M. Belkacem, V. Latora and A. Bonasera, Preprint LNS 28-09-94, Phys. Rev. C (1995), in press.
6. A. Bialas and R. Peschanski, Nucl. Phys. **B273**, 703 (1986); Nucl. Phys. **B308**, 857 (1988).
7. M. Ploszajczak and A. Tucholski, Phys. Rev. Lett. **65**, 1539 (1990); Nucl. Phys. **A523**, 651 (1991).
8. A. Bialas and R. C. Hwa, Phys. Lett. **B253**, 436 (1991).
9. H. R. Jaqaman and D. H. E. Gross, Nucl. Phys. **A524**, 321 (1991);
    D. H. E. Gross, A. R. DeAngelis, H. R. Jaqaman, Pan Jicai and R. Heck, Phys. Rev. Lett. **68**, 146 (1992);
    A. R. DeAngelis, D. H. E. Gross and R. Heck, Nucl. Phys. **A537**, 606 (1992).
10. M. Baldo, A. Causa and A. Rapisarda, Phys. Rev. C **48**, 2520 (1993).
11. N. Balazs, B. Schürmann, K. Dietrich and L. P. Csernai, Nucl. Phys. **A424**, 605 (1984).
12. P. Danielewicz, Phys. Lett. **B146**, 168 (1984).

7

A CLUSTERIZATION MODEL FOR BUU CALCULATIONS

E. J. Garcia-Solis and A. C. Mignerey

Chemistry Department
University of Maryland, College Park, MD

ABSTRACT

A clustering model that allows the recognition of mass fragments from dynamical simulations has been developed. Studying the evolution of a microscopic computation based on the nuclear-Boltzmann transport equation, a suitable time is chosen to identify the bound clusters. At this "stopping" time the number of binding surfaces for each test nucleon is found. Based on the number of nucleon bindings the interior nucleons are identified, and the cluster kernels are formed. An iterative routine is then applied to determine the coalescence of surrounding free nucleons. Once the fragment formation has been established, a statistical decay code is used to determine the final fragment distributions. Applications are given for the two systems ^{139}La on ^{27}Al, ^{139}La on ^{nat}Cu at 45 MeV/nucleon, with model predictions compared to experimental data. An overall agreement between the calculations and the experiment is found.

There are two main questions concerning the clusterization of the nucleons coming from a BUU-type calculation. One is when to stop the dynamical calculation; the other is how to proceed with actual clusterization given the output variables of the model. To deal with these questions the variables relevant to the computation must be addressed at two different levels.

At one level are the variables that describe the motion of the pseudo nucleons, or test particles, in a set of parallel systems. Provided that the pseudo nucleons obey the Newtonian equations of motion

$$\frac{d\mathbf{p}_i^{(n)}}{dt} = -\nabla_r U(\mathbf{r}_i^{(n)}; t) \text{ and } \frac{d\mathbf{r}_i^{(n)}}{dt} = \frac{\mathbf{p}_i^{(n)}}{m_i}, \tag{0.1}$$

where $\mathbf{p}_i$ is the particle momentum, $\mathbf{r}_i$ is the position, and m_i is the particle mass, the test particles will define a total phase-space distribution. This distribution is in turn a function of the second level variables that govern the evolution of the system as a collective ensemble via the Boltzmann–Uehling–Uhlenbeck (BUU) transport equation1

$$\left\{ \frac{\partial}{\partial t} + \frac{\mathbf{p}}{m} \cdot \nabla_r - \nabla_r U(\mathbf{r}; t) \cdot \nabla_p \right\} f(\mathbf{r}, \mathbf{p}; t) = \bar{I}[f],$$ (0.2)

where $\bar{I}[f]$ represents the average rate of change of the particle distribution due to two particle collisions, $U(\mathbf{r}; t)$ is the density dependent mean field potential, and $f(\mathbf{r}, \mathbf{p}; t)$ is the phase space distribution.

The collective ensemble described by Eq. (0.2) will follow a mean field trajectory, from which the average properties of the heavy-ion collision are calculated, and the parallel systems provide the fluctuations about these averages. Recently a consent is emerging that these fluctuations should play an essential role in the multifragmentation of the system at some point of the dynamical calculation (freeze out time) 6, 12, 13. Therefore, rather than using the mean field characteristics of the calculation, the properties of the parallel systems are compared to the experimental data 14.

In order to find the freeze-out or stopping time of the calculation, it is necessary to establish the point when the fluctuations around the mean field trajectory are more pronounced. Fig. 1 shows the evolution of the density function, in the plane of the reaction, as a function of time for the systems ^{139}La on ^{27}Al and ^{nat}Cu at $E/A = 45$ MeV, as calculated from the BUU using $\mathcal{M} = 300$ pseudo-particles. Distributions are shown for a range of input parameters and calculation times. For the smaller values of the impact parameter, where the projectile and target overlap substantially, a highly compressed system is formed. Then the system will start expanding up to a point when the process is repeated again. This is accompanied by emission of individual pseudo-particles in every cycle. At the end of the first compression, when the mean field applies an inward force toward the source that cancels the outward motion, the system will spend considerable time at a relatively low density and temperature, corresponding to an unstable region in the nuclear phase diagram. Under these conditions, it is expected that large fluctuations in the density will emerge 17, giving rise to the condensation of the system into a number of clusters. The description of this process resides outside the extent of the BUU, which deals only with the mean trajectory of the system, and is inadequate for descriptions of unstable evolutions. Therefore, the end of the first expansion indicates an approximate time to stop the dynamical calculation and proceed with the clusterization routine.

Natural variables which can be used to estimate the freeze-out time are the density and the averaged total kinetic energy of the system. The time at which to stop the BUU calculation t_{st} is then chosen as the time when these variables exhibit their absolute minima. The evolution as a function of time of the averaged total kinetic energy and the highest value of the density distribution computed at each time step for different impact parameters are displayed in Figs. 2 and 3 for the La on Al and Cu systems, respectively. At around 30 fm/c a highly compress system is formed, which subsequently expands, reaching a minimum in density at around 90 fm/c; therefore t_{st} is set at 90 fm/c for both systems.

1. THE MODEL

The common method to recognize cluster structures is to separate fragments according to the relative position in phase-space of the particles in a system 15, 12, 14. Using this approach, a nucleon belongs to the same cluster if it is sufficiently connected, that is

$$|\mathbf{r}_i - \mathbf{r}_i| < D_r$$ (1.1)

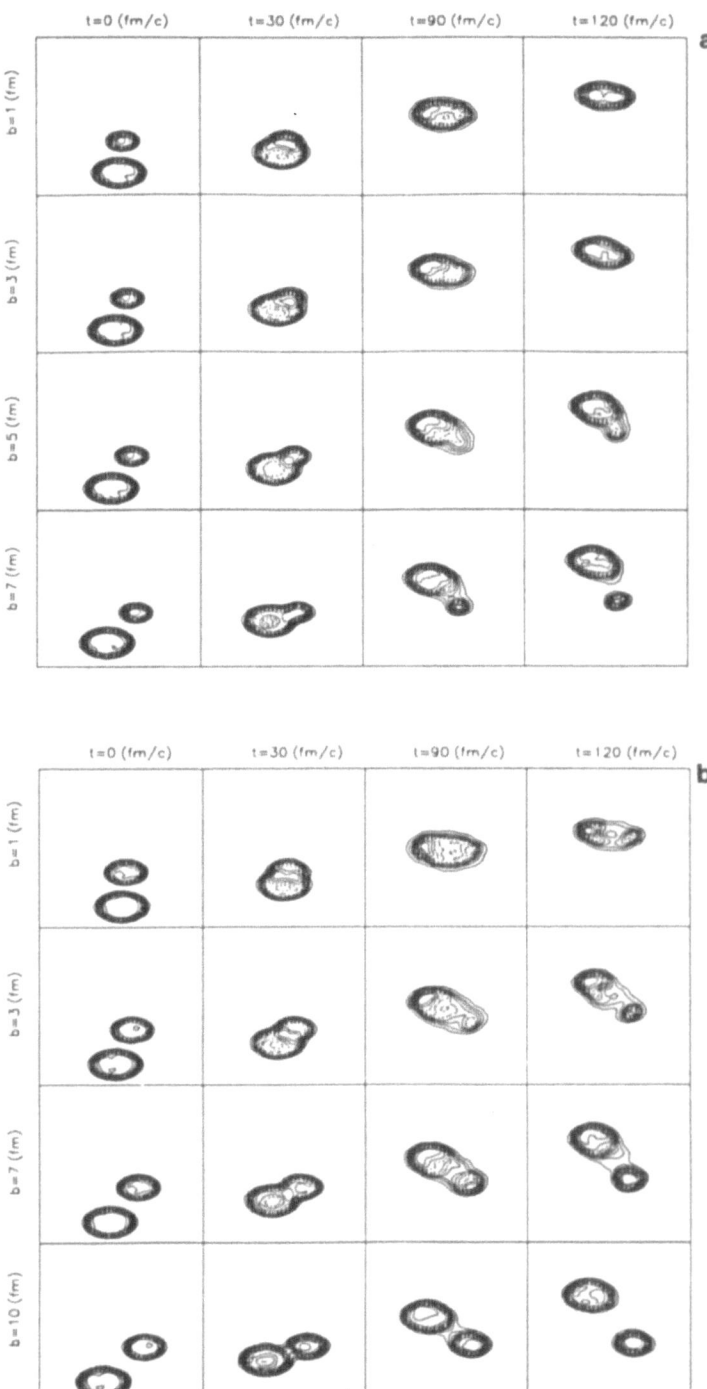

Figure 1. Evolution predicted by the BUU simulation of the density in the reaction plane, as a function of the time and impact parameter, for the systems ^{139}La on ^{27}Al and ^{nat}Cu at $E/A = 45$ MeV.

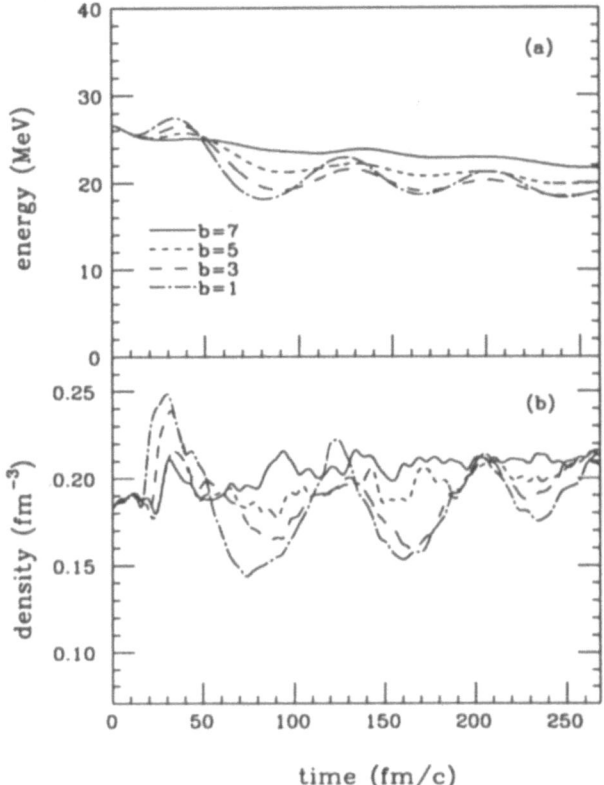

Figure 2. Predicted average total kinetic energy (a) and maximum value of the density function (b), as a function of time, for the system ^{139}La on ^{27}Al at $E/A = 45$ MeV. Each line represents the computation for different impact parameters.

or/and

$$|\mathbf{p}_i - \mathbf{p}_i| < D_p \qquad (1.2)$$

where $D_{r,p}$ is a parameter which is a function of the local density. Although this method is relatively simple and fast, it is not realistic if the clusters are not sufficiently well separated 3. For example, this method is not able to separate two structures that share one surface nucleon. Also strict application of Eqs. (3) and (4) will lead to the exclusion of energetic nucleons in the clusters, therefore miscalculating the amount of internal kinetic energy contained in the cluster.

An alternative approach to finding the clusters without these ambiguities, is by first labeling the interior particles in the parallel systems, then cluster them into fragment "seeds" using Eq. (1.1). To do this, suppose that at t_{st} we have a global particle distribution $\rho(\mathbf{r})$ for the ensemble. If the coordinate space is divided into cubic cells of side $2r_b$, every cell in the space can be defined as interior or exterior to a cluster by the following condition

Condition 1 *The cell i defined by its center at r_i will be interior if*
$\rho(r_i + r_b x_\mu) > \rho_0$ *for every* x_μ, $\mu = 1..6$

where $\mathbf{x}_\mu$ are six unitary vectors in the direction of the positive and negative axes of a coordinate space, with the positive $\mathbf{Z}$ axis directed in parallel to the beam velocity. Taking the

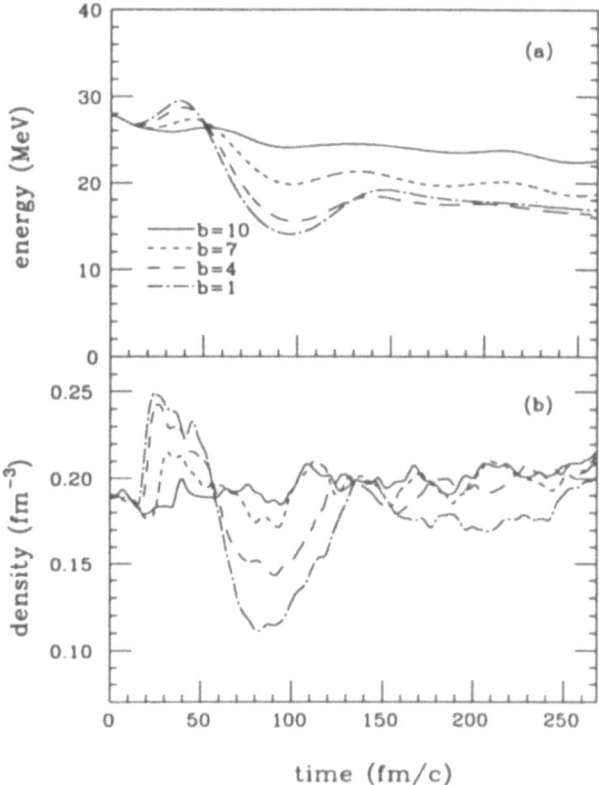

Figure 3. Predicted average total kinetic energy (b) and maximum value of the density function (a), as a function of time, for the system ^{139}La on ^{nat}Cu at $E/A = 45$ MeV. Each line represents the computation for different impact parameters.

position of every nucleon in each parallel system of the ensemble and applying condition 1, it is possible to determine which nucleons are, on the average, **entirely** surrounded by nuclear matter, tagging them as interior. Then, using the interior nucleons, it is possible to form the cluster "kernel" using a standard diagrammatic approach 18 based on Eqs. (3) and (4), with $D_r = r_b = 1.42$ fm, the average nuclear radius.

Note that a nucleon is defined as interior by the surrounding density of its cell, not by the density within the cell. To see how this affects the clustering procedure, take, for example, the system at the beginning of the dynamical calculation (second square of Fig. 1), if applying a common cluster routine Eq. (1.1), both nuclei touching each other would be considered a single fragment. On the other, hand if the interior nucleons are separated first, we would have, after clustering, two separated fragments, plus a number of exterior nucleons that can be treated differently.

Once the configuration of the cluster seeds for every parallel system is established, the corresponding surrounding nucleons tagged as exterior are tested by the condition

Condition 2 *The exterior nucleon i belongs to the cluster j if*
$$s_i \leq R_{cl}^j + r_b$$
and
$$\left| \mathbf{P}_{c.m.}^j - \mathbf{p}_i \right| \leq \sqrt{p(s_i)_{frm}^2 + 2m_i[BE + E_{co}^i]},$$

where $R_{cl}{}^j$ and $P_{c.m.}{}^j$ are the positions and momenta of the center of mass for the j cluster, respectively, and $p(s_i)_{frm}$, m_i, $\mathbf{p}_i$ and s_i are the Fermi moment, mass, momentum, and relative distance to the cluster center of the nucleon i, respectively. The nucleon binding energy is given by an average value of $BE = 8.0$ MeV and the neutron-incremented-energy $E_{co}{}^i = 5$ MeV for i neutrons and 0 MeV for protons. For nucleons that, by Condition 2, belong to two clusters, a random decision is taken. After the first pass using Condition 2, the procedure is repeated for the remaining exterior nucleons, computing the new values of the center-of-mass position and momentum at every iteration, until convergence to a constant mass of the clusters is achieved. The nucleons that, at this point, do not belong to any cluster are tagged as free. Finally, for those remaining free nucleons a coalescence check is done by using Eqs. (2) and (3).

When the configuration of the cluster is established, the collective properties for every cluster, such as translational kinetic energy, angular momentum, excitation energy, etc., are computed by using standard semi-classical formulas. Due to the instability of the cluster formed, is not possible to know the exact zero point of the potential energy 17. Thus, a parameter had to be introduced to calculate the excitation energy

$$E^* = E_{kin}{}^* - \chi E_{fermi}{}^* \tag{1.3}$$

where E_{kin}^* is the excitation energy due to the internal kinetic energy of the test particles, $E_{fermi}{}^*$ is the average fermi energy of the nuclei and χ is a parameter. This parameter is chosen in such way that the ground state $E^* = 0$ for the nuclei before the interaction.

2. APPLICATIONS

As an application of this clustering model, the reactions ^{139}La on ^{27}Al and ^{nat}Cu at $E/A = 45$ MeV are simulated. The results are then compared to experimental data 19 after running through the evaporation code GEMINI 16. The model cross sections and angular distributions, as a function of the detected charge, were calculated directly from the output of the evaporation code. On the other hand, the velocity distribution, and sum charge yield for different multiplicity gates were filtered according to the correspondent experimental set-up.

The results of the model calculation, together with the experimental data, are shown in Figs. 4 to 7. Figs. 4 and 5 show the angular distributions and integrated charge distributions for the Al and Cu target, respectively. For the Al target, the angular distributions (Fig. 4) for the heavier fragments $Z = 42$ down to $Z = 34$ are well reproduced by the calculations. On the other hand, for the range $Z = 34$ to $Z = 22$, the model overpredicts the yield of forward angles. After $\theta_{c.m.} = 50°$ the calculations again correspond to the experimental data. For Z values smaller than 20, the contrary trend is found; the model replicates well the data up to $\theta_{c.m.} = 50°$ and after this it underpredicts the yield.

In the bottom panel of Fig. 4 the experimental Z distribution in cross section is compared to the model predictions. The dashed lines in this figure represent the cross sections integrated for different intervals of impact parameter. In general, the cross sections agree reasonably well with the experimental data. However, for $Z = 7$ to $Z = 10$ the model underpredicts the yield, and it is possible to observe a bump around $Z = 30$. This bump grows proportionally to the interval of integration and may indicate that the fitting of the experimental angular distribution used to calculate the cross section was biased to exclude the fragments from more peripheral reactions, which are very forward peaked.

The angular distributions for the Cu target show the same general behavior as for the Al target, except for the overprediction of the calculation for fragments with $Z \geq 32$ at the

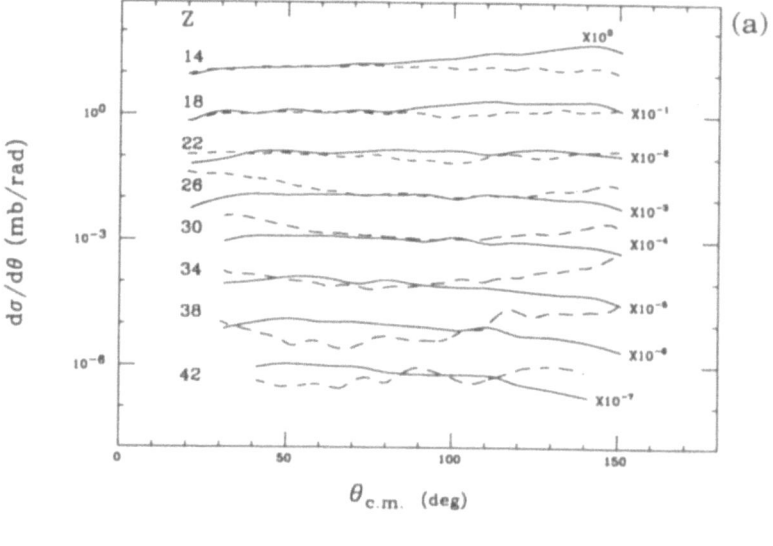

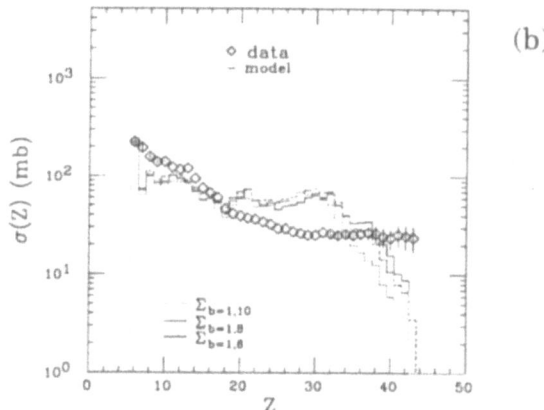

Figure 4. Angular distribution for selected Z values (a) and integrated (b) charge distributions for the reaction La + Al. For the angular distributions, the experimental (solid line) and calculated (dash lines) values are defined in the center of mass. For the total cross section, the experimental data are represented by diamonds, and the calculated distributions are represented by solid, dashed and dotted lines, for integration over impact parameters from 1 to 5 and 7 and 9 fm, respectively.

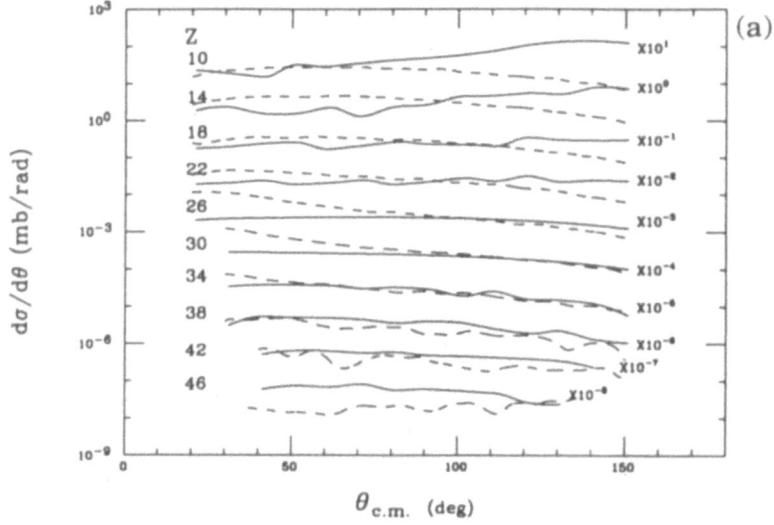

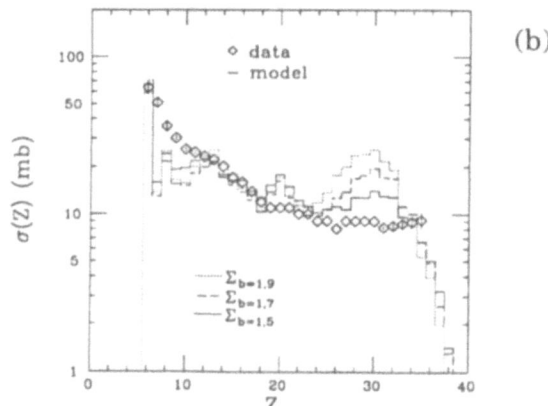

Figure 5. Angular distribution for selected Z values (a) and integrated (b) charge distributions for the reaction La+Cu. For the angular distributions, the experimental (solid line) and calculated (dash lines) values are defined in the center of mass. For the total cross section, the experimental data are represented by diamonds, and the calculated distributions are represented by solid, dashed and dotted lines, for integration over impact parameters from 1 to 5 and 7 and 9 fm, respectively.

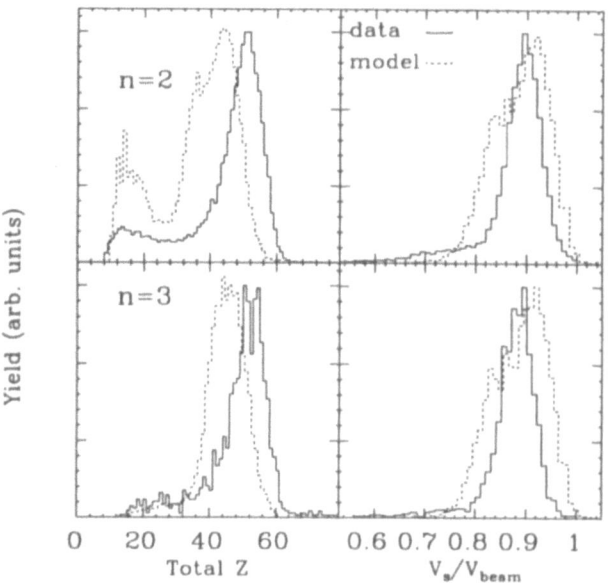

Figure 6. Experimental (solid lines) and calculated (dashed lines) sum of detected charge and source velocity distributions; expressed in velocity relative to the beam velocity V_s; for $n = 2$ and 3 events for the reaction La + Al.

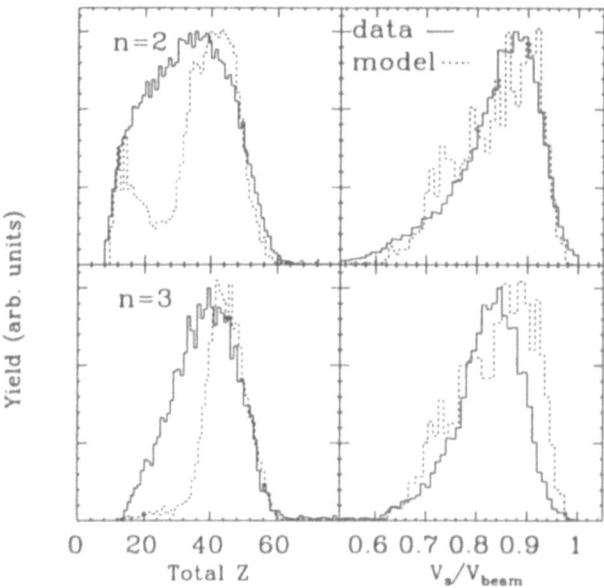

Figure 7. Experimental (solid line) and calculated (dashed line) sum of detected charge and source velocity distributions; expressed in velocity relative to the beam velocity V_s; for $n = 2$ and 3 events for the reaction La + Cu. The arrow indicates the velocity of the beam.

backward angles. For the smaller fragments, the calculation better fits the data for $\theta_{c.m.} \geq 100°$. The bump in the integrated cross sections around $Z = 32$ remains almost constant with increasing integration interval, and the cross section increases for larger integration intervals for fragments with $Z \geq 36$. This component of the spectra may again be interpreted as evaporated residues of the projectile-like fragments, coming from more peripheral reactions.

The sum detected charge and the source velocity distributions, for all coincidence events with multiplicity $n = 2$ and $n = 3$, are shown in Figs. 6 and 7. The dashed lines represent the calculated distributions and the solid lines the data. In general, for the $n = 2$ events, the peak in the Z_{tot} distribution is well reproduced by the calculation. However, the tail of the distribution is overpredicted for the Al target, and the central part of the distribution is underpredicted for the Cu target. For $n = 3$ events, the model is able to predict the peak of the distribution; however, it underestimates the width and the tale of the distribution. On the other hand, the model is able to reproduce the peaks of the V_S distributions within a few percent, also reproducing the width and the tail for $n = 2$ events for the Cu target. However, the width is overpredicted for the Al target. For $n = 3$ events, the velocity distribution widths are well reproduced, but the peak is shifted to higher velocities for both targets.

3. CONCLUSIONS

The results obtained confirm that it is possible to describe the fragment distributions produced in intermediate energy heavy-ion reactions considering statistical and dynamical features. In this work, a dynamic description of the nucleon collisions was coupled with a subsequent statistical decay of the primary source through a clustering subroutine. The specific clustering criterion is a reasonable approach, not only for dealing with dynamically separated clusters, but also for generally dense stages of the reaction. This feature makes a significant improvement to the predictions when compared with the results from other clustering models 2, 12, 14. Finally, it should be pointed out that, in spite of the reasonably good predictions that are obtained, this clustering routine is not optimal because it sharply stops the dynamical calculation. Further work in this area should focus on obtaining the clusterization directly from the dynamical evolution of the density distribution of the system. To this end, the formation of clusters from the ensemble systems is being investigated by increasing the fluctuations in the dynamical evolution, and performing a renormalization of the density distribution function by artificially inserting cells of empty space within the density volume.

REFERENCES

1. G. F. Bertsch and S. Das Gupta, *Phys. Rep.* **160** (1988) 190.
2. M. Colonna, M. DiToro, V. Latora and A. Smerzi, *Prog. Part. Phys.* **30** (1993) 17.
3. C. Dorso and J. Randrup, *Phys. Lett. B* **301** (1993) 328.
4. M. Colonna, P. Roussel-Chomaz, N. Colona, M. DiToro, L. G. Moretto and G. J. Wozniak, *Phys. Lett. B* **283** (1992) 180.
5. A. Ono, H. Horiuchi, T. Maruyama and A. Ohnishi, *Progress of Theoretical Physics* **87** (1992) 1185.
6. G. Batko and J. Randrup, *Nucl. Phys. A* **563** (1993) 97.
7. W. Cassing, V. Metag, U. Mosel and K. Niita, *Phys. Reports* **188** (1990) 363.
8. R. J. Charity, M. A. McMahan, G. J. Wozniak, R. J. McDonald, L. G. Moretto, D. G. Sarantites, L. G. Sobotka, G. Guarino, A. Pantaleo, L. Fiore, A. Gobbi and K. D. Hildenbrand, *Nucl. Phys.* **A483** (1988) 371.
9. H. Hauser and H. Feshbach, *Phys. Rev.* **87** (1952) 366.

10. L. G. Moretto, *Nucl. Phys.* **A247** (1971) 211.
11. Bao-An Li, W. Bauer and George F. Beretsch, *Phys. Rev. C* **44** (1991) 2095.
12. D. R. Bowman, C. M. Mader, G. F. Peaslee, W. Bauer, N. Carlin, R. T. de Souza, C. K. Gelbke, W. G. Gong, Y. D. Kim, M. A. Lisa, W. G. Lynch, L. Phair, M. B. Tsang, C. Williams, N. Colonna, K. Hanold, M. A. McMahan, G. J. Wozniak, L. G. Moretto, W. A. Friedman, *Phys. Rev. C* **46** (1992) 8834.
13. L. Phair, W. Bauer and C. K. Gelbke, *Phys. Lett. B* **314** (1993) 271.
14. J. P. Whitfield and N. T. Porile, *Phys. Rev. C* **49** (1994) 304.
15. M. Colonna, N. Colonna, A. Bonasera, and M. DiToro, *Nucl. Phys.* **A541** (1992) 295.
16. R. J. Charity, *Nucl. Phys.* **A483** (1988) 371; **A511** (1990) 59.
17. J. Randrup and G. Batko, *Prog. in Part. and Nucl. Phys.* **30** (1993) 117.
18. V. M. Kolybasov and Yu. N. Solokol'skikh, *Sov. J. Nucl. Phys.* **55** (1992) 1148.
19. P. Roussel-Chomaz, N. Colonna, Y. Blumenfeld, B. Libby, G. F. Peaslee, D. N. Delis, K. Hanold, M. A. McMahan, J. C. Meng, Q. C. Sui, G. J. Wozniak and L. G. Moretto, H. Madani, A. A. Marchetti, A. C. Mignerey, G. Guarino, N. Santoruvo, I. Iori, and S. Bradley, *Nucl. Phys.* **A551** (1993) 508.

8

ANALYSIS OF SMALL ANGLE PARTICLE–PARTICLE CORRELATIONS VIA CLASSICAL TRAJECTORY CALCULATIONS

N. N. Ajitanand

Department of Chemistry, SUNY at Stony Brook
Stony Brook, NY

ABSTRACT

A classical trajectory model, MENEKA, is described which addresses small angle correlations resulting from final state Coulombic interations between pairs of detected particles. The model takes particular care of the details of the detector array geometry and reproduces the observed singles and coincidence spectra. Three different forms of correlation functions are studied for their sensitivity to emission time, emission order and source size. The MENEKA analysis is applied to data from 34 MeV/A ^{40}Ar + ^{107}Ag to obtain the relative emission order for d–Li pairs. A similar analysis for the extraction of emission time information from particle energy-gated data for d–d pairs shows a wide range of emission times from $\sim$ 1000 fm/c at 10 MeV Ecm to $\sim$ 50 fm/c at 40 MeV Ecm.

1. INTRODUCTION

In general, heavy ion reactions pass through three important phases: the initial phase of collisions between nucleons of the projectile and target, the pre-thermal phase dominated by highly anisotropic emission of energetic particles following a small number of early collisions and a thermal phase in which the de-excitation can be best described in terms of a statistical model. At energies well above the Fermi energy this basic scenario is complicated by strong signals of collective effects such as flow and expansion. Since particle emission is a continuous feature of the evolving system, a detailed experimental study of the properties of these particles would be highly desirable. Near 4π detection systems are particularly suited for such studies and, indeed, in recent years, the use of such systems has resulted in detailed measurements of particle multiplicity, energy and angle over a wide range of bombarding energies.

It has been determined from studies of 7–34 MeV/A ^{40}Ar + ^{107}Ag that the detection of a pair of light charged particles at side angles is enough to trigger on high multiplicity

Advances in Nuclear Dynamics
Edited by Wolfgang Bauer and Alice Mignerey, Plenum Press, New York, 1996

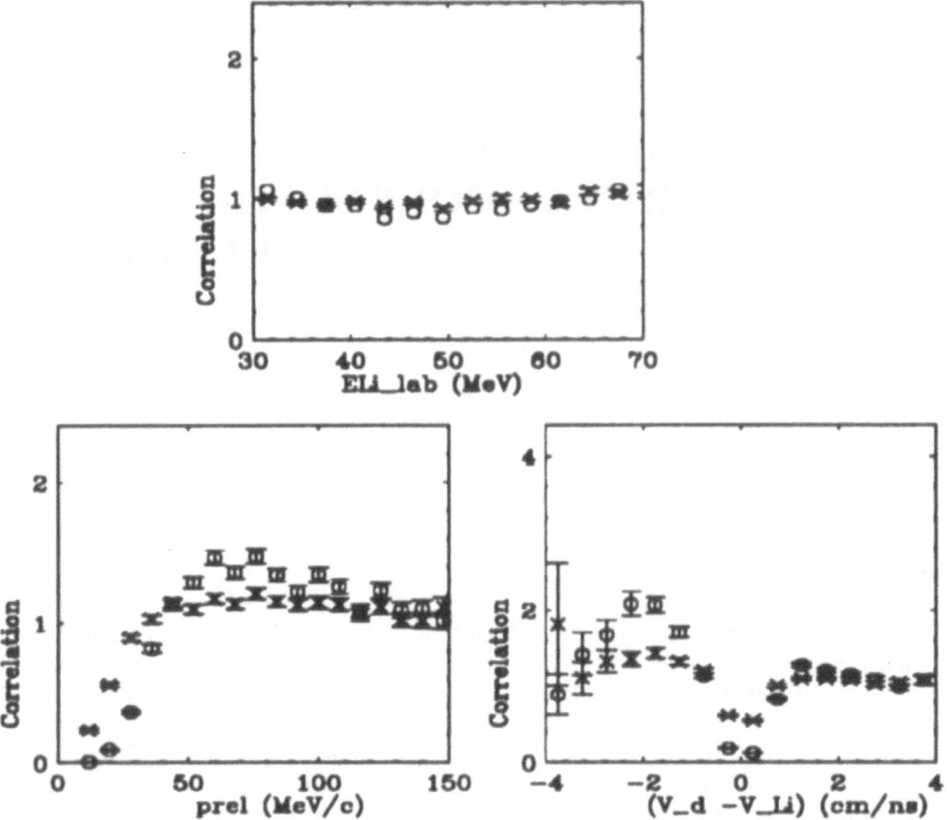

Figure 1. Dependence of calculated correlation functions for d–Li pairs from 27 MeV/A ^{40}Ar + ^{107}Ag on emission life time τ: open circles 60 fm/c, crosses 240 fm/c. Equal emission order, normal source size.

events.[1] By placing a small angle detection array at side angles one can then study small angle correlations between particle pairs from essentially central collision events. It is well known that the analysis of such data can yield important information such as emission time, and effective source size. The procedure followed is to compare the observed correlation with the results of a simulation model which incorporates the final state interations responsible for these correlations. Several such models have been developed in recent years but they have been lacking in terms of accounting for important experimental characteristics such as detector geometry, energy thresholds and observed singles and coincidence spectra.[2,3] In the following, we describe a simulation model MENEKA, which does address these effects.[4] The sensitivity of correlations to emission time, emission order and source size is illustrated for a specific case. New information obtained on the energy dependence of emission time through the application of this model, is also presented.

2. MENEKA — A CLASSICAL TRAJECTORY MODEL

In this model the emitter charge, mass and velocity are known quantities and the two particles of interest are emitted sequentially from the moving source. Briefly, the calculation

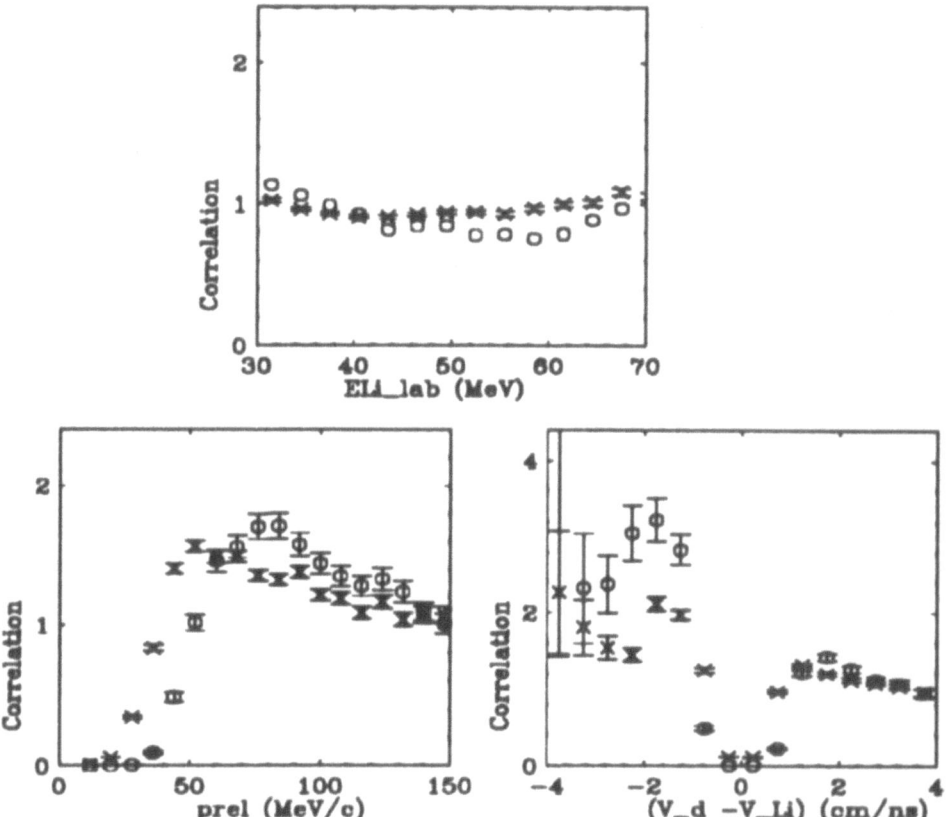

Figure 2. Dependence of calculated correlation functions on source size: open circles $r_0 = 1.2$ fm, crosses $r_0 = 2.4$ fm. $\tau = 6$ fm/c, equal emission order.

is performed in the following manner:

1) A lab direction is chosen for the first particle emission isotropically from a zone which overlaps the detector array.

2) The lab energy of the particle is chosen by natural Monte Carlo sampling of the observed singles spectrum for the detector array.

3) A transformation is made to obtain the velocity and direction of the particle in the emitter frame. The Jacobian of this transformation is termed the geometrical weight wg_1. By adding the residual nucleus recoil energy to the particle energy in the emitter frame one can obtain the channel energy e_1 for the first emission.

4) Corresponding to the energy e_1 one can obtain a grazing orbital angular momentum for emission lgr_1. A choice is then made for the orbital angular momentum l_1 from a triangular distribution going from 0 to lgr_1.

5) The hyperbolic trajectory of the particle corresponding to the chosen e_1 and l_1 (point charge approximation) can then be obtained. The azimuthal orientation of this hyperbolic trajectory is then chosen at random from 0 to 2π.

6) The time t at which the second particle is emitted with respect to the first is sampled from an an exponential emission time distribution characterized by a life time τ.

7) The first particle is released at the point where its hyperbolic trajectory crosses the

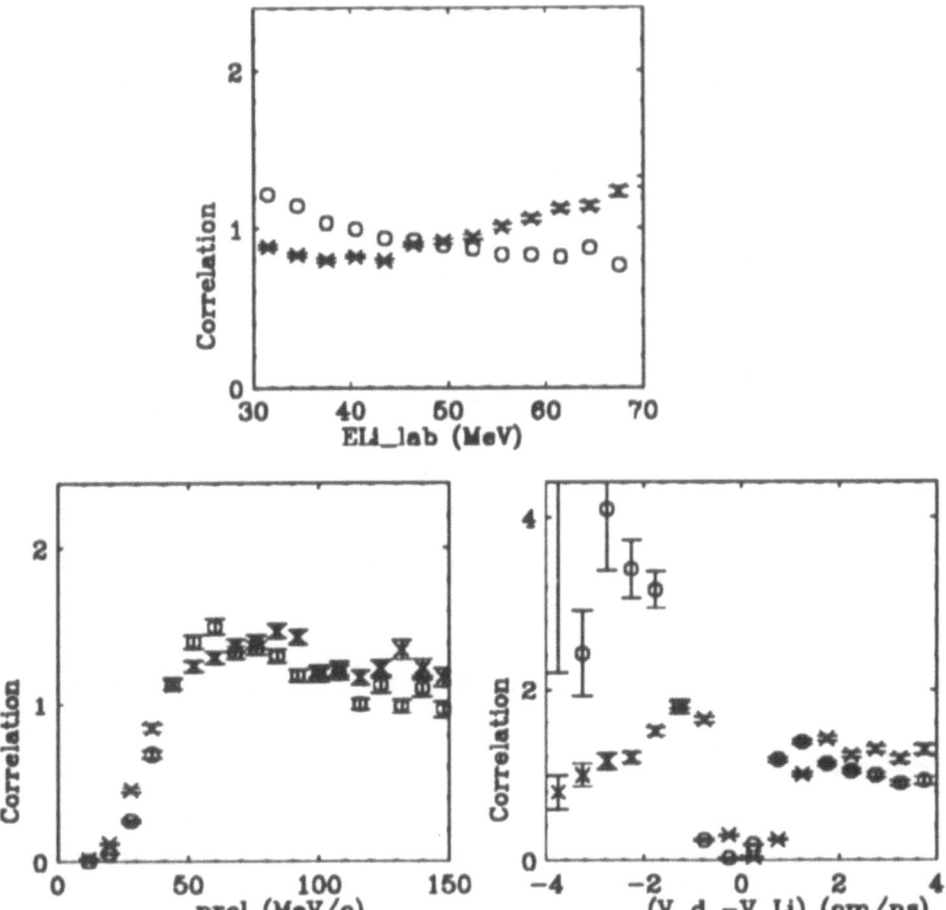

Figure 3. Dependence of calculated correlation functions on emission order: open circles d first, crosses Li first. $\tau = 60$ fm/c, normal source size.

barrier distance of the emitter and the hyperbolic trajectory is followed for a time t when the second particle is to be emitted.

8) The second particle lab direction and lab energy are chosen in the same manner as for the first. The direction in the emitter frame is then obtained which gives the corresponding geometric weight wg_2. As before the channel energy e_2, the orbital angular momentum l_2 and the corresponding hyperbolic trajectory for this emission are obtained. A choice is made for the azimuthal orientation for the second hyperbola which is closest to the orientation of the first hyperbola (a favored orientation for spin driven emission).

9) The second particle is released at the point where its hyperbola crosses the emitter barrier and the three body (two particles and residual nucleus) trajectory is computed to the asymptotic limit. All trajectory calculations are done assuming constant forces operating for a time step small enough to conserve the total energy within defined limits.

10) After checking for the detection of the two particles, the relative momentum (*prel*) of the two particles is calculated for the center points of the struck detectors and binned with a weight $wg_1 * wg_2$.

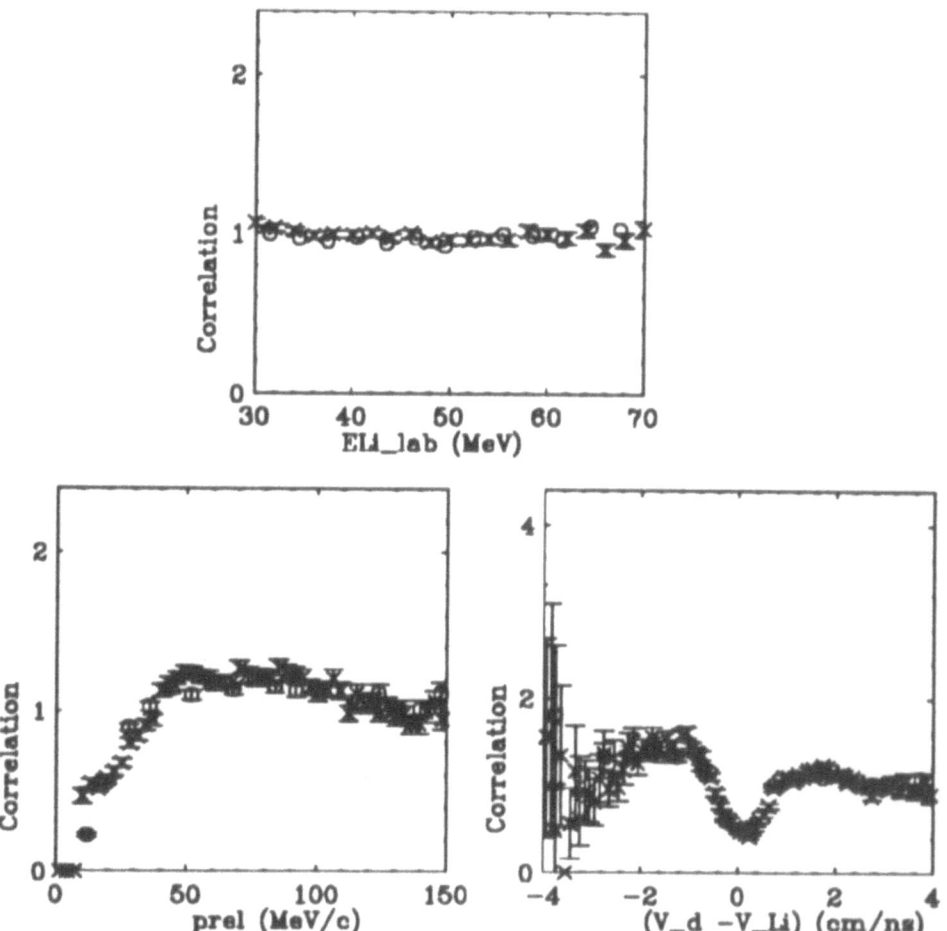

Figure 4. MENEKA fit to observed small angle correlation functions for d–d coincidences from 27 MeV/A ^{40}Ar + ^{107}Ag. τ = 240 fm/c, equal emission order and normal source size.

11) The real *prel* distribution is normalized to the fake *prel* distribution obtained by mixing different pair events and the correlation function (CF) is obtained by taking the ratio of the two distributions.

3. SENSITIVITY TESTS USING MENEKA

MENEKA has been used to calculate three different types of correlation functions between d and Li particles emitted from a system formed in 27 MeV/A ^{40}Ar + ^{107}Ag through 80% linear momentum transfer. These are *prel* correlation, *vdif* correlation and the lithium *elab* correlation. Here *vdif* stands for the difference between the magnitudes of the coincident deuteron and lithium lab velocities. The *elab* correlation is obtained by simply dividing the normalized coincidence and singles lab energy spectra of Li particles. In Figs. 1–3 we show

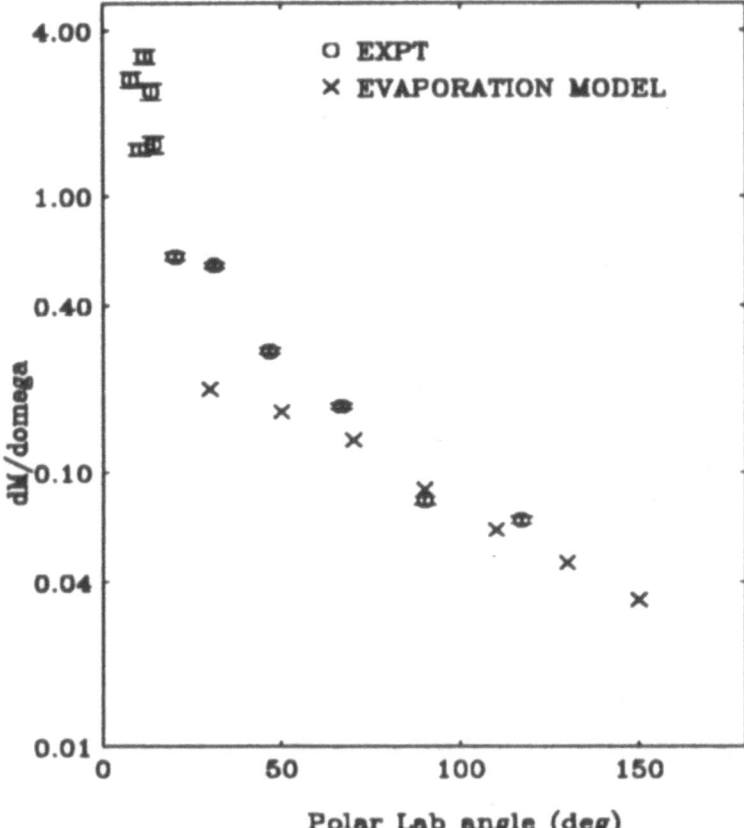

Figure 5. Lab angular distribution of deuterons from 34 MeV/A ^{40}Ar + ^{107}Ag.

the calculated correlation functions obtained in an array of 25 close packed detectors centered at 68° lab with a minimum center to center angle (γ) of 4°. Only coincident events with $\gamma \leq 7°$ are considered for all correlations.

Figure 1 shows the effect of changing emission life time τ from 60 fm/c to 240 fm/c. The order of emission is taken to be equal i.e. a deuteron is as likely to be emitted first as a lithium. It can be seen that the shorter emission life time results in stronger *prel* and *vdif* correlations but no noticeable change in the *elab* correlation. The effect on the *prel* CF is a reduction for low *prel* values and is understandable in terms of the increasing role of the repulsive interation. The effect on the *vdif* CF is seen as a reduction in the near zero *vdif* region and an enhancement in the negative *vdif* region. This behavior is the result of a mixture of the effects of emission time, relative velocity and emission order. A fast particle emitted first tends to escape the interation with the second particle. When the fast particle is emitted second it can catch up with the first and experience a retarding interation lowering its own speed while at the same time increasing the speed of the first particle. The interation can also scatter the particles out of the detection zone.

In Fig. 2 the effect of source size is explored. It was found that the source size change has a noticeable effect only for emission life times which are of the order of the particle

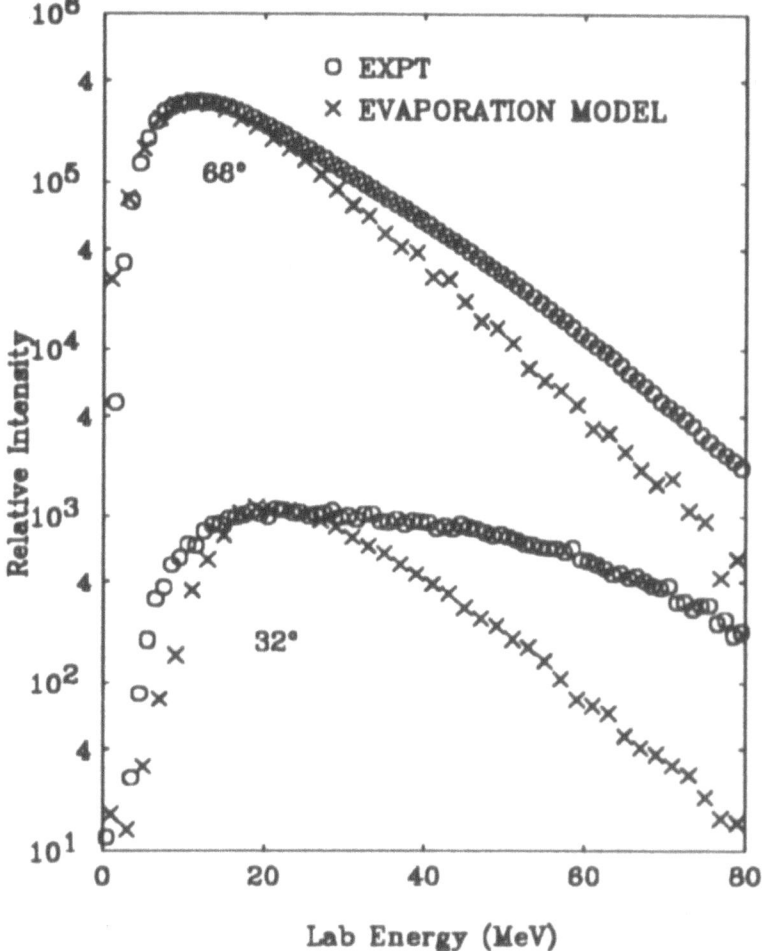

Figure 6. Lab energy distribution of deuterons from 34 MeV/A ^{40}Ar + ^{107}Ag.

traversal times across the emitting system. In Fig. 2 the emission life time has been kept at 6 fm/*c* and the emission order is equal. One sees that doubling the source diameter (i.e. reducing the density eight times) has a visible effect on the small *prel* part of the *prel* CF and the negative part of the *vdif* CF. A small effect can be noticed in the high energy region of the *elab* CF. Source size effects become negligible for emission life times in excess of 50 fm/*c*.

In Fig. 3 the source size is normal (i.e. radius parameter is 1.2 fm), emission life time is kept at 60 fm/*c* and the emission order is changed from deuterons first to lithium first. Whilst there is a strong effect on the negative region of the *vdif* CF, there is a negligible effect on the *prel* CF. There is also a strong effect on the *elab* CF of the heavy particle reflecting the interation induced recoil effect.

In Fig. 4 it is shown that a reasonable fit to the experimental data is obtained for all three correlation functions with τ = 240 fm/*c*, an equal emission order and a normal source size.[6]

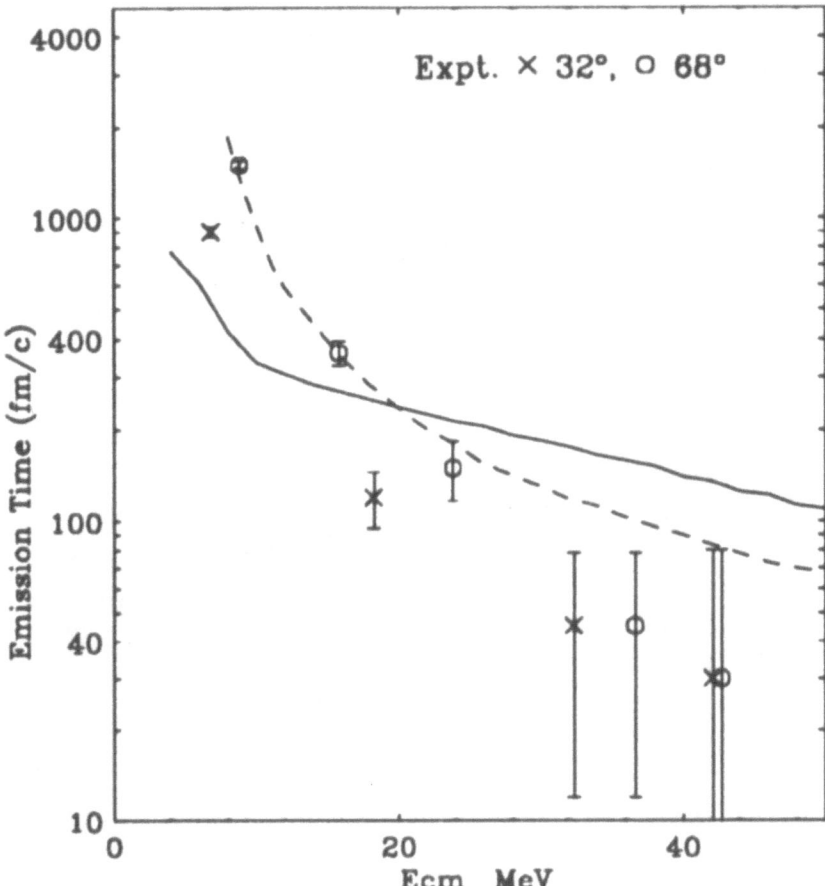

Figure 7. Average emission time as a function of deuteron energy from 34 MeV/A ^{40}Ar + ^{107}Ag. The data points are values inferred from d–d correlation data. The curves are from the EES model, continuous line: with expansion, dash: without expansion.

4. STUDY OF d–d CORRELATIONS USING MENEKA

Figure 5 shows the measured lab angular distributions of deuterons for the 34 MeV/A ^{40}Ar + ^{107}Ag reaction together with a calculated distribution for evaporated particles from a source resulting from 70% linear momentum transfer and a spin deposition of 0–120$\hbar$.[5] Fig. 6 shows the observed singles lab energy spectra of deuterons at 32° and 68° together with the corresponding evaporation calculations. From these figures it is apparent that the 32° spectrum is heavily mixed with pre-thermal emission and even the 68° spectrum is not free from such contamination. The small angle correlation data for coincident d–d pairs obtained at these two angles have been analyzed using MENEKA. The data were divided into different deuteron energy bins and emission life times were extracted for each bin using the MENEKA analysis.[7] The results are shown in Fig. 7. It can be seen that though the 32° data are in general showing smaller emission times, than the 68° data the overall trend is rather similar i.e. the emission times range from over a 1000 fm/c at 10 Mev Ecm to about 50 fm/c at 40 MeV Ecm. These data are compared with the Expanding Evaporating Source model calculations in the figure.[8] The model calculations are indeed able to reproduce the range of emission times however the inclusion of collective expansion produces a smaller energy dependence than observed.

5. CONCLUSIONS

We have used a classical trajectory model MENEKA, to test the sensitivity of different correlation functions (i.e. *prel*, *vdif* and *elab*) to emission time, emission order and source size. The *prel* CF is rather insensitive to emission order, while *vdif* CF shows sensitivity to both emission order and emission time. The single particle CF for *elab* of the heavy partner has very little dependence on emission time and source size but does show a considerable sensitivity to emission order. The MENEKA analysis of small angle *d–d* correlation data obtained in 34 MeV/A ^{40}Ar + ^{107}Ag shows a large variation of emission times with particle energy, a trend which is reproduced by an evaporation model.

This work was supported by the U. S. Department of Energy. Helpful discussions with J. M. Alexander and R. A. Lacey are appreciated.

REFERENCES

1. M. T. Magda et. al. Phys. Rev. C **45**, 1209 (1992).
2. S. E. Koonin, Phys. Lett. **70**, 43 (1977).
3. S. Pratt, Phys. Rev. Lett. **53**, 1219 (1984).
4. A. Elmaani et. Nucl. Instrum. Methods Phys. Res. A **313**, 401 (1992).
5. E. Bauge, thèse, UniversiteJoseph Fourier Grenoble-1 (1994)
6. C. J. Gelderloos, Private Communication (1995)
7. C. J. Gelderloos et. al. submitted to Phys. Rev. C
8. W. A. Friedman, Private Communication (1995)

PRODUCTION OF HOT NUCLEI WITH HIGH-ENERGY PROTONS, ^{3}He AND ANTIPROTONS

B. Lott

GANIL, BP 5027, 14021
Caen Cedex, France

ABSTRACT

Neutron and charged-particle multiplicity distributions have been measured for hot nuclei produced via the bombardment of energetic light projectiles or the annihilation of antiprotons at rest and in flight. The results show that fairly excited nuclei are formed under much better controlled experimental conditions than in the case of heavy-ion collisions. The data are reasonably well accounted for by INC simulations coupled with statistical models.

1. INTRODUCTION

In recent years, heavy-ion collisions have been widely exploited to produce hot nuclei with excitation energies approaching or even possibly exceeding their binding energies. However, several features complicate these experimental studies. The characteristics of these nuclei are indeed often poorly defined since the reaction mechanisms underlying the collisions are still not firmly established. Moreover, the strong excitation of collective modes (rotation, compression, deformation) makes it difficult to unravel the thermal effects from those induced by these modes in the decay pattern of the so-formed nuclei. Motivated by the above considerations, several experiments aiming at exploring alternative ways of producing hot nuclei have recently been undertaken. Both energetic light particles (p and ^{3}He) and antiprotons have been employed to bombard a variety of targets in experiments carried out at the Saturne (Saclay) and LEAR (CERN) facilities respectively. The most important issue addressed by these experiments was the determination of the distribution of excitation energy generated in these collisions. To this end, both neutron and light charged particle multiplicities were measured. In addition for fissile targets, the evolution of the fission probability with the excitation energy was studied as it provides interesting hints on the possible onset of more complex decay processes.

Advances in Nuclear Dynamics
Edited by Wolfgang Bauer and Alice Mignerey, Plenum Press, New York, 1996

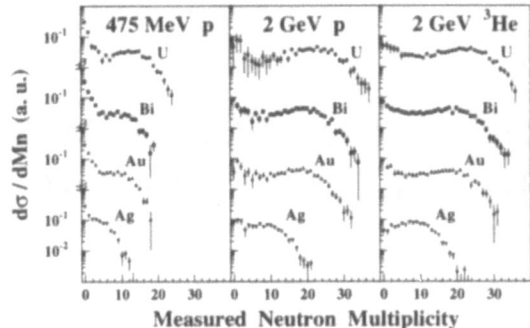

Figure 1. Inclusive multiplicity distributions observed for a variety of targets in 475 MeV p (left), 2 GeV p (middle) and 2 GeV ^{3}He (right) induced reactions.

2. SATURNE EXPERIMENT: 475 MeV, 2 GeV p AND 2 GeV ^{3}He

The collisions involving energetic light particles and nuclei are generally described in terms of a two-step scenario: a fast cascade process, in which the impinging nucleon(s) undergoes a series of incoherent collisions with the target nucleons thereby imparting energy to the nuclear medium, is subsequently followed by a statistical decay of the residual nucleus once the thermalization of the deposited energy is achieved. The first stage leads to the emission of energetic, forward-peaked nucleons and fragments whereas the second stage yields isotropically emitted products with thermal energy spectra. The measurement of the multiplicities of the latter particles allows to assess the amount of heat thermalized within the system after the first-stage nucleon–nucleon collisions have ended.

The neutron multiplicity was measured on an event-by-event basis by means of the 4 m^3 calorimetric detector ORION. This detector housed a reaction chamber accommodating a set of 10 silicon telescopes for the measurement of light charged particles as well as two multi-wire chambers devoted to detecting fission fragments.

Fig. 1 displays inclusive multiplicity distributions measured for different projectiles bombarding a variety of targets.[1] The measurement, performed in the inclusive mode, was triggered by requiring a coincidence between the prompt signal issued by Orion and a signal from a fast plastic positioned upstream from the target and used for tagging the incoming beam particles.

Three features are readily observable in Fig. 1. First, the average neutron multiplicity increases for heavier target nuclei. This behavior stems from two effects. On one hand, the heavier the target nucleus, the larger the number of collisions undergone by the projectile on its way through it, resulting in a higher deposited excitation energy. On the other hand, for a fixed excitation energy the heavier nuclei exhibit a larger propensity to emit neutrons instead of light charged particles because of the large Coulomb barrier experienced by the latter. Second, for all targets, the neutron multiplicities observed with the 2 GeV p beam are about twice as large as those measured with the 475 MeV p beam, indicating a very significant excitation energy increase. Finally, for a given target, the neutron multiplicity distributions for 2 GeV p and 2 GeV ^{3}He induced-reactions look strikingly alike. This finding corroborates the observation of similar mass distributions for residual nuclei produced by p and ^{3}He at the same total bombarding energy, reported in previous studies.[2]

To account for these data, theoretical simulations were performed using the intranuclear cascade model of J. Cugnon,[3] coupled with the statistical model Gemini.[4] In these

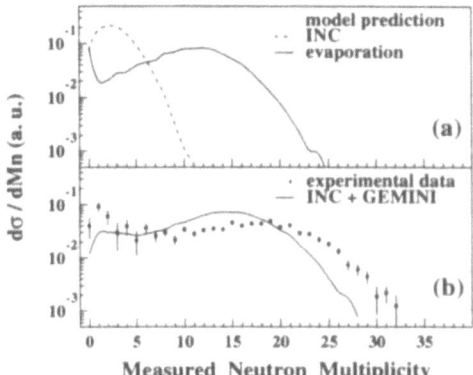

Figure 2. a) Theoretical multiplicity distributions of direct (dashed curve) and evaporated (solid curve) neutrons, folded with the detector efficiency. b) Comparison between the experimental neutron multiplicity measured for 2 GeV p + Au and that predicted by INC + GEMINI simulations.

simulations, the energy thermalization was assumed to be achieved after an elapsed time of 30 fm/c following the first interation. The neutron multiplicity predicted in these simulations was folded with the detection efficiency taking into account the angle and energy of each individual neutron. It is found that the resulting neutron multiplicity distribution reproduces the data reasonably well (Fig. 2b).

Fig. 2a compares the theoretical contributions to the neutron multiplicity corresponding to the two emission stages folded with the detection efficiency. The evaporative component prevails over the direct one very significantly, proving the sensitivity of the measured multiplicity to the excitation energy in the system. The calculations reproduce the differential neutron multiplicity measured for the five individual sectors of ORION equally well, giving confidence in the model reliability in predicting the relative magnitudes of these two neutron components.

To seek further confirmation of the validity of the simulations, the characteristics of the

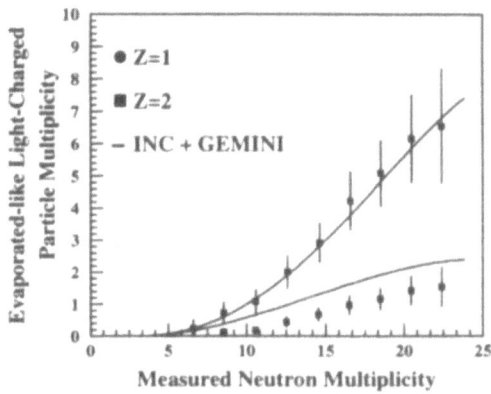

Figure 3. Experimental evaporated multiplicities of H and He as function of the neutron multiplicity, together with simulation predictions (solid lines).

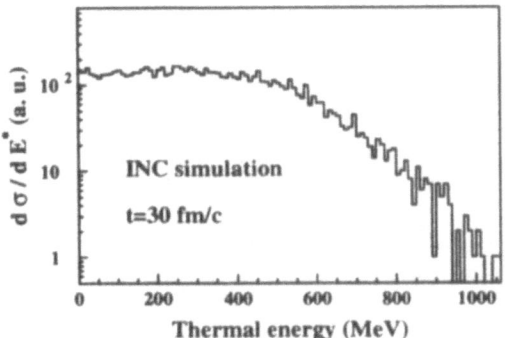

$\frac{d\sigma}{dE^*}$ (a. u.)

INC simulation

t=30 fm/c

Thermal energy (MeV)

Figure 4. Theoretical excitation energy distribution generated in the interation of a 2 GeV p with a Au target.

detected light charged particles were investigated. At backward angles, the α-particle energy spectra gated on large neutron multiplicities were found consistent with an evaporation from a emitter with a temperature around 5 MeV, whereas at more forward angles, the spectra exhibit an increasingly prominent high-energy tail due to direct particles. This temperature of 5 MeV is consistent with the average energy of 500 MeV predicted by the model for the same neutron multiplicity range. The dependence on the measured neutron multiplicity of the observed average H and He multiplicities assessed for the evaporative component is displayed in Fig. 3, and compared with the theoretical predictions.

From the fair agreement between data and results of simulations, the theoretical excitation energy distribution (Fig. 4) can be considered as mirroring the physical one in a realistic fashion. About 15% of the total reaction cross section would thus be associated with excitation energies larger than 500 MeV at the end of the INC stage.

2.1. Decline of Fission

What is the prevalent decay channel at high excitation energy? To shed light on this important issue, the probability of fission was investigated as function of the neutron multiplicity, M_n. Fission is known to be a slow process and observing it is an unambiguous evidence of the system survival as a selfbound object throughout the deexcitation stage. For a given neutron multiplicity, the fission probability was inferred from the yield of coincident fission fragments measured with the set of MWC, corrected for the limited solid angle coverage and normalized to the total reaction yield associated with the same neutron multiplicity, measured in the inclusive mode. Fig. 5 depicts the evolution of the fission probability for increasing neutron multiplicity for 475 MeV p + U and 2 GeV ^{3}He + U (for the former system the probability has been arbitrarily set to 1 for $M_n = 8$). At 475 MeV, the probability rises at low multiplicity and saturates for a large range of neutron multiplicities: as observed in previous studies,[5] fission exhausts the total yield associated with sufficiently large excitation energy. For the 2 GeV ^{3}He-induced reaction, a similar behavior is observed at low multiplicity but the fission probability drops significantly when one moves to large multiplicities. This observation corroborates the earlier finding, obtained with radiochemical methods, that the fission cross section decreases for increasing bombarding energies. This finding was interpreted as an indication for the onset of multifragmentation. The data presented here are not conclusive in this respect. The average IMF multiplicity estimated for events associated with the largest

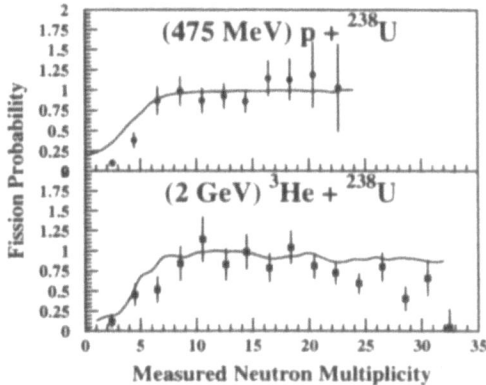

Figure 5. Probability of fission as function of the neutron multiplicity for 475 MeV p + U (top) and 2 GeV ³He + U (bottom). The solid lines are the results of the simulations described in the text.

neutron multiplicities measured in the present experiment is fairly low, about 0.2, seeming to rule out multifragmentation as an important factor in the behavior of the fission probability shown in Fig. 5.

An alternative explanation is that the system ends up as a cold, heavy residue. Its fissility could indeed be strongly reduced through the large mass and charge losses it suffers during both the direct and evaporation stages prior to reaching the saddle point. The production of heavy residues has actually already been observed[6] to gradually replace fission as one of the preeminent decay channels for systems formed in very dissipative heavy ion collisions. Theoretical simulations are currently underway to provide hints on which of these two alternatives is the most credible.

2.2. Conclusion

Using light energetic projectiles seems to be a promising way to produce hot nuclei. In a near future, the products multiplicities and hence the excitation energy generation will be investigated at still higher bombarding energies, for which some interesting data already exist.[7]

It is worth stressing the relevance of the present study to the problems of nuclear waste transmutation and electric power production via the recently proposed[8] hybrid systems.

3. LEAR EXPERIMENTS: ANTIPROTONS AT REST AND 2 GeV/c

3.1. Introduction

The idea motivating the use of antiprotons for producing hot nuclei consists in taking advantage of part of the 1.9 GeV of energy released in the annihilation of the antiproton with one nucleon of the nucleus to heat up the latter. This naive expectation is actually supported by intranuclear cascade models,[9,10] which predict that antiproton annihilation into a nucleus allows to form quite hot systems associated with a moderate loss of direct nucleons and a weak excitation of rotation (average spin lower than $10\hbar$) and compression ($\rho_{max} < 1.3\rho_0$) modes. The antiproton–nucleus interaction proceeds via different mechanisms depending upon the

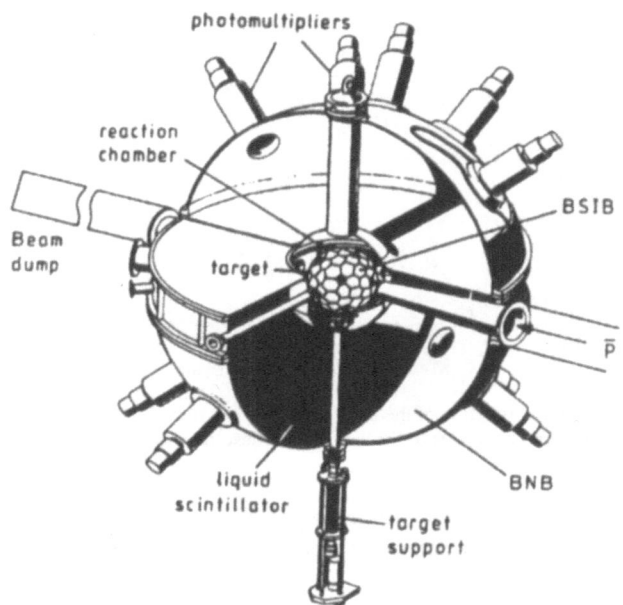

Figure 6. Sketch of the experimental setup used at LEAR

antiproton energy. At rest ($E_{kin} \simeq 10$ keV), the antiproton is captured in remote orbitals, leading to the formation of an antiprotonic atom. The antiproton then cascades inward via a series of X-rays transitions until it annihilates on the surface of the nucleus because of its very short mean free path into the nucleus at that energy. In this process, five pions are released on the average with mean kinetic energy of 210 MeV. This energy makes the pions very suitable to interact with nucleons to form Δ resonances, which is a very efficient way to impart energy to the nucleus. However, the annihilation site is located so far out of the nucleus that the solid angle spanned by the latter as viewed by the pions is very limited. Consequently, most of them actually miss the nucleus and the average excitation energy deposited is very moderate, about 200 MeV as already observed.[11]

At higher bombarding energy, the interation does not give rise to the formation of an antiprotonic atom. The expected energy deposition is larger than that at rest because a larger number of pions interact with the nucleus as the annihilation takes place deeper into the nucleus due to a lower annihilation cross-section at high energy. In addition, the pions are more forward focused because of the Lorentz boost. However, the pion kinetic energy moves progressively away from that of the Δ resonance, causing eventually a decrease in the excitation energy generated within the target nucleus. The optimum bombarding energy for the production of hot nuclei is predicted to lie around 2 GeV.

3.2. Experimental Method

The above arguments motivated a recent experiment at the Low-Energy Antiproton Ring which provides high-quality antiproton beams with energy between 20 MeV and 1.2 GeV (2 GeV/c). A preliminary account of this experiment has already been given elsewhere.[12]

Three runs were performed at rest (by degrading the 20 MeV beam), 635 MeV and 1.2 GeV of bombarding energy. The intensity was limited to $500\bar{p}/s$ for the first run and reached

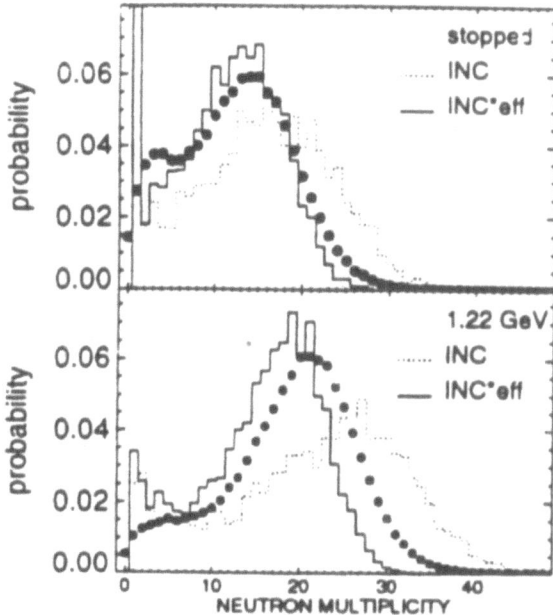

Figure 7. Neutron multiplicity distributions (uncorrected) measured for the annihilation of stopped (top) and 1.22 GeV (bottom) antiprotons on a Gold target together with results of simulation calculations. The solid (dotted) histograms correspond to multiplicities folded (not folded) with the detector efficiency.

about $10^5 \bar{p}/s$ for the two others. The experimental setup comprised mainly two 4π devices, devoted to detecting neutrons, charged particles and fragments (Fig. 6). The Berlin Neutron Ball (BNB) with a volume of 1.5 m³ housed the Berlin Silicon Ball (BSIB), consisting of 157500μm-thick Si-detectors. The BNB measured the neutron multiplicity on an event-by-event basis while the BSIB provided energy and identification for stopped particles with $Z < 2$ as well as a rough identification for heavier fragments. In addition, a set of 16 CsI detectors positioned at backward angles allowed to measure light particles over a large energy range.

3.3. Preliminary Results

Preliminary neutron multiplicity distributions measured for the two extreme bombarding energies with the Au target are displayed in Fig. 7 together with results of theoretical simulations. The experimental neutron multiplicities shown here are uncorrected for spurious contributions related to pion decay or pion secondary interations which could amount to one or two units. Two features of the distributions in Fig. 7 are worth mentioning: there is a significant increase in the average multiplicity when the annihilation takes place "in flight" as compared with the situation at rest, as expected from the above arguments and for both bombarding energies the shapes of the distributions are gaussian-like, contrasting strongly with those observed in heavy-ion or light particle-induced reactions which exhibit a low-multiplicity peak (see Fig. 1). This difference can be ascribed to the much weaker dependence of the energy deposition on the impact parameter for antiproton-induced reactions, since once annihilation occurs a sizable energy is left into the nucleus even for very grazing collisions.

In Fig. 7, a good agreement is observed between the preliminary data and results of

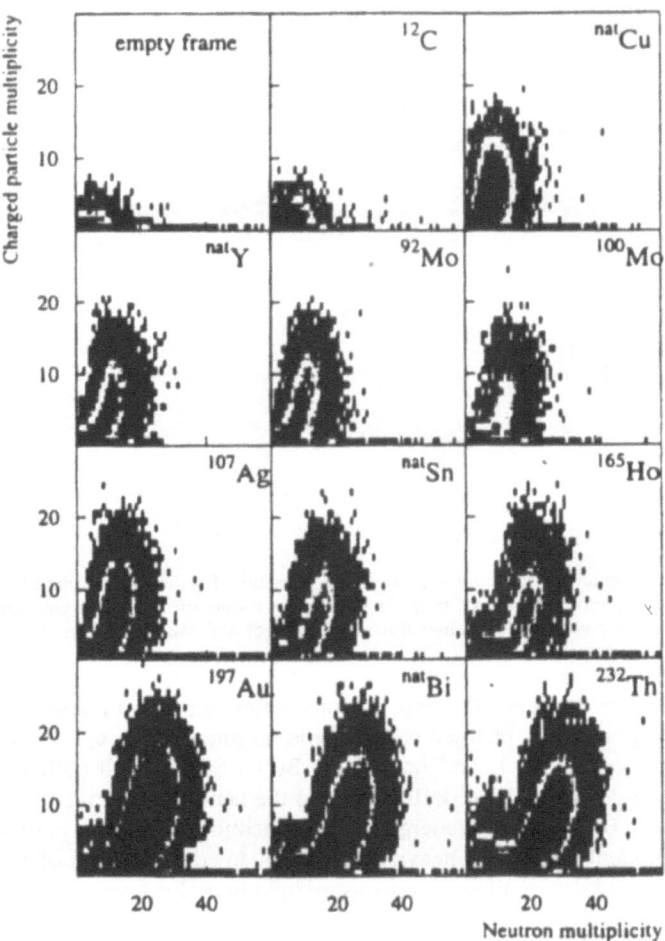

Figure 8. Joint multiplicity distributions measured in the annihilation of 1.2 GeV antiprotons on a variety of targets.

INC simulations,[10] indicating that the exciting expectations raised by numerous theoretical works may actually well be fulfilled.

Fig. 8 displays joint multiplicity distributions measured at 1.2 GeV for a variety of targets. The measured multiplicities increase with the target mass because of the larger number of nucleons the pions can interact with, giving rise to a larger excitation energy in the nucleus. The patterns of these joint distributions are very similar to those previously observed[13] in heavy-ion reactions in the Fermi-energy domain which were shown to be strongly dominated by evaporated particles. For heavy targets, indeed, light charged particle emission sets in only for events associated with a large number of neutrons. This Coulomb-barrier hindrance points thus to evaporation as the dominant emission process. One must emphasize again that in the present case the emitting source is much better defined than in the case of heavy-ion collisions.

Another interesting finding concerns the light charged particle multiplicities observed for the Copper target, which reach values up to 14. As the total charge carried by nucleons is 28 and both p and α-particles are observed, this means that the system basically vaporizes into light fragments for some part of the events. Further analysis is going on to explore whether or not these events exhibit a critical behavior.

3.4. Conclusion

The preliminary results obtained in the LEAR runs seem to confirm the interesting prospects put forward by previous theoretical works. Although the total available energy in the system is much less than in heavy ions collisions, this drawback is largely offset by cleaner experimental conditions making the analysis more straightforward.

ACKNOWLEDGMENTS

The data presented here and their interpretation result from a collective work. The Saturne experiment was performed by a GANIL-HMI Berlin-Saturne-IPN Orsay-KVI Groningen-U. Liège collaboration involving H. G. Bohlen, J. Cugnon, H. Fuchs, J. Galin, D. Guerreau, D. Hilscher, D. Jacquet, U. Jahnke, X. Ledoux, S. Leray, M. Morjean, A. Péghaire, L. Pienkowski, G. Röschert, H. Rossner, R. H. Siemssen and C. Stéphan.

The LEAR PS208 experiment was carried out by a collaboration between HMI Berlin, GANIL, TU-München, U. Warsaw, FZ-Rosendorf, CERN and INR-Moscow. Members of this collaboration are W. Bohne, J. Eades, T. von Egidy, P. Figuera, H. Fuchs, J. Galin, F. Goldenbaum, Ye. Golubeva, K. Gulda, F. J. Hartmann, D. Hilscher, A. S. Iljinov, U. Jahnke, J. Jastrzebski, W. Kurcewicz, M. Morjean, S. Neumaier, G. Pausch, A. Péghaire, L. Pienkowski, D. Polster, S. Proschitzki, B. Quednau, H. Rossner, S. Schmid, W. Schmid and P. Ziem.

REFERENCES

1. L. Pienkowski et al., Phys. Lett. B **43**, 1804 (1994).
2. A. I. Warwik et al., Phys. Rev. C **27**, 1083 (1983).
3. J. Cugnon, Nucl. Phys. A **462**, 751 (1987).
4. R. Charity et al., Nucl. Phys. A **483**, 391 (1988).
5. M. H. Simbel, Z. Phys. A **333**, 177 (1989).
6. E. Schwinn et al., Nucl. Phys. A **568**, 169 (1994).
7. K. Kwiatkowski et al., Contribution to this conference.
8. F. Carminati et al., CERN preprint CERN/AT/93-47 (ET).
9. J. Cugnon et al., Nucl. Phys. A **484**, 542 (1988).
10. Ye.S. Golubeva et al., Nucl. Phys. A **483**, 539 (1988).
11. J. P. Bocquet et al., Z. Phys. A **342**, 183 (1992).
12. D. Hilscher, Invited contribution to the LEAP'94, Third Biennal Conference on Low-Energy Antiproton Physics, Bled, Slovenia, September 12–17, 1994.
13. J. Töke et al., Phys. Rev. Lett. **75**, 2920 (1995).

HARD PHOTON INTENSITY INTERFEROMETRY IN HEAVY-ION COLLISIONS AT INTERMEDIATE ENERGIES

A. Badalà,[1] R. Barbera,[1,2] A. Palmeri,[1] G. S. Pappalardo,[1] F. Riggi,[1,2]
A. C. Russo,[1] G. Russo,[1,3] and R. Turrisi[1]

[1]Istituto Nazionale di Fisica Nucleare, Sez. di Catania
Corso Italia, 57 — I 95129 Catania, Italy
[2]Dipartimento di Fisica dell'Università di Catania
Corso Italia, 57 — I 95129 Catania, Italy
[3]Istituto Nazionale di Fisica Nucleare, Laboratorio Nazionale del Sud
Via S. Sofia, 44 — I 95123 Catania, Italy

ABSTRACT

The technique of intensity interferometry has been applied to the pairs of high-energy photons coming from the ^{36}Ar + ^{27}Al reaction at 95 MeV/nucleon. For the first time, the experimental correlation distributions $C(q_{rel})$ and $C(q_0)$, as functions of the relative momentum and energy of the two detected photons, have been separately analyzed in order to extract both the spatial size and lifetime of the emitting source. The found values are in agreement with dynamical approaches based on the *bremsstrahlung* radiation picture from first-chance proton–neutron collisions.

1. INTRODUCTION

The main features of the huge existing phenomenology concerning hard-photon (E_γ > 25–30 MeV) production in heavy-ion collisions at intermediate energies can be explained within the framework of an incoherent *bremsstrahlung* radiation picture from first-chance single proton–neutron collisions at the early stage of the reaction.[1,2] The chaoticity of the photon source and its strong space–time localization, foreseen by these models,[2] allow and justify, in principle, the use of the intensity interferometry technique[3] as a method of investigation which could provide valuable information, both on the geometry and dynamics of the collision, in an almost model-independent way. Notwithstanding the absence of final state effects, which makes photons ideal probes to study interference effects, very few investigations has been made in this field[4,5,6,7,8,9,10] and the conclusions are far from being clear.

Advances in Nuclear Dynamics
Edited by Wolfgang Bauer and Alice Mignerey, Plenum Press, New York, 1996

This is due to the practical difficulties concerning the experimental detection of high-energy $\gamma - \gamma$ coincidence events, such as the low high-energy-photon production rates and the large background due to the hard component of the cosmic radiation and to the main decay channel of the neutral pion. Furthermore, a complete and unambiguous understanding of the results is strongly dependent on the correct choice and treatment of the correlation observables and on the effects of the experimental filter on the correlation function.

2. THE EXPERIMENT

In a recent work[11] we have established and described, through full GEANT3[12] simulations, a new analysis method of photon intensity interferometry experiments with multidetectors where, for the first time, both the relative momentum, $q_{rel} = |\vec{p}_{\gamma_1} - \vec{p}_{\gamma_2}|$, and energy, $q_0 = |E_{\gamma_1} - E_{\gamma_2}|$, of the two detected photons, are simultaneously used to extract the spatial size and lifetime of the emitting source.

In this contribution we report on an experiment where this new technique has been applied to the study of the correlations at small relative momenta between high-energy gamma rays (E_{γ_1}, $E_{\gamma_2} > 30$ MeV) emitted in the ^{36}Ar + ^{27}Al reaction at 95 MeV/nucleon. Photons have been detected by the BaF$_2$ ball of the MEDEA multidetector consisting of 144 modules of barium fluoride (20 cm thick) arranged in six rings covering the whole solid angle between $\theta = 40°$ and $\theta = 140°$. A detailed description of this multidetector can be found in Ref. 13. Photons have been separated from charged particles by means of the usual pulse-shape analysis technique and from neutrons using the time-of-flight information. For the energy reconstruction of the recorded photons, it has been used a detection technique which considers, besides the two modules having the maximum deposited energy (the two "most touched" detectors), two cluster configurations of nine packed crystals (the eight around each of them) in order to recover the sideward leakages and, hence, to improve the collection process of the electromagnetic shower. The properties of the BaF$_2$ ball of the MEDEA multidetector as a photon spectrometer have been already extensively studied in Refs. 13, 14 and experimentally tested in Refs. 11, 13, 15, 16.

Particular care has been devoted to the identification of neutral pions and to the estimation of possible background sources as cosmic rays and e^+e^- pairs coming from γ-conversion. Neutral pions are recorded through the simultaneous detection of the couples of photons coming from their main decay mode. These photons are separated by imposing severe conditions on the experimental distributions of the relative angle, θ_{12}, and invariant mass, m_{inv}, as functions of the total energy, $E_1 + E_2$, of the two detected photons (see Fig. 1, upper panel and lower panel, respectively). The cuts reported in Fig. 1 select those photons coming from π^0 decay and derive from the results of full GEANT3 simulations performed to determine the detector efficiency. A complete study of the MEDEA multidetector as a neutral pion spectrometer is reported in Ref. 14 and first experimental results are reported in Refs. 17, 18.

Taking into account the absolute cosmic ray intensity at sea level,[19] the expected cosmic rate for the BaF$_2$ ball of MEDEA is about 200 Hz (without correction for the geomagnetic latitude of the experimental site). In order to reduce this large number we imposed several conditions in the off-line analysis of two-photon events. Only those events where the two photons were detected in coincidence with at least two charged particles have been kept for further analysis. Moreover, the two photons must be recorded by two different clusters of detectors, i.e. $\theta_{12} \geq 15°$. Considering (i) that the beam current was limited to have not more than 10^4 reactions per second (in order to avoid undesired pile-up effects with BaF$_2$ detectors

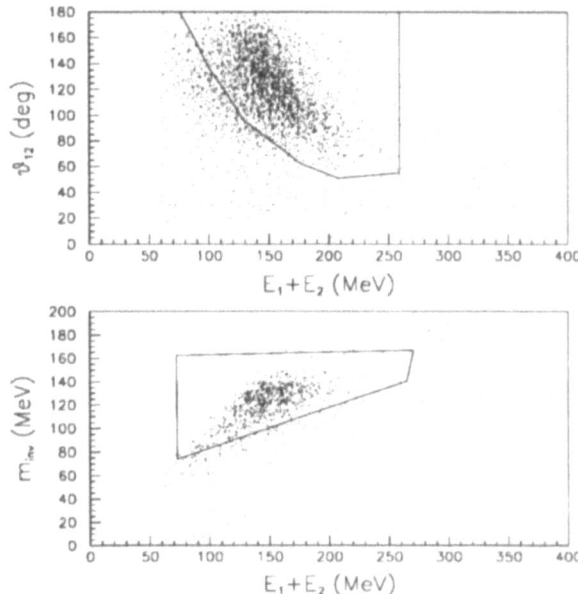

Figure 1. Relative angle (upper panel) and invariant mass (lower panel) versus total energy distributions of the pairs of photons detected in the ^{36}Ar + ^{27}Al reaction at 95 MeV/nucleon. In both plots, the contours define those pairs of photons coming from π^0 decay.

which have typical recovery times in the order of 1 μs), (ii) that the coincidence window was set to about 100 ns, (iii) that, because of the time resolution of the BaF$_2$ modules, all photons coming from the target are gathered in about 2 ns, and (iv) that the global efficiency of the BaF$_2$ ball for cosmic muons originating two distinct showers, each having an energy larger than 30 MeV, is 25% on the average,[11] the expected cosmic rate decreases to about $3 \cdot 10^{-4}$ Hz which means, taking into account the total run time, less than 10 two-photons events generated by cosmic rays. This number has to be compared with 16888 two-high-energy-photon total recorded events. In any case, as it was shown in Ref. 11 by means of full GEANT3 simulations, these special pairs of photons usually concern the high part of the relative momentum spectrum and, hence, can not influence the photon–photon correlation signal. Even less important is the background induced by the e^+e^- pairs coming from γ-conversion. Due to the fact that the MEDEA multidetector operates in a vacuum, the only possible sources of γ-conversion are the target and a thin layer of teflon and aluminized Mylar, 200 μm thick, which wraps each module.[13] The pair-production cross section[12] for photons having an energy greater than 60 MeV (the minimum one to produce two 30-MeV showers) takes values in the range of 0.01–1 μb for all atomic numbers. This means that, taking into account the thickness of the used target (1.6 mg/cm^2) and the measured yield of more-than-60 MeV photons, the probability that a e^+e^- pair can be produced inside the target is about 10^{-8}. The number of e^+e^- pairs generated in the layers of teflon and aluminized Mylar could be considerably larger but, since the two particles are produced very close to one (and only one) scintillator, the aforementioned condition on the minimum relative detection angle reduces the probability that γ-conversion could originate two distinct showers near to zero (see Refs. 9, 10).

Before to study two-γ events, inclusive photon data have been submitted to a moving source analysis in order to compare our results with the existing systematics. Hard-photon

$(E_\gamma > 30$ MeV) spectra at different laboratory angles, cleaned of neutral pion contamination (by GEANT3 simulations), have been simultaneously fitted with the usual parameterization[20] assuming an exponential energy spectrum and an isotropic + dipolar angular distribution in the source frame. The fitting procedure gives a photon source velocity $\beta = 0.19 \pm 0.01$ and a dipolar term $\alpha = 0.23 \pm 0.01$ (in fair agreement with the *p–n bremsstrahlung* picture), and an inverse slope parameter $E_0 = 28.9 \pm 0.3$ MeV. The total measured hard-photon cross section is $\sigma_\gamma = 1.22 \pm 0.01$ mb[16] corresponding to an hard-photon probability per *p–n* collision $P_\gamma = \sigma_\gamma/(\sigma_R \langle N_{pn} \rangle_b) = (1.37 \pm 0.01) \cdot 10^{-4}$.

3. THE PHOTON–PHOTON CORRELATION FUNCTION

In order to extract from two-photon data the values of the spatial size and lifetime of the hard-photon emitting source, we have adopted, as correlation observables, the relative momentum, q_{rel}, and energy, q_0, of the two detected photons. Up to now, because of the Lorentz-invariant definition of the correlation function,[21,22] all other authors[4,5,6,7,8,9,10] have used the invariant transferred momentum, $q_{inv} = \sqrt{q_{rel}^2 - q_0^2}$, as correlation observable. As it has been already observed,[11,22] q_{inv} behaves as an effective relative momentum between the two photons only in their mutual center-of-mass frame. The spatial parameter conjugate to q_{inv} does not represent the radius of the emitting source, but only the width of the distribution of relative separations in the source itself. This parameter necessarily contains an ensemble average over all the distribution of the pair velocities used to measure the correlation function and, hence, the relationship with the real source size is dependent on the distribution of the accepted pairs for a given experimental set-up (usually the extracted parameter is larger than the real one[22]).

The relative momentum distribution of all detected two-photon events is reported in Figure 2. Filled dots refer to those couples of photons coming from π^0 decay, i.e. standing inside the contours drawn in Fig. 1, while open dots refer to those couples of photons standing outside the contours drawn in Fig. 1. These latter pairs of photons, producing the large bump centered around 80–90 MeV/c, are mainly ($\sim 90\%$) relative to "badly" detected pions and, with an estimated relative probability of about 10%, to $\pi^0 - \gamma$ events (where a photon coming from the emitting source and only one of the two photons coming from the π^0 decay are detected) as it has been foreseen in Ref. 11 by GEANT3 simulations. True $\gamma - \gamma$ events are expected to place themselves in the lowest part of the relative momentum spectrum. The absence of values of q_{rel} lower than about 10 MeV/c is due to the limitation imposed both in the minimum energy and relative angle of the two detected photons.

In order to evaluate the correlation function, the relative momentum distribution plotted in Fig. 2 for two-photons events has been divided by that obtained folding the inclusive photon events through the event mixing technique.[23] The normalization factor takes explicitly into account the different run-times of the two sets of events. The correlation distribution $1 + C(q_{rel})$, as a function of the relative momentum, is reported in the upper panel of Fig. 3. Three contributions are visible. A peak, due to neutral pions (centered around 140 MeV/c), a large bump, due to those "badly" measured pion-events discussed before (between 50 and 150 MeV/c), and the photon–photon correlation signal (for $q_{rel} < 45$ MeV/c). The continuous line passing through the points refers to the result of a best-fit procedure using the sum of three gaussian functions ($c = 1$):

$$C(q_{rel}) = \lambda_{R'} \, e^{-\frac{q_{rel}^2 R'^2}{\hbar^2}} + A_1 \, e^{-\frac{(q_{rel}-q_1)^2}{2\sigma_1^2}} + A_2 \, e^{-\frac{(q_{rel}-q_2)^2}{2\sigma_2^2}} \ . \tag{3.1}$$

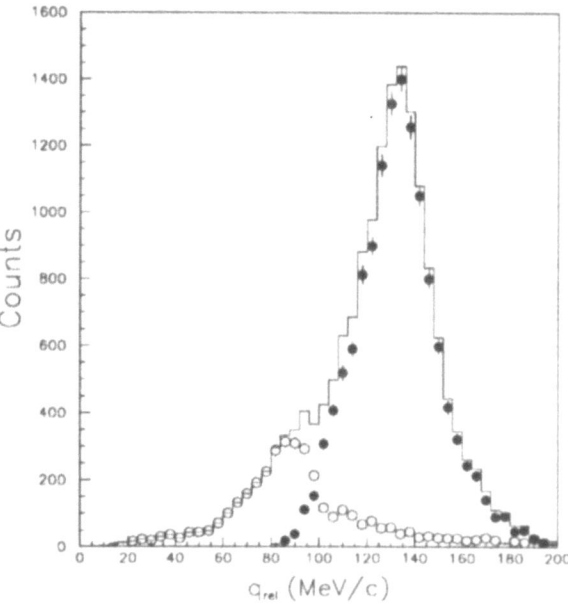

Figure 2. Relative momentum distribution of two-photon events detected in the ^{36}Ar + ^{27}Al reaction at 95 MeV/nucleon (continuous histogram). Filled dots refer to those pairs of photons coming from π^0 decay, i.e., standing inside the contours drawn in Fig. 1, while open dots refer to those pairs of photons standing outside.

one for each component. R' is the width of the spatial correlation signal.[11,21,22] Other curves in the same figure refer to each of the three contributions. For the sake of correctness, the first two points in the correlation distribution, having $q_{rel} < 15$ MeV/c, have been not included in the fitting because the evaluated detector efficiency for this correlation function, reported in the lower panel of Fig. 3, shows a steep fall-off to zero below that value of the relative momentum.

Using the parameterization of Eq. (1), it has been possible to estimate a two-hard-photon total cross section $\sigma_{\gamma\gamma} = 1.19 \pm 0.24$ μb in fair agreement with the expected value $\sigma_R \langle N_{pn} \rangle_{\gamma\gamma} P_\gamma^2 = 0.93 \pm 0.02$ μb where, for our system, $\langle N_{pn} \rangle_{\gamma\gamma} \equiv \langle N_{pn}(N_{pn} - 1) \rangle_b = 17.2$.[26] This value of the two-hard-photon cross section, coupled with that of σ_γ, makes negligible any eventual spurious contribution coming from internal γ-conversion (or virtual *bremsstrahlung*) where a photon is produced by the interation of protons coming from the reaction with nucleons of the target and/or of the layer of teflon and aluminized Mylar. In fact, taking into account that the mean kinetic energy of detected protons is about 40–50 MeV and that the corresponding *bremsstrahlung* cross section σ_γ^p is less than 1 μb,[27] the expected cross section of the process where one photon comes from the reaction and the other one is created by a proton is $10^3 - 10^4$ times smaller than $\sigma_{\gamma\gamma}$ (the probability of the process where both photons are created by protons is still smaller).

Upper panel of Figure 4 shows the correlation distribution $C(q_0)$, as a function of the relative energy, plotted only for those couples of photons having $q_{rel} < 45$ MeV/c. The continuous curve is the result of a best-fit procedure with the function ($c = 1$):

$$C(q_0) = \lambda_{\tau'}\, e^{-\frac{q_0^2 \tau'^2}{\hbar^2}}$$

(3.2)

where τ' is the width of the temporal correlation signal.[11,21,22] In this case all the points have

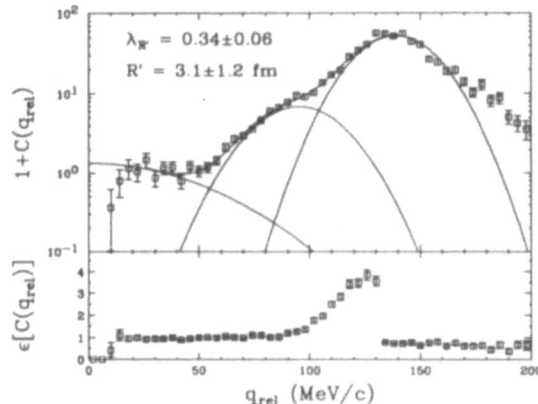

Figure 3. Upper panel: correlation distribution $1 + C(q_{rel})$, as a function of the relative momentum, for the pairs of photons detected in the ^{36}Ar + ^{27}Al reaction at 95 MeV/nucleon. Continuous curves are relative to the result of a best-fit procedure (see text). Lower panel: corresponding detector efficiency.

been included in the fitting due to the shape of the evaluated detector efficiency for this correlation function, reported in the lower panel of the same figure. It is worth noting that the extracted values of the so-called λ-parameters are not so far from that of 0.5 expected for a completely incoherent emitting source,[21] even if we are aware that (i) the true physical meaning of such experimental quantity is still an open question in the framework of quantum statistics,[24,25] and (ii) the interference between the two photon-polarization vectors always acts to reduce the values of $C(q_{rel} = 0)$ and $C(q_0 = 0)$.[11]

Furthermore, the distortion effect due to the interference between the polarizations of the two emitted photons, always present in the correlation function,[21] and that due to the detector efficiency, make R' and τ' different from the real source parameters. As it has been pointed out in Ref. 11, these effects are also not simply additive and operate at the same time in a very complicated way both depending on the source parameters and detector geometry. Only by complete computer simulations it is possible to find the relations between those widths and the true R and τ. These simulations, relative to the BaF$_2$ ball of the MEDEA multidetector, have been performed in Ref.[11] using GEANT3, and the result was that it is possible to successfully parameterize the dependence of R' and τ' on R and τ through simple relations $R' = a(\tau)R + b(\tau)$ and $\tau' = c(R)\tau + d(R)$, where the coefficients a and b are linear functions of τ, and c and d are linear functions of R. Inverting these equations and substituting the values of the widths R' and τ', extracted from the best-fit procedure, we found $R = 1$ fm and $\tau = 4$ fm/c with a global statistical error of about 50%. As it has already been pointed out,[11,22] R and τ represent only the σ-values of the assumed space–time source distribution and they are connected to their rms-values by the relations $\sqrt{\langle \vec{r} \cdot \vec{r} \rangle} = \sqrt{3}R$ and $\sqrt{\langle t^2 \rangle} = \tau$. Using the geometrical estimation[26] of the average number of participant nucleons producing two hard-photons, $\langle N_{pn} \rangle_{\gamma\gamma}$, we get a gaussian radius[28] of the two-nuclei overlap zone of $R_{ov} = \sqrt{2/5}\,(1.2\langle N_{pn} \rangle_{\gamma\gamma}^{1/3}) = 1.96$ fm which is in good agreement with the experimentally extracted spatial rms-value of 1.73 ± 0.86 fm. These small values of both R and τ can also be explained within microscopic dynamical phase-space approaches[2] which predict hard-photon production by first-chance proton–neutron collisions in the overlap zone of the two interacting nuclei at the early stage of the reaction.

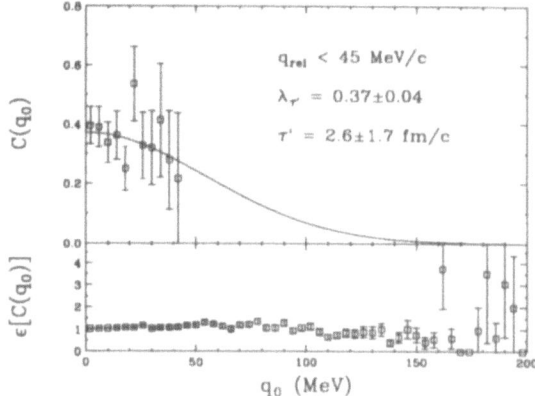

Figure 4. Upper panel: correlation distribution $C(q_0)$, as a function of the relative energy, for the pairs of photons detected in the ^{36}Ar + ^{27}Al reaction at 95 MeV/nucleon and having $q_{rel} \lesssim 45$ MeV/c. Continuous curve is relative to the result of a best-fit procedure (see text). Lower panel: corresponding detector efficiency.

REFERENCES

1. H. Nifenecker and J. A. Pinston, Prog. Part. Phys. **23**, 271 (1989) and references therein.
2. W. Cassing, V. Metag, U. Mosel, and K. Niita, Phys. Rep. **188**, 363 (1990) and references therein.
3. For a review, see, e.g.: *Proceedings of the Interational Workshop on Particle Correlation and Interferometry in Nuclear Collisions*, Nantes, France, 1990, edited by D. Ardouin (World Scientific, Singapore, 1990); D. H. Boal, C. K. Gelbke, and B. J. Jennings, Rev. Mod. Phys. **62**, 553 (1990), and references therein; W. Bauer, C. K. Gelbke, and S. Pratt, Annu. Rev. Nucl. Part. Sci. **42**, 77 (1992), and references therein.
4. R. Ostendorf, in *Proceedings of the Workshop on Physics Related to TAPS*, Schiermonnikoog, Denmark, 1990, edited by W. Kühn and H. Löhner, GSI Report GSI-91-21, 1991, p. 127.
5. R. Ostendorf *et al.*, Nouvelles du GANIL **38**, 16 (1991).
6. M. Marques, P. Lautridou, G. Martinez, T. Matulewicz, R. Ostendorf, J. Quebert, and Y. Schutz, Nouvelles du GANIL **47**, 33 (1993).
7. Y. Schutz, in *Proceedings of the XXIII Mazurian Lakes Summer School on Nuclear Physics*, Piaski, Poland, 1993, GANIL Preprint P 93 26.
8. R. W. Ostendorf *et al.*, Nuclear Physics at GANIL, Compilation 1992–1993, edited by M. Bex and J. Galin, 1994, p. 183.
9. M. Marques *et al.*, Nuclear Physics at GANIL, Compilation 1992–1993, edited by M. Bex and J. Galin, 1994, p. 186.
10. M. Marques *et al.*, Phys. Rev. Lett. **73**, 34 (1994).
11. A. Badalà, R. Barbera, A. Palmeri, G. S. Pappalardo, F. Riggi, A. C. Russo, G. Russo, and R. Turrisi, Nucl. Instrum. Methods A **351**, 387 (1994).
12. CERN Application Software Group, *GEANT: Detector Description and Simulation Tool* (CERN, Geneva, 1993); CERN Program Library Long Writeups W5013.
13. E. Migneco *et al.*, Nucl. Instrum. Methods A **314**, 31 (1992).
14. A. Badalà, R. Barbera, A. Palmeri, G. S. Pappalardo, F. Riggi, and A. C. Russo, Nucl. Instrum. Methods A **306**, 283 (1991).
15. E. Migneco *et al.*, Phys. Lett. B **298**, 46 (1993).
16. A. Badalà, R. Barbera, A. Palmeri, G. S. Pappalardo, F. Riggi, A. C. Russo, G. Russo, and R. Turrisi, in preparation.
17. A. Badalà *et al.*, Phys. Rev. C **47**, 231 (1993).
18. A. Badalà *et al.*, Phys. Rev. C **48**, 2350 (1993).
19. O. C. Allkofer, K. Carstensen, and W. D. Dau, Phys. Lett. B **36**, 425 (1971).
20. R. Merrouch *et al.*, Nouvelles du GANIL **38**, 4 (1993).
21. D. Neuhauser, Phys. Lett. B **182**, 289 (1986).

22. W. A. Zajc, in *Proceedings of a NATO Advanced Study Institute on Particle Production in Highly Excited Matter*, IL Ciocco, Italy, 1992, edited by H. H. Gutbrod and J. Rafelski (Plenum Press, New York, 1993).
23. D. Drijard, H. G. Fischer, and T. Nakada, Nucl. Instrum. Methods A **225**, 367 (1984).
24. R. M. Weiner, Phys. Lett. B **232**, 278 (1989).
25. W. A. Zajc, Nucl. Phys. A **525**, 315c (1991).
26. H. Nifenecker and J. P. Bondorf, Nucl. Phys. A **442**, 478 (1985).
27. M. Schäfer, T. S. Biró, W. Cassing, U. Mosel, H. Nifenecker, and J. A. Pinston, Z. Phys. A **339**, 391 (1991).
28. W. A. Zajc *et al.*, Phys. Rev. C **29**, 2173 (1984).

11

HEAVY RESONANCE PRODUCTION IN ULTRARELATIVISTIC NUCLEAR COLLISIONS

D. Seibert

Department of Physics
McGill University
Montreal, QC H3A 2T8, CANADA

ABSTRACT

Are heavy quarks produced thermally or only by hard parton collisions? What is the probability that a produced heavy quark or antiquark is observed in a given resonance?

1. INTRODUCTION

In this talk, I discuss the problems of heavy quark production and of heavy resonance production, i.e., are heavy quarks produced thermally or only by hard parton collisions, and what is the probability that a produced heavy quark or antiquark is observed in a given resonance? My collaborators are Tanguy Altherr at CERN,[1] who has since died tragically in a climbing accident, and George Fai at Kent State University.[2]

We use very simple ultrarelativistic nuclear scenarios, basically as proposed long ago by Bjorken.[3] We assume that the collisions can be divided into the following periods:

1. hard parton collisions, in which essentially all of the final-state entropy is produced, for which the typical time scale is about 0.1 fm (we use the standard high energy conventions that $\hbar = c = k_B = 1$);

2. thermal equilibration,[4] which ends after 0.2–0.3 fm, and possibly even sooner[5];

3. expansion of thermally equilibrated gluon plasma[4] (GP), or possibly quark gluon plasma (QGP);

4. phase transition, when the temperature $T \simeq 1$ fm^{-1};

5. resonance gas (RG) expansion; and

6. freezeout, after which the particles leave the hot matter and flow conveniently to the detectors.

Advances in Nuclear Dynamics
Edited by Wolfgang Bauer and Alice Mignerey, Plenum Press, New York, 1996

Of course, it is possible that the process of thermal equilibration will coincide with some of the later equilibrium processes, but we believe that this will produce only small changes to most of our results. It is also possible that there is no phase transition but instead some fast but smooth change to the entropy and energy densities as a function of T, but again that should not produce large changes in our results.

We make the following approximations to model the dynamics. First, we assume that the hot matter is cylindrically symmetric, so

$$f(p_x, p_y, p_z) = f(p_T = \sqrt{p_x^2 + p_y^2}, p_z), \tag{1.1}$$

where f is any distribution function.

Second, we assume approximate boost-invariance as observed in proton collisions,[3]

$$f(\tau, p_T, y) = f(\tau, p_T). \tag{1.2}$$

Finally, we assume that the hot matter does not expand transversely during the collision, so that the volume at proper time τ is

$$\frac{dV(\tau)}{dy} = \pi R^2 \tau, \tag{1.3}$$

where R is the nuclear radius.

We also occasionally assume approximate entropy conservation,

$$\pi R^2 \tau s \simeq 3.6 dN/dy, \tag{1.4}$$

where s is the entropy density and dN/dy is the number of produced particles per unit rapidity. In that case, we treat the hot matter as an ideal gas with zero chemical potential, so

$$s(T) = \sum_i \frac{g_i}{(2\pi)^3 T} \int d^3 p \frac{E_i(p) + k^2/3E_i(p)}{e^{E_i(p)/T} \pm 1}. \tag{1.5}$$

Combining Eqs. (1.4) and (1.5), we then estimate T as a function of τ for given dN/dy.

2. HEAVY QUARK PRODUCTION

In this section I review the theory of heavy quark production. There have been three types of calculations of heavy quark production in ultrarelativistic heavy ion collisions. The initial production cross sections have been estimated from perturbative quantum chromodynamics (QCD), either by simply taking cross sections for proton collisions[6] and scaling by A^2, where A is the number of nucleons in the nucleus, or in a more sophisticated manner by using the nuclear (instead of the bare nucleon) parton structure functions.[7,8] The results obtained with these two methods do not differ very much.

Thermal production cross sections have been calculated by taking the production matrix elements, convoluting with the four-volume element and the thermal distribution functions, and integrating over assumed collision histories.[9,10,11,12,1] These thermal calculations have grown in sophistication with time, but the results have not changed very much. Our major addition to these calculations has been the inclusion of heavy quark production through thermal gluon decay, which has been previously neglected but is the dominant term in the weak coupling limit,[1] due to the anomalously large thermal gluon width.[13] For typical values of the strong coupling constant probed in nuclear collisions, this new production term is comparable to those previously calculated, although it does not dominate.[14]

Finally, there is the parton cascade model,[5,15,16] which attempts to include both of the previous calculations by using a perturbative QCD cascade, in which interations are more or less arbitrarily cut off at some lower momentum so that cross sections remain finite. This model seems to work reasonably well for strange (s) quark production, which at least seems to be correctly predicted for proton collisions and scales with projectile and target in a reasonable manner. Results for charm (c) and bottom (b) production are probably not reliable, since proton collision results are not reproduced and the cross sections scale as $A^{5/3}$ in proton-nucleus collisions, instead of A as observed by experimenters.

In my opinion, the best density estimates and dominant quark production mechanisms for an Au + Au collision at RHIC energy ($\sqrt{s} = 200$ GeV/nucleon) are:

s: Pure thermal production gives $dN_s/dy \simeq 100$, the parton cascade model gives $dN_s/dy \simeq 50$, so I expect that $dN_s/dy \simeq 100$, mostly from thermal collisions.

c: Thermal production gives $dN_c/dy \simeq 1$, perturbative QCD gives $dN_c/dy \simeq 1$, so I expect that $dN_c/dy \simeq 1$ with roughly equal production from hard parton collisions and thermal collisions.

b: Thermal production gives $dN_b/dy < 10^{-3}$, perturbative QCD gives $dN_b/dy \simeq 0.02$, so I expect $dN_b/dy \simeq 0.02$ mostly from hard parton collisions.

3. FREEZEOUT CONDITIONS

Once we have our heavy quarks, the next question is, "When do they freeze out?" We attempted to solve this problem with a simple model of heavy quark production and dynamics in an ultrarelativistic nuclear collision.[2] We first assume that all heavy quark–antiquark pairs are produced at time $\tau = 0$; this is reasonable for c and b quarks even if they are produced thermally, since thermal production is also concentrated at early times, although it is not such a good assumption for s quarks. We then assume that these quarks quickly thermalize, and that their subsequent trajectory is a random walk in a thermal bath, with collision frequencies taken from Pisarski[13]; the bath temperature is estimated by assuming entropy conservation. Finally, we say that the quarks have frozen out when either (i) their mean free path is larger than the mean distance to leave the hot matter, or (ii) their random walk has taken them out of the hot matter.

This is of course complicated by the fact that we expect a phase transition to RG during the evolution of the hot matter. We take this into account by assuming that the interation cross section per unit entropy is approximately the same in the two phases, and that in the RG phase the quarks are contained in mesons which we model as $q\bar{q}$ pairs to estimate the mesonic mean free paths. We thus obtain different results by varying the transition temperature, T_c, and the ratio of the number of degrees of freedom in the two phases, ν.

I show results in Table 1 for S + S collisions at SPS energy ($\sqrt{s} = 20$ GeV/nucleon), and for Au + Au collisions at SPS, RHIC, and LHC ($\sqrt{s} = 7$ TeV/nucleon) energies, averaged over $T_c = 150$ and 200 MeV, and over $\nu = 5$ and 10. Here T_f is the freezeout temperature, τ_f is the proper time, r_Q is the distance the quark has moved from its point of creation, and n_f is the mean number of collisions before freezeout. This last is the most important number — when $n_f \simeq 1$ (S + S collisions at SPS), statistical recombination models should not be expected to

Table 1. Freezeout Parameters, Adapted from Ref. 2

A	dN/dy	Q	T_f/T_c	τ_f (fm/c)	r_Q (fm)	n_f
32	85	s	1.0c	2.0	2.5	1
		c	1.0c	3.6	2.5	1
		b	0.8c	8.3	2.7	1
197	1000	s	0.9b	15	8.8	4
		c	0.9b	26	8.7	5
		b	0.7b	56	8.6	5
197	2000	s	1.0w	21	9.3	7
		c	0.9w	37	9.3	8
		b	0.7w	78	9.3	10
197	3500	s	1.0w	26	9.3	10
		c	1.0w	45	9.3	11
		b	0.8w	97	9.3	15

work very well, while when $n_f \gg 1$ (Au + Au collisions at all energies considered) we expect that statistical models should describe heavy resonance data very well.

4. STATISTICAL RECOMBINATION

Our statistical recombination model[2] is very similar to those used by recent authors.[18,19,20] The main difference is that, in addition to ensuring that the quark and antiquark densities, respectively ρ_Q and $\rho_{\bar{Q}}$, are conserved at freezeout, we ensure that the local quark–antiquark density, $\rho_{Q\bar{Q}}^{(2)}(x, x)$, is also conserved. This is not trivial, as typically $\rho_{Q\bar{Q}}^{(2)}(x, x) \gg \rho_Q \rho_{\bar{Q}}$. We thus include three chemical potentials:

1. quark, μ_Q;

2. antiquark, $\mu_{\bar{Q}} \neq -\mu_Q$, since the heavy quarks are not in chemical equilibrium;

3. pair, $\mu_{Q\bar{Q}}$.

The chemical potential for resonance i is

$$\mu_i = \sum_Q \left(k_i^{(Q)} \mu_Q + k_i^{(\bar{Q})} \mu_{\bar{Q}} + k_i^{(Q\bar{Q})} \mu_{Q\bar{Q}} \right). \tag{4.1}$$

Here $k_i^{(Q)}$ ($k_i^{(\bar{Q})}$) is the number of quarks (antiquarks) of flavor Q, and $k_i^{(Q\bar{Q})}$ is the number of pairs (the smaller of $k_i^{(Q)}$ and $k_i^{(\bar{Q})}$). We include all confirmed meson and baryon resonances[21] in our statistical recombination model.

We estimate the two-particle density (and thus $\mu_{Q\bar{Q}}$) from the freezeout conditions obtained from our simulation.

$$\frac{\rho_{Q\bar{Q}}^{(2)}}{\rho_Q} = \rho_{\bar{Q}} + \frac{3}{4\pi} \left(\frac{3}{5 r_Q^2} \right)^{3/2}. \tag{4.2}$$

Results are shown in Table 2, again averaged over T_c and ν. Although the corrections to s resonance production from the pair chemical potential are less than order unity, the corrections

Table 2. Freezeout Densities, Adapted from Ref. 2

A	dN/dy	Q	dN$_Q$/dy	ρ_Q (fm^{-3})	$\rho_{Q\bar{Q}}^{(2)}/\rho_Q\rho_{\bar{Q}}$
32	85	s	5	0.06	1
		c	0.03	2×10^{-4}	40
		b	0.0003	9×10^{-7}	7000
197	1000	s	25	0.01	1
		c	0.3	8×10^{-5}	3
		b	0.003	4×10^{-7}	500
197	2000	s	50	0.016	1
		c	1	2×10^{-4}	2
		b	0.02	2×10^{-6}	80
197	3500	s	250	0.06	1
		c	5	7×10^{-4}	1
		b	0.4	3×10^{-5}	6

to c and b resonance production are huge, giving orders of magnitude changes in fugacities. Thus, this pair chemical potential (or its equivalent) must be included to calculate c and b resonance production.

Finally, this recombination model can be used to gain insight into predicted phenomena such as suppression of J/ψ resonance production in ultrarelativistic nuclear collisions.[17] Predictions for the fractions of $c\bar{c}$ and $b\bar{b}$ pairs that freeze out as $c\bar{c}$ and $b\bar{b}$ mesons are given in Tables 3 and 4. These quantities serve as surrogates for the J/ψ and Υ suppression, as these resonances are the most common sources for observed J/ψ and Υ resonances. It is worth noting that the heavy mesons are all predicted to freeze out in the RG in ultrarelativistic nuclear collisions, so their suppression probably depends most strongly on (and thus carries the most information about) their interations with the RG, and not with the QGP; thus, their dynamics is only weakly dependent on the presence or absence of QGP. Otherwise, it is obvious from the strong dependences on T_c and ν that the theoretical uncertainty in these calculations is huge, so much work remains to be done on the problem of how produced heavy quarks emerge as observed heavy resonances.

5. CONCLUSIONS

We have calculated heavy quark production; s quarks are mostly produced thermally, b quarks are mostly produced by hard parton collisions, whil c quarks are produced by both

Table 3. $c\bar{c}$ Meson Fractions, from Ref. 2

T_c (MeV):		150		200	
A	dN/dy	ν: 5	10	5	10
32	85	0.066	0.075	0.029	0.021
197	1000	0.021	0.002	0.008	0.001
197	2000	0.009	0.004	0.004	0.001
197	3500	0.009	0.010	0.005	0.002

Table 4. $b\bar{b}$ Meson Fractions, from Ref. 2

| T_c (MeV): | | 150 | | 200 | |
A	dN/dy	ν: 5	10	5	10
32	85	0.62	0.16	0.36	0.072
197	1000	0.50	0.053	0.20	0.016
197	2000	0.20	0.013	0.070	0.005
197	3500	0.075	0.008	0.027	0.002

mechanisms. We have attempted to estimate when freezeout will occur for the various heavy quark species, and whether there are enough collisions that statistical recombination models should be reliable guides to heavy resonance production; the latter seems to be true for Au + Au collisions at SPS energy and above, but not necessarily for S + S collisions at SPS. We found that heavy quarks typically freeze out from the RG, and not from the QGP, so the relative numbers of the various heavy resonances are only weakly dependent on the presence or absence of QGP. Finally, we found that it is necessary to include a pair chemical potential in statistical recombination models, or to otherwise allow for the fact that heavy quarks and antiquarks are always produced in coincidence.

This work was supported in part by the Natural Sciences and Engineering Research Council of Canada, and in part by the FCAR fund of the Québec government. It is based on earlier work supported in part by the U. S. Department of Energy under Grant No. DOE/DE-FG02-86ER-40251, and in part by the North Atlantic Treaty Organization under a Grant awarded in 1991.

REFERENCES

1. T. Altherr and D. Seibert, Phys. Lett. B **313**, 149 (1993); Phys. Rev. C **49**, 1684 (1994).
2. D. Seibert and G. Fai, Phys. Rev. C **50**, 2532 (1994).
3. J. D. Bjorken, Phys. Rev. D **27**, 140 (1983).
4. E. Shuryak, Phys. Rev. Lett. **68**, 3270 (1992).
5. K. Geiger and J. I. Kapusta, Phys. Rev. D **47**, 4905 (1993).
6. P. Nason, S. Dawson, and R. K. Ellis, Nucl. Phys. **B327**, 49 (1989); W. Beenakker, W. L. van Neerven, R. Meng, G. A. Schuler, and J. Smith, Nucl. Phys. **B351**, 507 (1991).
7. I. Sarcevic and P. Valerio, Phys. Lett. **B338**, 426 (1994).
8. Z. Lin and M. Gyulassy, Columbia University preprint CU-TP 638 (December 1994).
9. J. Rafelski and B. Müller, Phys. Rev. Lett. **48**, 1066 (1982).
10. T. Matsui, B. Svetitsky, and L. McLerran, Phys. Rev. D **34**, 783 (1986).
11. A. Shor, Phys. Lett. B **215**, 375 (1988).
12. T. Biró, E. van Doorn, B. Müller, M. Thoma and X. Wang, Phys. Rev. C **48**, 1275 (1993).
13. R. Pisarski, Phys. Rev. D **47**, 5589 (1993).
14. N. Bilić, J. Cleymans, I. Dadić, and D. Hislop, University of Cape Town preprint UCT-TP 213/94 (August 1994), unpublished.
15. K. Geiger, Phys. Rev. D **47**, 133 (1993).
16. K. Geiger, Phys. Rev. D **48**, 4129 (1993).
17. T. Matsui and H. Satz, Phys. Lett. B **178**, 416 (1986); A. Capella *et al.*, Phys. Lett. B **206**, 354 (1988); J. Ftanik, P. Lichard, and J. Pisut, Phys. Lett. B **207**, 194 (1988); S. Gavin, M. Gyulassy, and A. Jackson, Phys. Lett. B **207**, 257 (1988); R. Vogt *et al.*, Phys. Lett. B **207**, 263 (1988); J. P. Blaizot and J. Y. Ollitrault, Phys. Lett. B **217**, 386 (1989).
18. K. Redlich, J. Cleymans, H. Satz, and E. Suhonen, Nucl. Phys. **A566**, 391c (1994).
19. J. Letessier, A. Tounsi, U. Heinz, J. Sollfrank, and J. Rafelski, Phys. Rev. D, in press.
20. P. Lévai and J. Zimányi, Phys. Lett. B **304**, 203 (1993).
21. Particle Data Group, Phys. Rev. D **45**, 1 (1992).

12

SIGNATURES OF STATISTICAL DECAY

D. Horn,[1] G. C. Ball,[1] D. R. Bowman,[1] A. Galindo-Uribarri,[1] E. Hagberg,[1]
R. Laforest,[2]* J. Pouliot,[2]† and R. B. Walker[1]

[1]AECL, Chalk River Laboratories
Chalk River, Ontario
K0J 1P0, Canada
[2]Laboratoire de Physique Nucléaire
Université Laval
Ste-Foy, Québec
G1K 7P4, Canada

ABSTRACT

The partition of decay energy between the kinetic energy of reaction products and their Q-value of formation is obtained in a statistical derivation appropriate to highly excited nuclei, and is shown to be in a constant ratio. We measure the kinetic energy fraction, $R = \Sigma E_{kin}/(\Sigma E_{kin} + \Sigma Q_0)$, over a wide range of excitation energy for well-defined systems formed in the $^{35}Cl + ^{12}C$ reaction at $35A$ MeV. Relationships between excitation energy, charged-particle multiplicity, and intermediate-mass-fragment multiplicity, observed in this work and in recent experiments by a number of other groups, follow from the derivation of the average kinetic energies and Q-values.

1. INTRODUCTION

A number of scaling phenomena and correlations between observables have recently been discovered in the deexcitation of highly excited nuclei. These include:

- the correlation of N_{IMF}, the number of intermediate-mass fragments, with N_c, the total number of charged products,[1]

- the correlation of N_{IMF} with Z_{bound}, the total amount of charge contained in fragments heavier than hydrogen,[2]

---

*Present address: Laboratoire de Physique Corpusculaire, ISMRA et Université de Caen, Blvd. du Maréchal Juin, F-14050 Caen, France.
†Present address: Hôtel-Dieu de Québec, Département de Radio-Oncologie, Québec, Canada.

Advances in Nuclear Dynamics
Edited by Wolfgang Bauer and Alice Mignerey, Plenum Press, New York, 1996

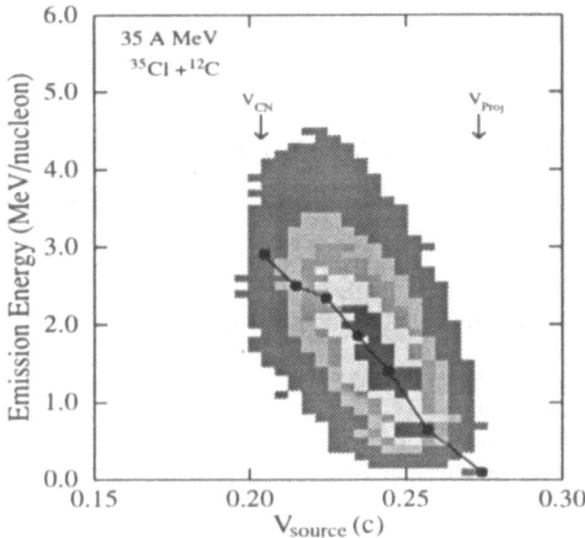

Figure 1. Emission energy ($E_{kin} + Q_0$) per nucleon in the moving frame plotted as a function of the velocity of the center of mass of the detected fragments for $35A$ MeV ^{35}Cl+ ^{12}C. Points represent a massive-transfer simulation, filtered by the experimental acceptance; arrows indicate the velocities of the projectile and of the projectile-target center of mass.

- the approximate proportionality of N_{IMF} to E_t, the measured transverse energy,[3]

- the linear relationship between N_c and T, the nuclear temperature,[4] and

- the scaling of IMF multiplicity yield ratios with $\frac{1}{\sqrt{E_t}}$.[5,6]

    Each of these observations has been taken as indicative of statistical decay. We here introduce a schematic derivation for the partitioning of the decay energy between the kinetic and the mass-excess degrees of freedom. Because of its relatively small width at moderate multiplicities, the ratio of the kinetic energy of the products to the total available decay energy is proposed as an event-by-event signature for statistical decay. Further consequences of the derivation include the dependence of the mean multiplicities, $\langle N_c \rangle$ and $\langle N_{IMF} \rangle$, on excitation energy and the prediction of multiplicity yield ratios. The derived quantities are confronted with our experimental results for the decay of light nuclear systems with measured velocities, sizes, and excitation energies.

2. DERIVATION OF STATISTICAL OBSERVABLES

    The probability of emitting a particle with kinetic energy, E_{kin}, from a state of excitation energy, E, is

$$P(E, E_{kin}, Q_0) \propto E_{kin}\sigma_{inv}(E_{kin})\frac{\rho_f(E - E_{kin} - Q_0)}{\rho_i(E)}, \tag{2.1}$$

where ρ_f and ρ_i are the final- and initial-state level densities and $\sigma_{inv}(E_{kin})$ is the cross section for the inverse reaction. The mean kinetic energy may be derived by integrating E_{kin} with the probability distribution of equation (1):

$$\langle E_{kin} \rangle = \frac{\int_0^{E-Q_0} E_{kin} P(E_{kin}) dE_{kin}}{\int_0^{E-Q_0} P(E_{kin}) dE_{kin}}. \tag{2.2}$$

In the limit of high excitation energy and negligible emission barriers, this gives the traditional[7] results for the mean kinetic energy,

$$\langle E_{kin} \rangle = 2\sqrt{E/a} = 2T, \tag{2.3}$$

and its variance,

$$\sigma^2_{E_{kin}} = 2E/a = 2T^2, \tag{2.4}$$

where a is the nuclear level-density parameter. These results are appropriate for neutron emission and for charged-particle emission from light systems, where Coulomb barriers are very low. The many exit channels accessible to states of high excitation energy permit one to approximate the density of available ground-state Q-values as a continuous function, $f(Q_0)$. A similar integral may then be performed for Q_0, also giving a proportionality to $\sqrt{E/a}$, with the proportionality constant depending on $f(Q_0)$. Thus,

$$\langle Q_0 \rangle = 2\sqrt{E/a} = 2T \tag{2.5}$$

if the density of available exit channels is a linear function of Q_0. For systems in the mass range studied experimentally here, this is a reasonable approximation. Whatever the proportionality factor, the kinetic energy fraction,

$$R = \frac{\Sigma E_{kin}}{(\Sigma E_{kin} + \Sigma Q_0)}, \tag{2.6}$$

should then be a constant, independent of excitation energy, with the specific case of $\langle Q_0 \rangle = 2T$, giving $\langle R \rangle = 0.50$.

For statistical processes, the observed width depends on the number of samplings of the parent distribution. In our case the number of samplings is the number, N, of detected charged particles. For $\langle R \rangle = 0.50$, that width is

$$\sigma_R = \frac{1}{4\sqrt{2N}}. \tag{2.7}$$

We now examine the relative *rates* for barrier-dominated and barrier-independent emission. If the probability per unit time for emitting n particles is approximately proportional to $\exp(-n\langle \Delta E \rangle/T)$, where $\langle \Delta E \rangle$ is the average deexcitation energy per particle emitted, then the particle emission *rate* is

$$\langle n \rangle = \frac{\int dn\, n \exp(-n\Delta E/T)}{\int dn \exp(-n\Delta E/T)} \propto \frac{T}{\Delta E}. \tag{2.8}$$

For light-ion emission from a light, highly excited nucleus, we neglect the Coulomb barrier and obtain

$$\langle \Delta E \rangle = \langle E_{kin} \rangle + \langle Q_0 \rangle = 4T. \tag{2.9}$$

However, for emission of heavier fragments, barrier effects may dominate, so that $\langle \Delta E \rangle \approx B$. A fixed or generic value of B, previously demonstrated[5] to be appropriate for IMF emission, would imply the rates for the two processes to be in the ratio,

$$\frac{\langle n_{IMF} \rangle}{\langle n \rangle} \propto \frac{T/B}{T/4T} \propto T. \tag{2.10}$$

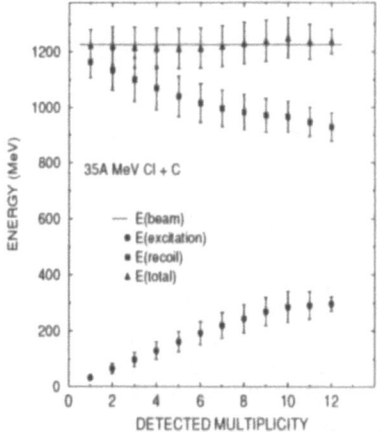

Figure 2. Energy sums and distribution widths (bars) for 35A MeV ^{35}Cl+ ^{12}C as a function of the number of detected charged particles. The solid line indicates the beam energy.

If the average deexcitation per particle is $\langle \Delta E \rangle = 4T$, the mean number of particles emitted would be $\langle N \rangle = E/4T$, giving

$$N \propto \sqrt{E}, \tag{2.11}$$

and variance,

$$\sigma_N^2 = N/4. \tag{2.12}$$

From the ratio of rates,

$$N_{IMF} \propto E. \tag{2.13}$$

If the proportionality of transverse energy to excitation energy is assumed, then equations (11) and (13) consolidate the first four points of the introduction.

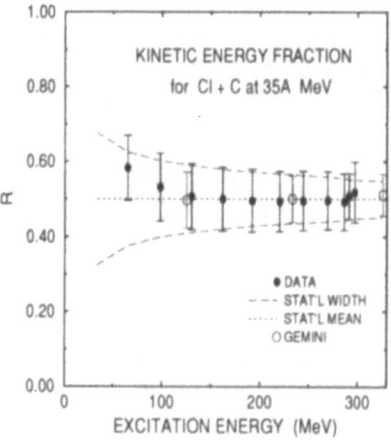

Figure 3. Centroids and widths of kinetic energy fraction, R, for ^{35}Cl + ^{12}C at 35A MeV. Data for multiplicities 2 to 12 are plotted.

3. EXPERIMENTAL DETERMINATION OF SOURCE PROPERTIES

A systematic comparison of the derived quantities with experiment requires the event-by-event determination of the size, velocity and excitation energy of the decaying system. One type of reaction which lends itself to the isolation and measurement of an excited source is the massive transfer mechanism.[8] A beam of 35A MeV ^{35}Cl ions, provided by Chalk River's TASCC facility, collided with a 2-mg/cm^2 carbon target, producing a variety of reaction products. The kinematics of the reaction served to focus the reaction products into the 6° to 25° angular range of our detector array, a close-packed assembly of 40 phoswich counters.[9] All ions were identified by atomic number and isotopic distributions were measured with a set of high-resolution Si/CsI detector telescopes.[10] Our selection of massive transfer events was facilitated by the thresholds and limited angular acceptance of the array, which reduced our sensitivity to target-like "spectator" matter.

To ensure that no major component of an event went undetected, we analyzed only events in which at least 15 units of charge were identified. The velocity of the center of mass of all detected ions was computed, and the kinetic energy of each product was calculated within the moving frame. Based on the average mass excess deduced from the isotope distributions for each detected ion, ΣQ_0 was added to ΣE_{kin} for each event, and the resulting deexcitation energy per detected nucleon was plotted as a function of source velocity. Fig. 1 shows the relationship between emission energy per nucleon and source velocity, starting at projectile-like values of source velocity and zero emission energy in the lower right portion of the figure, and extending to compound-nucleus velocity and large emission energy in the upper left. The entire range of massive transfer phenomena, as previously observed at lower energies,[11] is evident in the figure. The points superimposed on the figure represent centroids of the distributions in energy and velocity for the simulated massive transfer of zero to twelve nucleons from the carbon target to the heavier projectile. Decay of the excited system is simulated[12] by a Monte Carlo event generator and filtered by the experimental acceptance. Quantitative agreement with the data has been demonstrated[13] for simultaneous projection of the same set of calculated results on both the energy and velocity axes.

A massive transfer reaction has the useful feature that the mass of the recoiling system may be deduced from its velocity, allowing a model-dependent estimate of total excitation energy. The validity of this estimate can then be tested by the consistency check demonstrated in Fig. 2, where the excitation energy, recoil energy, and deduced total energy are plotted as a function of the number of detected charged particles. Here, the multiplicity-dependence of the momentum transfer and energy deposition is evident. Significantly, the energy sums yield the initial projectile energy, indicating that the efficiency corrections have been properly applied. The behavior of any unobserved component must then be consistent with that of the detected fragments.

4. COMPARISON OF DERIVED AND MEASURED STATISTICAL OBSERVABLES

Fig. 3 shows the kinetic energy fraction, R, plotted as a function of excitation energy for events in which two to twelve charged products were detected. The excitation energy attributed to each point is the mean for a given multiplicity. Filled circles are the centroids of the R distributions, with bars indicating the distribution widths, σ_R. The $\langle R \rangle = 0.50$ value and associated widths from equation (7) are indicated by the dashed curves. To investigate the effects of a more comprehensive statistical treatment, a calculation was performed with

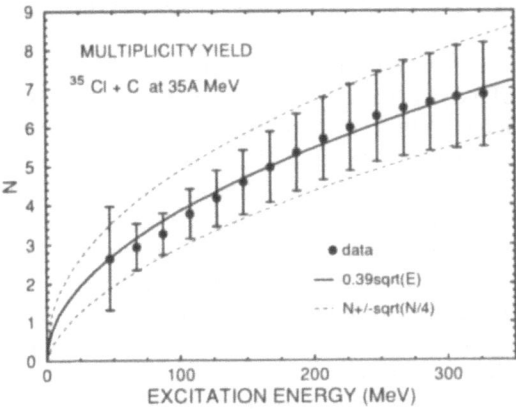

Figure 4. Centroids and widths of multiplicity distributions for ^{35}Cl + ^{12}C at $35A$ MeV, plotted as a function of reconstructed excitation energy. The proportionality constant for the solid line, $k = 0.39$, was chosen to fit the data.

the code, GEMINI,[14] with average gamma-ray emission energies obtained from a modified version of PACE,[15] and the results for three excitation energies plotted as open circles. The data are obviously in good agreement with both the schematic calculation and the statistical code results, though some bias due to detector acceptance may be apparent at the lowest multiplicities. The statistical nature of the decay is reflected in the R distributions over the range of excitation energies from 2 to nearly 7 MeV per nucleon. The narrowness of the distribution supports the prediction of equation (7) and indicates that, even for relatively low multiplicities, R may be useful as an event-by-event indicator of statistical decay. Indeed, the technique has recently been applied to a series of projectile fragmentation experiments,[16] giving $\langle R \rangle$ values from 0.50 to 0.57.

The mean number of charged fragments emitted should, according to equation (11), be proportional to the square root of the excitation energy. Fig. 4 shows the centroids and

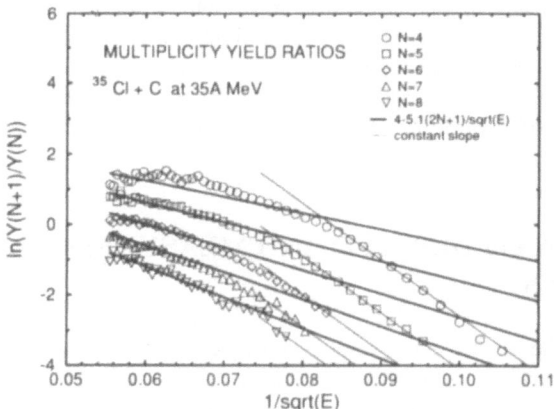

Figure 5. Natural logarithm of multiplicity yield ratios for ^{35}Cl + ^{12}C at $35A$ MeV, plotted as a function of $E^{-1/2}$. Heavy lines represent ratios based on "barrier-independent" calculations of N and σ; light lines represent constant slope.

widths of the measured multiplicity distributions as a function of excitation energy. The solid line represents a square-root proportionality, and the widths determined by equation (12) are indicated by the dashed lines. The centroids and widths predicted by our schematic statistical derivation are in good agreement with experiment for the higher excitation energies, but at lower energies, where the Coulomb barrier might impose upper limits on the number of charged particles, the experimental widths are narrower.

Moretto et al.[5] have demonstrated that for IMF emission, which may be dominated by barrier effects, the natural logarithm of the multiplicity yield ratio is a linear function of $E^{-1/2}$. We plot this ratio for emission of all ions in Fig. 5. Note that the data do not, except at the lowest excitation energies, have the linear dependence upon $E^{-1/2}$ expected from the systematics of Ref. 5. Based on the behavior of the widths in Fig. 4, it would instead be reasonable to expect agreement at higher excitations with multiplicity yield ratios predicted for barrier-independent emission. In this case, the assumption of gaussian multiplicity distributions gives

$$\ln \left(\frac{Y(N+1)}{Y(N)} \right) = 4 - \frac{2(2N+1)}{k\sqrt{E}}, \tag{4.1}$$

where k is the proportionality constant of equation (11). These "statistical" ratios, indicated by the heavy lines in Fig. 5, do, in fact, approximate the data at high excitation energies. For lower excitation energies, the constant slopes of the barrier-dominated systematics may be more appropriate.

5. CONCLUSIONS

- A schematic calculation showed, for $E \gg B$, that the partition of decay energy between kinetic energy of emission and Q_0 should be in a constant ratio for statistical decay.

- The kinetic energy fraction, $R = \Sigma E_{kin}/(\Sigma E_{kin} + \Sigma Q_0)$, and its width were measured for a well-determined reaction mechanism.

- The predicted mean and distribution width for R were observed in the data and reproduced in calculations with a well-known statistical decay code.

- At sub-vaporization excitation energies, emission at the barrier has a different energy dependence than emission well above the barrier.

- The relationship between the two processes is exemplified by many of the correlations observed between charged-particle production, IMF production, and excitation energy.

REFERENCES

1. D. R. Bowman et al., Phys. Rev. C46 (1992) 1834.
2. J. Hubele et al., Z. Phys A340 (1991) 263.
3. L. Phair, private communication.
4. N. Porile, Proc. XI Winter Workshop on Nuclear Dynamics, Key West, Florida, 1995 Feb. 11–17.
5. L. G. Moretto, D. N. Delis, and G. J. Wozniak, Phys. Rev. Lett. 71 (1993) 3935.
6. L. G. Moretto et al., Phys. Rev. Lett. 74 (1995) 1530.
7. P. Morrison in experimental Nuclear Physics, Vol. II, ed. E. Segré (John Wiley and Sons, New York, 1953), p.173.
8. P. J. Siemens et al., Phys. Lett. 36B (1971) 24.
9. C. Pruneau et al., Nucl. Inst. and Meth. in Phys. Res. A297 (1990) 404.

10. D. Horn *et al.*, Nucl. Inst. and Meth. in Phys. Res. **A320** (1992) 273.
11. N. Colonna *et al.*, Phys. Rev. Lett. **62** (1989) 1833.
12. D. Horn *et al.*, PR-TASCC-4: 3.1.7; AECL-10545.
13. D. Horn *et al.*, Proc. Int. Workshop on Heavy-Ion Fusion, Padova, Italy, 1994 May 24–27, editors A. M. Stefanini *et al.* (World Scientific, 1994) 208.
14. GEMINI code: R. J. Charity *et al.*, Nucl. Phys. **A483**(1988)371.
15. PACE2 code: A. Gavron, Phys. Rev. **C21** (1980) 230, modified by J. R. Beene.
16. R. Roy *et al.*, to be published in Proc XXXIII Interational Winter Meeting on Nuclear Physics, Bormio, Italy, 1995 Jan. 23–28.

13

PROBING THE DEGREES OF FREEDOM IN HOT COMPOSITE NUCLEI VIA CHARGED PARTICLE EMISSION STUDIES

Measuring Too Many Degrees of Freedom Puts a Strain on the Statistical Model of Nuclear Evaporation

Morton Kaplan,[1] Craig M. Brown,[1] Joanna B. Downer,[1]* Zoran Milosevich,[1] Emanuele Vardaci,[1+] James P. Whitfield,[1] Craig Copi,[2‡] and Paul DeYoung

[1]Carnegie Mellon University
Pittsburgh, PA
[2]Hope College
Holland, MI

ABSTRACT

There are many experimental observables whose measurement can provide information on evaporative particle emission from highly excited composite nuclei. Simultaneous observations of several degrees of freedom in such hot systems allows a stringent test of theoretical models by requiring that the calculations reproduce the multiple characteristics with a unique set of model input parameters. We make such comparisons for two data sets, one involving a relatively heavy (high Z) system and the other referring to a relatively light (low Z) system. For the former case, reasonably good agreement is achieved between statistical model calculations and experimental data, whereas for the latter example, significant inconsistencies are found.

In this paper our objective is to discuss the utilization of light-charged-particle emission as a probe of hot nuclear matter, with particular emphasis on the importance of measuring several observables simultaneously. We shall consider nuclear evaporation as a statistical process, in the context of its widespread applications in the emission of charged particles from heavy-ion-induced nuclear reactions. Thus we assume that the detected particle emissions arise from the de-excitation of highly excited, compound-nucleus-like systems involving substantial amounts of angular momentum. Experimentally, this implies that the particle data have already been subjected to, and passed, the usual criteria for signatures of compound nuclei (CN) in thermal equilibrium, namely invariant cross section maps, forward-backward symmetry in the CN center-of-mass, and roughly Maxwellian

*Present address: Washington University, St. Louis, MO 63130.
+Present address: University of Napoli, 80125 Napoli, Italy.
‡Present address: University of Chicago, Chicago, IL 60637.

energy spectra characterized by relatively low apparent temperatures. Although specific limits are difficult to define, what we have in mind are systems with initial excitation energies $E^* \sim 50 - 250$ MeV, initial temperatures $T \sim 1.0 - 4.0$ MeV, and initial spins $J \sim 0 - 70\hbar$.

It is worthwhile to remind the reader of the basic ingredients which are in the statistical model and their relationship to system properties. The probability P of equilibrium emission of a particle of type i, with channel energy ε and orbital angular momentum l, into an angle θ, from a nuclear system of atomic number Z, mass number A, excitation energy E^*, and spin J is given by two major factors [1-3]:

$$P_{i,\varepsilon,l,\theta}(Z, A, E^*, J) \; \alpha \; \begin{bmatrix} \text{Transmission} \\ \text{Coefficient} \end{bmatrix} \times \begin{bmatrix} \text{Level} \\ \text{Density} \end{bmatrix}$$

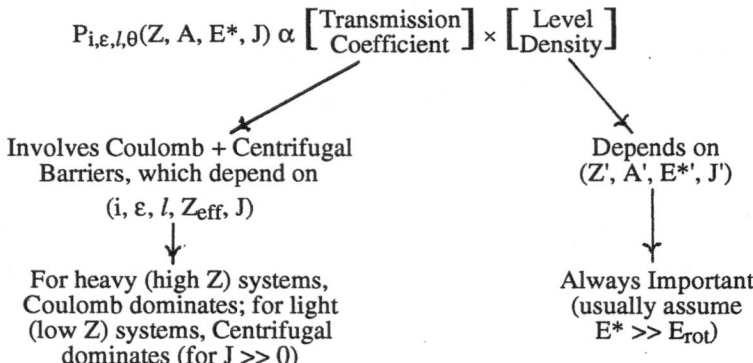

Involves Coulomb + Centrifugal
Barriers, which depend on

$(i, \varepsilon, l, Z_{eff}, J)$

For heavy (high Z) systems,
Coulomb dominates; for light
(low Z) systems, Centrifugal
dominates (for $J \gg 0$)

Depends on
$(Z', A', E^{*'}, J')$

Always Important
(usually assume
$E^* \gg E_{rot}$)

As the angular momentum is quite important for the cases of interest here, one should note that the spin always enters through the parameter β_2 (or some equivalent form):

$$\beta_2 = \frac{\hbar^2 J^2}{2IT} \left[\frac{\mu R^2}{I + \mu R^2} \right]; \text{ i.e., } \frac{\text{Rotational Energy}}{\text{Temperature}};$$

where I is the moment-of-inertia and μ is the reduced mass of the system at the instant of particle emission [4,5].

What can we measure experimentally, and how do such observables depend on the parameters of the statistical model? With modern, but relatively straightforward, detection systems, it is not difficult to obtain:

(1)   Charged particle energy spectra (depend strongly on effective barriers, and on T; weakly on J).

(2)   Particle angular distributions (depend strongly on J, weaker on T).

(3)   Particle evaporation cross sections (multiplicities); these depend on the same parameters as in (1) and (2) above.

(4)   Coincidences and correlations with fragments (helps the characterization of emission sources).

(5)   Particle-particle coincidences and correlations,

proton-proton
proton-alpha   }   a) Integrated cross sections.
alpha-alpha   b) In-plane vs. out-of-plane correlations.

p-p-p
p-p-α }   c) Integrated cross sections.
p-α-α
α-α-α

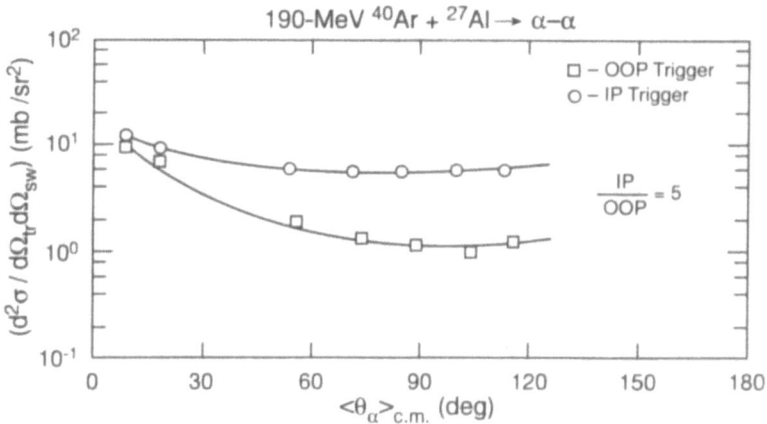

Figure 1. Angular distributions of ^{4}He in coincidence with another ^{4}He particle in a trigger detector. The two trigger detectors are each at ~90° c.m., but differ in their planarity with respect to the other detectors.

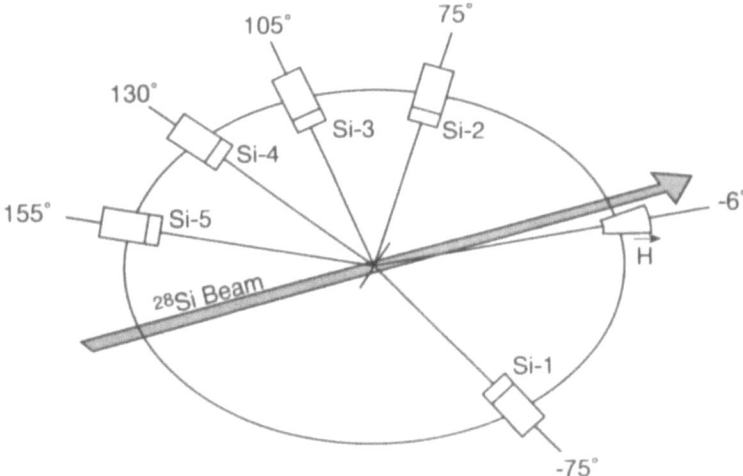

Figure 2. Schematic diagram of the experimental setup showing five Si light particle telescopes and the magnetic spectrometer (H) for ER detection.

To have confidence in our understanding of the statistical model and its underlying physics, the theoretical simulations and calculations should be able to predict all of the above properties in reasonable agreement with experiment.

In the present report, we shall not dwell on the generally accepted observation that the effective proton and alpha evaporation barriers are substantially lower than the corresponding (time-reversed) fusion barriers [6,7]. A systematic study of this phenomenon has been published earlier [7], and the statistical model computer codes used here have been modified to incorporate the empirical barrier lowerings.

The angular momentum associated with heavy ion reactions can profoundly affect the angular distributions of evaporated light particles. To illustrate this sensitivity, we show in Fig. 1 the α-α angular correlations measured [8] in the reaction 190-MeV ^{40}Ar + ^{27}Al. As

the shapes of these distributions are primarily spin driven, the large difference in anisotropy (a factor ~5) between the in-plane (IP) trigger and the out-of-plane (OOP) trigger re-
flects the efficiency of the α-α correlations in identifying the spin orientation of the
emitter. From Fig. 1 it can be seen that IP vs. OOP α-α correlations can give spin signals
which are significantly stronger than the more traditional α-γ coincidence measurements
[9].

Let us look at some charged particle results for a heavy system, ^{149}Tb*, formed at an
excitation energy of 240 MeV in the reaction 344-MeV ^{28}Si + ^{121}Sb. The experiment was
carried out as indicated schematically in Fig. 2, using five silicon telescopes to detect
charged particles in coincidence with evaporation residues (ER) separated by a magnetic
spectrometer (H) at -6° to the beam. Examples of the light particle energy spectra are
shown in Figs. 3 and 4 for protons and alphas, respectively. The data are compared with
statistical model simulations using the code Lilita_N95, an extensively modified version
[10] of the LILITA program originally developed by J. Gomez del Campo [11].

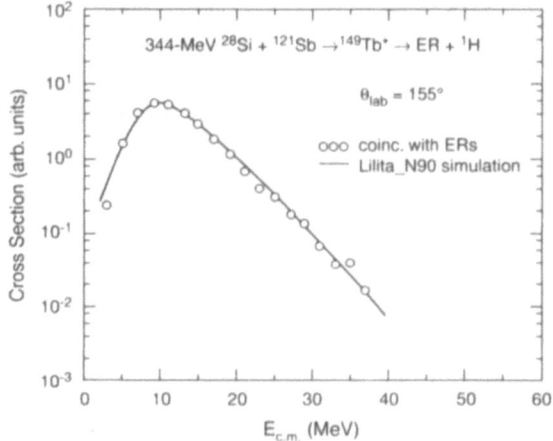

Figure 3. Comparison of calculated (line) and experimental (points) energy spectra for protons in coincidence with evaporation residues for the reaction 344-MeV ^{28}Si + ^{121}Sb.

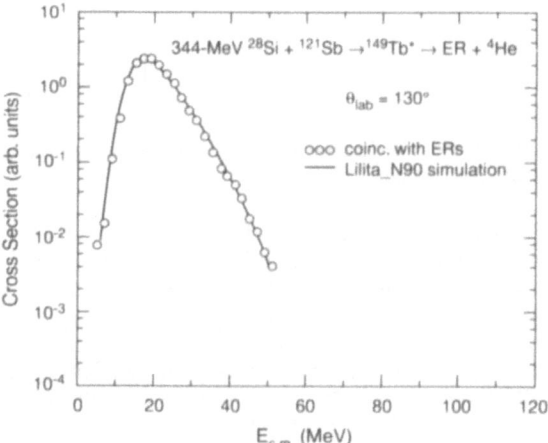

Figure 4. Comparison of calculated (line) and experimental (points) energy spectra for alphas in coincidence
with evaporation residues for the reaction 344-MeV ^{28}Si + ^{121}Sb.

Table 1. 344-MeV ^{28}Si + ^{121}Sb $\rightarrow$ $^{149}_{65}$Tb* (E* = 240-MeV)

	M_n	M_p	M_α	M_α/M_p	$<E_p>$ (MeV)	$<E_\alpha>$ (MeV)	η_α	$M_{\alpha\alpha}/M_{pp}$	$M_{\alpha\alpha}/M_{p\alpha}$
Lilita_N95 (J_{max} = 60)	8.10	3.97	2.72	0.69	11.92	18.63	1.36	0.50	0.72
Modgan (J_{max} = 60)	8.32	3.79	2.31	0.61	12.02	18.96	1.41		
Experiment:				0.72				0.61	0.80

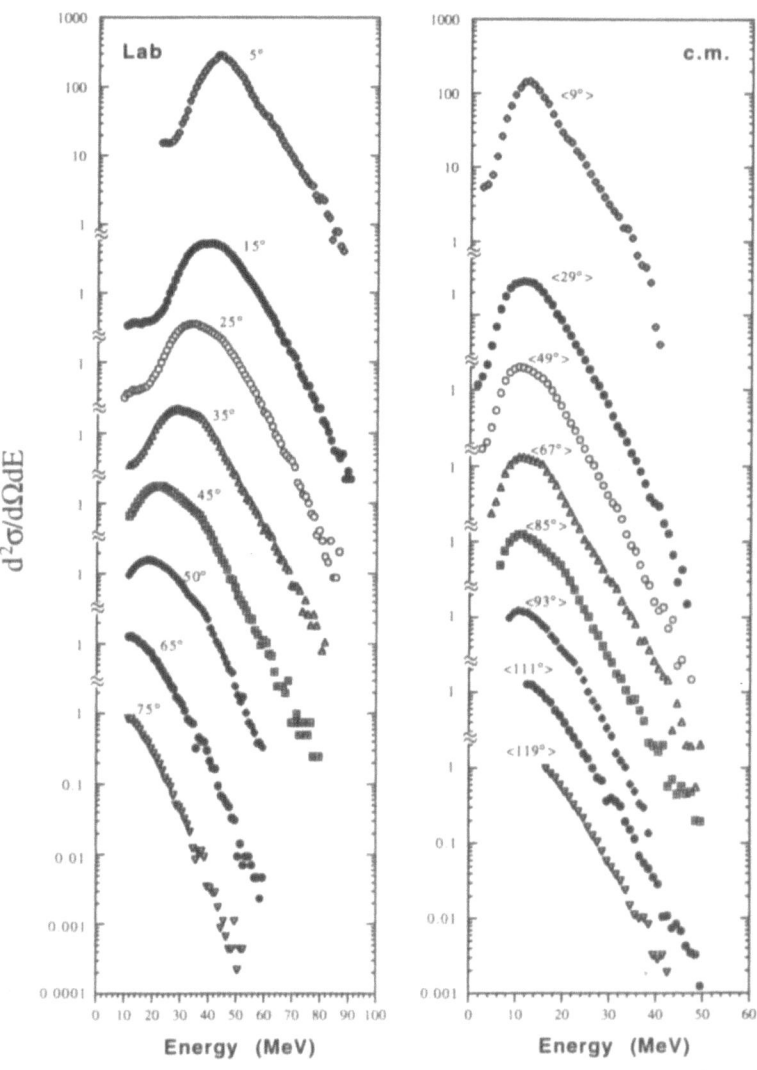

Figure 5. Alpha energy spectra in both the laboratory (left) and center-of-mass (right) reference frames.

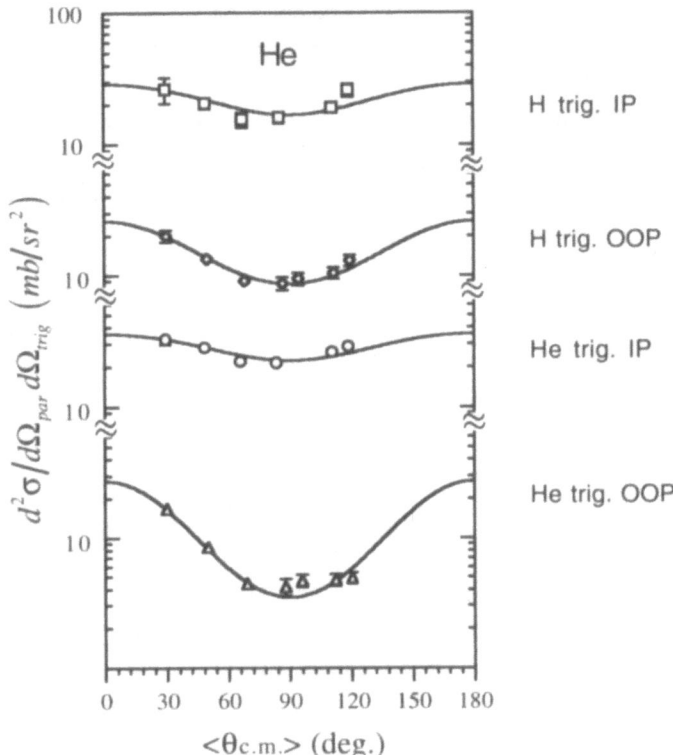

Figure 6. Exclusive angular distributions of alphas in coincidence with protons and alphas as trigger particles.

The calculations were performed using a triangular spin distribution from 0 to $J_{max} = 60\hbar$, a value roughly consistent with ER cross-section estimates from systematics. As can be seen, the statistical model predictions with realistic angular momenta and empirically-fitted barriers are in excellent agreement with the experimental data for both ^{1}H and ^{4}He. In Table 1 we give values for various additional properties predicted by the Lilita_N95 simulations, using the same set of input parameters as for the energy spectra.

Along with the average proton ($<E_p>$) and alpha ($<E_\alpha>$) energies, which have already been effectively compared with experiment in Figs. 3 and 4, Table 1 also lists the particle multiplicities M_n, M_p, and M_α for neutrons, protons, and alpha particles, respectively, and the anisotropy η_α for the alpha c.m. angular distribution. To compare with the Lilita_N95 predictions, some corresponding results are shown from the MODGAN statistical-model program [12], using identical input parameters. Within the statistical precision of the two Monte Carlo simulation codes, the agreement between the predicted sets of average energies, particle multiplicities, and η_α is excellent. From an experimental viewpoint, estimation of evaporative particle multiplicities requires a knowledge of the ER cross section, which is not directly available. Hence to minimize unnecessary uncertainties, we have chosen to compare calculated with experimental multiplicities in terms of the multiplicity ratios, which eliminates the need for absolute ER data. Table 1 presents the inclusive ratio M_α/M_p, and the particle-particle coincidence multiplicity ratios, $M_{\alpha\alpha}/M_{pp}$ and $M_{\alpha\alpha}/M_{p\alpha}$. The agreement between experiment and Lilita_N95 model simulations is quite good, indicating that a single set of statistical model input parameters can reproduce both the

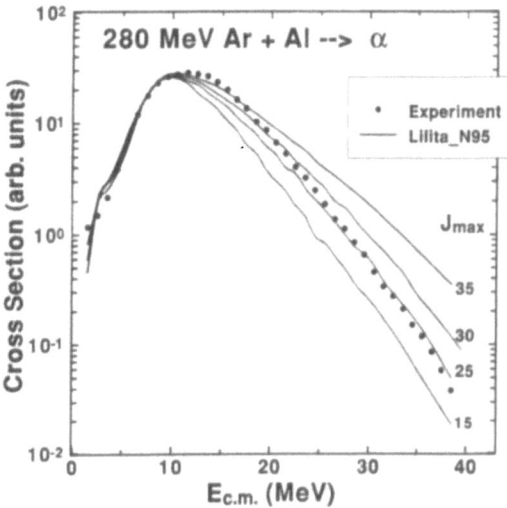

Figure 7. Experimental c.m. alpha energy spectrum compared to Lilita_N95 simulations for several values of the spin as indicated.

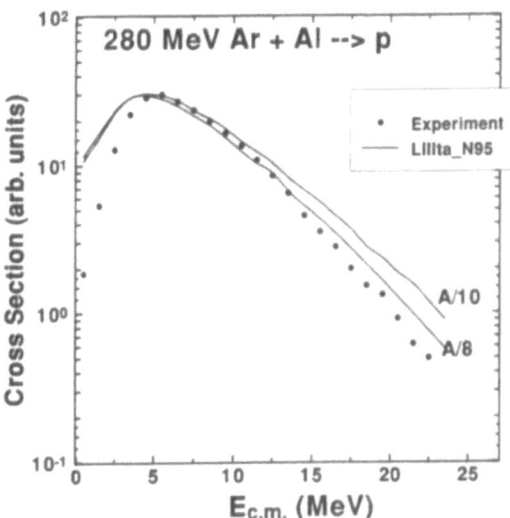

Figure 8. Experimental c.m. proton energy spectrum compared to Lilita_N95 simulations for two values of the level density parameter.

particle energy spectra and the several integrated particle multiplicity ratios. In our view, this is a significant accomplishment.

Let us now turn our attention to a relatively light (low Z) system, ^{67}Ga*, at an excitation energy of 127 MeV. Experimentally, we have extensively studied this system via two different entrance channels, 280-MeV ^{40}Ar + ^{27}Al and 670-MeV ^{55}Mn + ^{12}C, which systematics would suggest are formed with significantly different angular momentum distributions. Calculationally, of course, the properties predicted by the statistical model depend on the input spin distribution, and would be identical for the two channels if the same spins were used.

Figure 5 shows a series of inclusive alpha-particle energy spectra measured for the reaction 280-MeV ^{40}Ar + ^{27}Al. For this same reaction, the ^{4}He exclusive angular distributions are given in Fig. 6, with the selected coincidence trigger requirements as indicated (trigger particle either ^{1}H or ^{4}He and located either IP or OOP). From these figures, it can be seen that: (a) the c.m. energy spectra are roughly Maxwellian in shape and essentially independent of c.m. emission angle; and (b) the ^{4}He exclusive angular distributions exhibit strong angular-momentum effects with the out-of-plane triggers yielding enhanced anisotropies.

A representative c.m. alpha energy spectrum is shown in Fig. 7, together with calculated spectra for several values of J_{max}, as selected for the triangular spin distribution in the CN from 0 to J_{max}. The predicted ^{4}He spectra are clearly sensitive to J_{max}, and although no detailed attempt has been made to optimize a fit to the experimental data, one must conclude that the required J_{max} will be near to 25$\hbar$. This is substantially lower than the 46 - 56$\hbar$ range for J_{max} that would be derived from estimates of fusion cross section systematics. Recall, however, that J_{max} is only a convenient parameter for adjusting within the model calculation and that an *increase* in the moment-of-inertia (away from spherical) would yield the same result as a *decrease* in the spin.

From the statistical theory of nuclear evaporation, the predictions for proton energy spectra are not very sensitive to the spin distribution and thus are poor indicators of J_{max}. Figure 8 presents proton data together with Lilita_N95 simulations using $J_{max} = 25\hbar$ and two values of the level density parameter a, as indicated.

From Fig. 8 it seems that minor adjustments of the level density and the barrier penetrability would improve the already reasonable agreement between calculation and experiment. Note that Fig. 8 is on an expanded scale compared to Fig. 7, and thus the apparent

Table 2. 280-MeV ^{40}Ar + ^{27}Al → $^{67}_{31}$Ga* (E* = 127-MeV)

		Lilita_N95 Calculations:			
J_{max}	M_{α}/M_{p}	η_{α}	$M_{\alpha\alpha}/M_{pp}$	$M_{\alpha\alpha}/M_{p\alpha}$	
15	0.35	1.25	0.13	0.34	
25	0.45	1.61	0.21	0.42	<Spectra
30	0.53	1.83	0.29	0.50	
35	0.63	2.11	0.41	0.60	
38	0.70	2.22	0.51	0.67	
42	0.81	2.48	0.67	0.78	<M; η
45	0.90	2.51	0.83	0.88	

	Experiment (280-MeV ^{40}Ar + ^{27}Al):		
M_{α}/M_{p}	η_{α}	$M_{\alpha\alpha}/M_{pp}$	$M_{\alpha\alpha}/M_{p\alpha}$
1.17	2.70	0.75	0.41

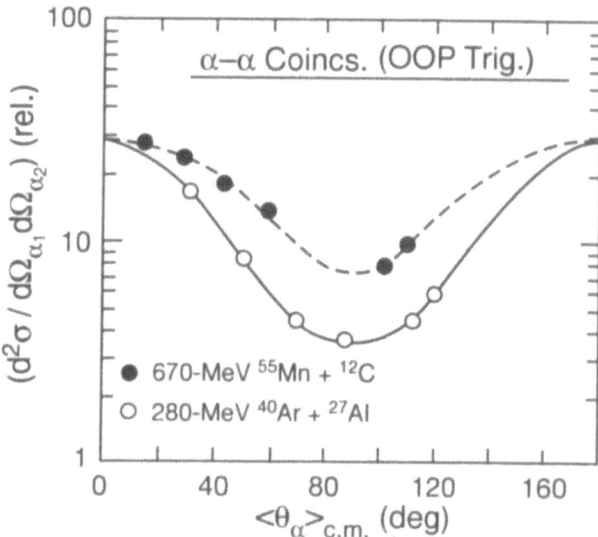

Figure 9. Coincidence α-α angular correlations with OOP triggers for the two reactions leading to the same compound nucleus, ^{67}Ga, at $E^* = 127$ MeV.

deviations are not so large. The Lilita_N95 calculations also yield predictions for the particle multiplicities and the angular anisotropies, and these are shown as functions of J_{max} in Table 2. As expected from the statistical theory, the multiplicity ratios M_α/M_p, $M_{\alpha\alpha}/M_{pp}$, and $M_{\alpha\alpha}/M_{p\alpha}$ all increase with increasing spin, as does the alpha anisotropy η_α. Also shown in Table 2 are the experimental data for these quantities. Comparison between the measurements and calculations, and allowing for some uncertainties in the data, indicates that relatively high spins seem to be required to agree with the particle multiplicities and the alpha anisotropy. In Table 2 we indicate roughly by the arrow and symbol M, η where the "acceptable" spins might begin. In contrast, we recall from the comparisons of energy spectra, particularly Fig. 7, that J_{max} close to $25\hbar$ was required to fit the data. This is indicated in Table 2 by the arrow labeled "Spectra". Thus we find a significant contradiction, or inconsistency, between experiment and the statistical model calculations: For this light (low Z) system, the spin distribution (or moments-of-inertia) needed to reproduce the particle multiplicities and alpha anisotropy leads unavoidably to alpha energy spectra which are much too broad. As noted earlier in this paper, it is for such light nuclear systems that spin (or centrifugal) effects play a more dominant role in governing de-excitation observables.

Experimentally, we find that the particle energy spectra in the two entrance channels, 280-MeV ^{40}Ar + ^{27}Al and 670-MeV ^{55}Mn + ^{12}C, are nearly identical in shape when transformed to the c.m. system. Thus there is no indication from the spectra that different spin distributions are involved in the two entrance channels. Are we, therefore, to infer that the spins are actually very similar? This question can be addressed by examination of Fig. 9, which shows the measured out-of-plane α-α coincidence correlations for the two reactions. Clearly, the observed anisotropies differ by a factor of ~2, and this is positive indication of the spin differences for the emitting compound nuclei in the two reactions.

In conclusion, we have found that there are important differences between heavy and light hot nuclear systems, in terms of the relative role played by angular momentum in the de-excitation processes. The message we have attempted to convey is that simultaneous measurements of many degrees of freedom provide severe constraints on the models of even simple, "well understood", nuclear processes. We believe that this should serve as incentive for strengthening the models by incorporating better (more realistic) physics and fewer approximations.

ACKNOWLEDGMENT

This work was supported at Carnegie Mellon University by the Division of Nuclear Physics, Department of Energy, and at Hope College by the Physics Division, National Science Foundation.

REFERENCES

1. T. Ericson and V. Strutinski, *Nucl. Phys.* **8**, 284 (1958).
2. T. Ericson, *Adv. Phys.* **9**, 425 (1960).
3. T. Dossing. Licentiat Thesis, Unversity of Copenhagen, 1977.
4. G. L. Catchen *et al.*, *Phys. Rev.* **C21**, 940 (1980).
5. N. N. Ajitanand *et al.*, *Phys. Rev.* **C34**, 877 (1986).
6. G. La Rana *et al.*, *Phys. Rev.* **C35**, 373 (1987).
7. W. E. Parker *et al.*, *Phys. Rev.* **C44**, 774 (1991).
8. M. Kaplan *et al.*, unpublished data.
9. N. G. Nicolis *et al.*, *Phys. Rev.* **C41**, 2118 (1990).
10. G. La Rana and M. Kaplan, private communication. The code has been renamed to distinguish it from several other versions of LILITA.
11. J. Gomez del Campo and R. G. Stokstad, Report ORNL TM-7295, 1981; J. Gomez del Campo *et al.*, *Phys. Rev.* **C19**, 2170 (1979).
12. Program MODGAN, N. N. Ajitanand *et al.*, SUNY at Stony Brook, private communication 1995. This code contains essentially the same physics ingredients as Lilita_N95, although the calculational details of the two codes are substantially different.

14

ASSESSING THE EVOLUTIONARY NATURE OF MULTIFRAGMENT DECAY

E. Cornell,[1] T. M. Hamilton,[1] D. Fox,[1]* Y. Lou,[1] R. T. de Souza,[1]
M. J. Huang,[2] W. C. Hsi,[2] C. Schwarz,[2]† C. Williams,[2] D. R. Bowman,[2]*
J. Dinius,[2] C. K. Gelbke,[2] D. O. Handzy,[2] M. A. Lisa,[2]‡ W. G. Lynch,[2]
G. F. Peaslee,[2]§ L. Phair,[2]† M. B. Tsang,[2] G. VanBuren,[3]¶ R. J. Charity,[3]
L. G. Sobotka,[3] and W. A. Friedman[4]

[1]Department of Chemistry and Indiana University Cyclotron Facility
Indiana University, Bloomington, IN
[2]National Superconducting Cyclotron Laboratory and Department of Physics
and Astronomy
Michigan State University, East Lansing, MI
[3]Department of Chemistry
Washington University, St. Louis, MO
[4]Department of Physics
University of Wisconsin, Madison, WI

ABSTRACT

Multifragment decays of central collisions in ^{84}Kr + ^{197}Au at $E/A = 35$, 55, and 70 MeV are studied. The dependence of the extracted emission time on the velocity of the fragment pair is investigated. More energetic pairs manifest a stronger Coulomb interaction indicating emission from a source of smaller spatial–temporal extent than less-energetic pairs. This trend can be understood in the context of a statistical model which allows the source characteristics to evolve as the fragments are emitted.

1. INTRODUCTION

Large, highly excited nuclear systems are observed to undergo the process of multi-fragmentation, i.e., they decay into a relatively large number of intermediate mass nuclear

---

*Present address: CRL, Chalk River, Ontario K0J 1J0, Canada.
†Present address: LBL, University of California, Berkeley, CA 94720.
‡Present address: GSI, D-64220, Darmstadt, Germany.
§Present address: Department of Chemistry, Hope College, Holland, MI 49423.
¶Present address: Department of Physics, MIT, Cambridge, MA 02139.

Advances in Nuclear Dynamics
Edited by Wolfgang Bauer and Alice Mignerey, Plenum Press, New York, 1996

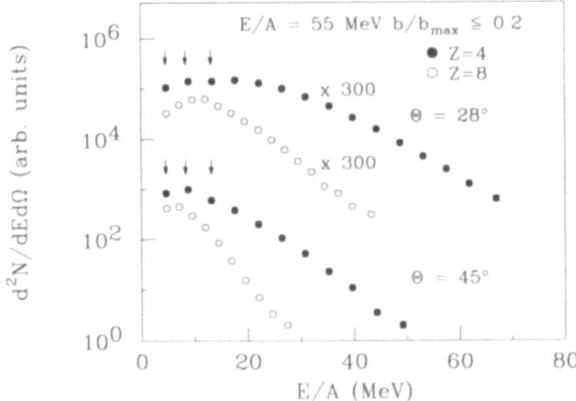

Figure 1. Inclusive kinetic energy spectra for Be and O fragments (closed and open symbols) emitted in central collisions. The arrows indicate velocity cuts of v_{min} = 3, 4, and 5 cm/ns.

fragments (IMFs: $3 \leq Z \leq 20$).[1-15] Current evidence suggests that these fragments are produced from the decay of systems at low density.[4,5] A crucial open question regarding this process is whether these complex fragments arise from a well defined freeze-out condition, or whether they are formed during the dynamic evolution of the system.

The inclusive kinetic energy spectra of fragments originating from central collisions provides little indication of an answer to this question. At a given angle these spectra are smooth, relatively featureless distributions which can be described by simple Boltzmann-like functions involving a single temperature, Coulomb barrier, and source velocity, and in some cases collective expansion energy. Typical spectra are shown in Fig. 1.

To explore the general systematics of multifragmentation, we have previously studied the dependence of fragment multiplicity on incident energy for the $^{84}Kr+^{197}Au$ system in the range E/A = 35–400 MeV.[16] Light charged particles and IMFs produced in the collisions were detected in the angular range $5.4° \leq \theta_{lab} \leq 160°$ by the MSU Miniball/Washington University Miniwall 4π detector array. Experimental details have been previously described.[16,17]

2. FREEZE-OUT VERSUS EVOLUTION

In this analysis, our attention is specifically directed at the question of whether the multifragment final state arises from a "freeze-out" scenario or whether it has an "evolution-ary" nature. In order to preferentially select events from a single equilibrated source, we select on central collisions. We have related the charged particle multiplicity to an impact parameter scale following a geometrical prescription[18] and selected events which correspond to $b/b_{max} \leq 0.2$. In our definition, b_{max} refers to the maximum interaction radius for which two charged particles are emitted. These central collisions have charged particle multiplicities corresponding to $N_C \geq$ 24, 33, and 38 at E/A = 35, 55, and 70 MeV respectively. For these central collisions, multifragment decay describes the average behavior of the excited system. For $b/b_{max} \leq 0.2$ the average multiplicity of IMFs, $\langle N_{IMF} \rangle \approx$ 4, 5, 6 at E/A = 35, 55, and 70 MeV respectively.[16]

Our central question can be addressed by relating more exclusive observables to specific portions of the one-body energy spectrum. In particular, we examined the fragment–

fragment velocity correlations for different portions of the one-body velocity distribution. If the multifragmentation process were to involve a sharp freeze-out then one would expect little dependence of these fragment correlations on different portions of the energy spectrum. On the other hand, if the yield were to arise during the evolution of the system, then different components of the spectra may arise from different conditions which could, in turn, provide different fragment–fragment correlation signals.

3. CORRELATION FUNCTIONS AS A FUNCTION OF IMF VELOCITY

Fragment–fragment velocity correlations are a powerful tool for extracting information about the spatial–temporal dimensions of the emitting source.[19-26] This technique utilizes the mutual Coulomb repulsion of the fragments as a probe of the emitting system. The Coulomb repulsion results in a reduction of the probability for observing fragments at low relative velocity. Velocity correlation functions, $R(v_{red})$, were constructed, using procedures previously employed,[22] by relating the coincidence yield Y_{12} to the product of the single particle yields Y_1 and Y_2:

$$\sum Y_{12}(v_1, v_2) = C[1 + R(v_{red})] \sum Y_1(v_1)Y_2(v_2)$$

where v_1 and v_2 are the laboratory velocities of the fragments, the reduced velocity , $v_{red} = (v_1 - v_2)/(Z_1 + Z_2)^{1/2}$,[21] and C is a normalization constant determined by the requirement that $R(v_{red}) \to 0$ at large relative velocities where the Coulomb repulsion is small. The single particle yields were constructed by selecting fragments from different events which satisfy the same constraints as the coincidence yield. The use of the reduced relative velocity allows summation over different charge combinations.[21]

As Fig. 2a-c clearly indicates, the fragment–fragment correlation functions depend strongly on the kinetic energy of the fragment pairs. For each of the correlation functions shown, the fragments were selected on the basis of the velocity of the slowest fragment of the pair. This minimum velocity, v_{min} is defined as the velocity of the less energetic fragment. All the correlation functions shown in this work were summed over all pairs $4 \leq Z_1, Z_2 \leq 9$ emitted in the angular range $25° \leq \theta_{lab} \leq 50°$. The normalization constant was determined from the inclusive correlation function in the range $0.05c \leq v_{red} \leq 0.08c$. For orientation, the kinetic energies which correspond to these minimum velocities are depicted as the arrows on the energy spectra in Fig. 1. For the $E/A = 35$ MeV data the $v_{min} = 5$ cm/ns correlation function was not shown because for this cut there were non-negligible momentum conservation effects. All of the correlation functions exhibit a "Coulomb hole," a strong suppression of pairs of low relative velocity. As the minimum velocity of the pair is increased, a fairly dramatic increase in the width of the Coulomb hole is observed at all three incident energies.

To quantify this effect sufficiently to pursue qualitative observations, we have extracted the width of the Coulomb hole in the correlation function at half its asymptotic value (HWHM) for each cut of fragment energy. In Fig. 3a, the values of the widths of the Coulomb holes are plotted against the velocity cut-off (minimum velocity) used to construct the correlation function. An increase in the HWHM with increasing v_{min} is clearly evident. For the v_{min} cuts shown, the correlation functions are affected by negligible momentum conservation effects.

We next consider the possible implications of this trend. The shape of the correlation function is associated with the space–time structure of the fragment emission process. In general, the wider the Coulomb hole the smaller the separation in space–time between the

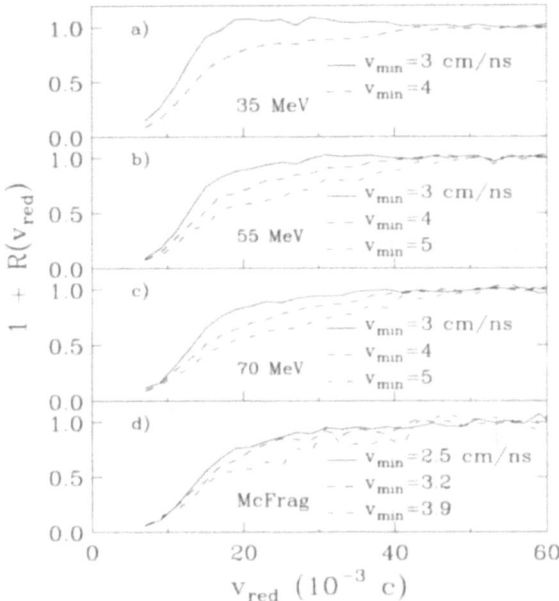

Figure 2. a-c) Experimental correlation functions at E/A = 35, 55, 70 MeV with different restrictions on v_{min}. d) Correlation functions constructed from the predictions of a microcanonical ensemble model for different restrictions on the minimum velocity.

emission of contributing fragments pairs. Thus, the dependence of the strength of the Coulomb interaction suggests that different space–time situations are associated with the emission related to different parts of the spectra. Specifically, the higher energy fragments are emitted with smaller space–time separation between fragments. This result is a clear indication of an evolutionary process.

In earlier studies of proton-proton correlations,[28] similar investigations found that different effective source sizes for proton emission were associated with different proton momenta. This trend was interpreted as due to contributions to the one-body energy distribution from protons originating both from an early dynamical stage, as well as, a late evaporative stage. This temporal behavior for proton emission is not surprising however, since proton emission does not require attainment of a low density phase.

To further examine the general trends observed in the correlation functions, we have performed 3-body Coulomb trajectory calculations in which we assume fragment emission from the surface of a source of fixed initial size. The source was assumed to be a nucleus with $Z = 40, A = 92$ and a radius of 7 fm. The distribution function for the time between emissions was assumed to have the form $\exp(-t/\tau)$ and be characterized by a single characteristic time constant, τ. Correlations functions, characterized by v_{min}, were constructed from the 3-body Coulomb trajectory calculations. The observed inclusive correlation function, at 55 MeV/A, is well described by the decay of a $R = 7$ fm source with a characteristic emission time of τ of 100 fm/c. Within the context of our 3-body model, the dependence of the width of the Coulomb hole on the minimum velocity for different values of τ is shown as solid lines in Fig. 3a.

While the reference 3-body calculations for a given size and decay rate show a dependence of the predicted widths on velocity cut-off, the trend is opposite the trend observed in the data, namely, one finds a decrease in width with increasing minimum velocity. This decrease

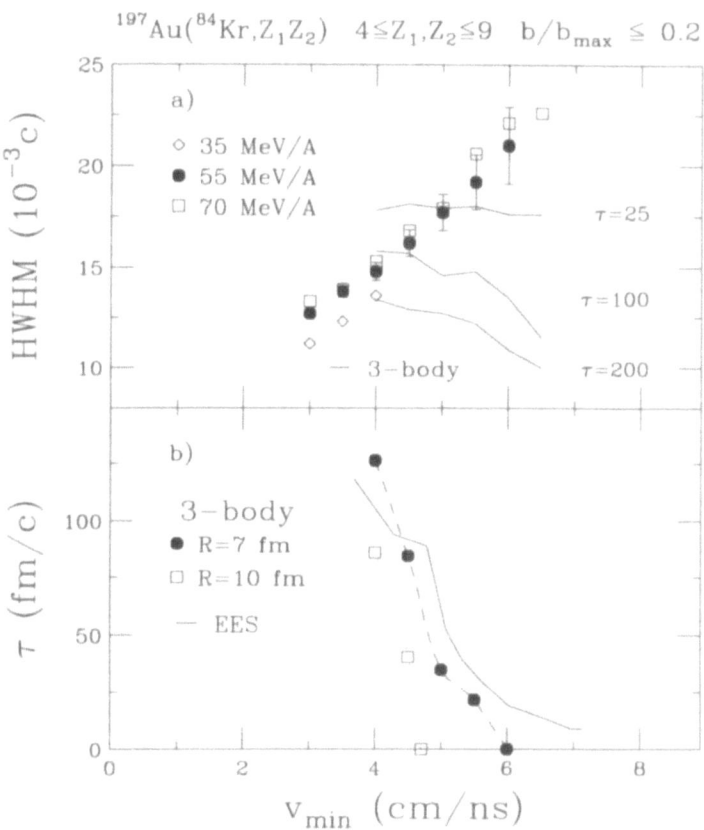

Figure 3. a) Dependence of the Coulomb interaction (HWHM) between fragments on the minimum fragment velocity. Diamonds, circles and squares represent the experimental data at $E/A = 35, 55$, and 70 MeV. Solid lines indicate the results of 3-body Coulomb trajectory calculations for $\tau = 25, 100$, and 200 fm/c. b) Dependence of the extracted emission time on the minimum velocity of the fragment pair. The solid circles and open squares correspond to the experimental data at $E/A = 55$ MeV.

may simply be associated with an increase in the initial spatial separation of the members of the pair due to the increasing velocity of the first emitted fragment. Furthermore, if one were to assume that variation in time separation is more significant than the variation in spatial separation, and if one makes some reasonable assumption of source size, then through comparison with the 3-body calculations each observed hole width can be associated with a given emission time constant. This procedure has been used to obtain the points displayed in Fig. 3b, which relate mean emission times to the minimum velocity cut-offs in the spectra. The two different sets of points arise from the use of two different values for the source radius in the reference trajectory calculations.

4. MODEL COMPARISONS

We have examined the predicted relationship between emission rates and spectral velocities for two specific models of multifragmentation, one representative of the *freeze-out* scenario and one based on an *evolutionary* scenario.

The first model we have examined is the Berlin microcanonical statistical model (McFrag).[29] This model has been used to investigate the inclusive correlation function,[30] but it fails to reproduce the fragment kinetic energy spectra, providing a very narrow range of fragment velocities. The source was assumed to have $Z = 79, A = 197$ with an excitation energy of 2400 MeV. These parameters roughly reproduce the experimentally measured IMF and charged particle multiplicities at $E/A = 55$ MeV. [25] The dependence of the correlation function on the fragment velocity predicted by the McFrag model is shown in Fig. 2d. The lines shown span the entire dynamic range in the energy spectrum. The strong dependence of the correlation function on the fragment velocity exhibited in the data is not predicted by the model. Due to the limited range of the spectra provided by the model, the velocity cut-offs have been scaled (based on the fraction of the velocity distribution) to the velocity cut-offs for the 55 MeV/A experimental data. The near independence of the Coulomb interaction on the fragment velocity apparent in Fig. 2d is not unexpected, since this model assumes a simultaneous emission of all the fragments and contains no evolutionary aspects. The only dependence of the correlation function on final spectral velocity comes through the initial spatial configurations of the decay fragments. It appears that sampling different initial spatial configurations does not result in a strong enough dependence of the correlation function on the fragment kinetic energy.

For the second model we examined the predictions of the EES model[9] which explicitly incorporates dynamical evolution within the emission process. While this model predicts smooth singles spectra for the fragment energy spectra, these spectra are composed of contributions arising from different times in the expansion and decay of the source. In this model, the source temperature changes with time due to both de-excitation due to particle emission and adiabatic changes in density. The Coulomb acceleration also changes with time. Therefore, the portion of the spectra associated with each instant in time is provided by instantaneous properties. A schematic calculation was done for an initial source of $A = 197$ and $Z = 79$, with an initial temperature of 12 MeV. This calculation predicts an average multiplicity of about 4 IMFs. Due to the changing conditions there is a variation of the mean separation times associated with the velocity of the fragments. The calculation shows a decreasing mean emission (separation) time with increasing fragment energy. This result can be understood as follows: The most energetic part of the spectra is populated for the highest temperatures and Coulomb energies. These conditions only exist early in the evolution of the system. After some time has passed fragments of these kinetic energies are rarely emitted. This confined "window" of opportunity for high energy fragments affects the mean separation time associated with their emission. The model suggests that the mean separation time increases with decreasing fragment energy until the vicinity of the yield peak where it levels off. For the very lowest velocity fragments, conditions of low temperature and small source charge confine the emission to the latest stages of the process where the emission is slowest and the separation between fragments grows sharply. An attempt has been made to compare the qualitative predictions of the EES model to the trends shown in Fig. 3b. This result is depicted as the solid line in that panel of the figure. This curve was constructed from the predicted relationship between mean emission times and fragment velocities in the source frame for pairs of ^{7}Li fragments. The velocities of fragments predicted in the model were transformed from the center-of-mass to the laboratory using the experimentally determined center-of-mass velocity. These transformed results have been compared with the experimental data in Fig. 3b. The schematic model calculations do not reproduce the exact dependence of τ on v_{min}, indicating that the precise relationship between expansion and fragment emission dynamics is not yet fully understood. The calculations do, however, predict a trend which is very similar to that observed experimentally, indicating that expansion and fragment formation occur on commensurate time scales.

5. SUMMARY

In summary, we have examined the relationship between the fragment–fragment velocity correlation functions and the velocity of the fragments from which they are constructed, for central collisions in the reaction ^{84}Kr+^{197}Au at $E/A = 35$, 55, and 70 MeV. In each of these cases there is an increase in the width of the Coulomb hole of the correlation function with increasing fragment velocity. This trend is opposite to that obtained with trajectory calculations which assume a single decay constant and source size. The data appear inconsistent with a single freeze-out condition, and are not predicted by a model based on such a scenario. A statistical model which incorporates dynamics into the fragment emission phase qualitatively predicts the trend observed in the data. The strong relationship between the fragment velocities and the mean emission times suggests that there are changes in the character of the source on a time scale concurrent with the fragment emission, and thus that the mechanism of multifragmentation is evolutionary.

We would like to acknowledge the valuable assistance of the staff and operating personnel of the K1200 cyclotron at Michigan State University for providing the high quality beams which made this experiment possible. One of the authors R. D. gratefully acknowledges the support of the Sloan Foundation through the A. P. Sloan Fellowship program. W. F. acknowledges the support of the National Institute for Nuclear Theory, University of Washington, Seattle. This work was supported by the U. S. Department of Energy under DE-FG02-92ER40714 and the National Science Foundation under Grant No. PHY-89-13815, PHY-90-15957, and PHY-92-14992.

REFERENCES

1. J. W. Harris et al., Nucl. Phys. **A471**, 241c (1987).
2. J. Finn et al., Phys. Rev. Lett. **49**, 1321 (1982).
3. C. A. Ogilvie et al. Phys. Rev. Lett. **67**, 1214 (1991).
4. R. T. de Souza et al., Phys. Lett. **B268**, 6 (1991).
5. D. R. Bowman et al., Phys. Rev. Lett. **67**, 1527 (1991).
6. G. Bertsch and P. J. Siemens, Phys. Lett. **B126**, 9 (1983).
7. D. H. E. Gross et al., Phys. Rev. Lett. **56**, 1544 (1986).
8. J. Bondorf et al., Nucl. Phys. **A444**, 460 (1985).
9. W. A. Friedman Phys. Rev. C **42**, 667 (1990).
10. H. W. Barz Nucl. Phys. **A448**, 753 (1986).
11. J. Aichelin et al., Phys. Rev. C **37**, 2451 (1988).
12. G. Peilert et al., Phys. Rev. C **39**, 1402 (1989).
13. D. H. Boal and J. N. Glosli, Phys. Rev. C **37**, 91 (1988).
14. W. Bauer et al., Phys. Rev. Lett. **58**, 863 (1987).
15. L. G. Moretto and G. J. Wozniak, Ann. Rev. Nucl. Part. Sci. **43**, 379 (1993).
16. G. F. Peaslee et al., Phys. Rev. C **49**, R2271 (1994).
17. R. T. de Souza et al., Nucl. Instr. Meth. **A295**, 109 (1990).
18. C. Cavata, et al., Phys. Rev. C **42**, 1760 (1990).
19. R. Trockel et al., Phys. Rev. Lett. **59**, 2844 (1987).
20. Y. D. Kim et al., Phys. Rev. Lett. **67**, 14 (1991).
21. Y. D. Kim et al., Phys. Rev. C **45**, 338 (1992).
22. D. Fox et al., Phys. Rev. C **47**, R421 (1993).
23. E. Bauge et al., Phys. Rev. Lett. **70**, 3705 (1993).

24. T. C. Sangster et al.,Phys. Rev. C **47**, R2457 (1993).
25. D. Fox et al., Phys. Rev. C **50**, 2424 (1994).
26. T. Glasmacher et al., Phys. Rev. C **50**, 952 (1994).
27. R. Bougault et al., Phys. Lett. **B232**, 291 (1989).
28. W. G. Gong et al., Phys. Rev. C **43**, 1804 (1991).
29. X. Z. Zhang et al., Nucl. Phys. **A461**, 641 (1987);**A461**, 668 (1987).
30. O. Schapiro et al., Nucl. Phys. **A568**, 333 (1994).

15

REACTION MECHANISMS OF THE MOST VIOLENT ^{24}Mg + ^{12}C COLLISIONS AT 25A AND 35A MeV*

Y. Larochelle,[1] G. C. Ball,[2] L. Beaulieu,[1] B. Djerroud,[1†] D. Dore,[1‡]
A. Galindo-Uribarri,[2] P. Gendron,[1] E. Hagberg,[2] D. Horn,[2] E. Jalbert,[1]
R. Laforest,[1§] J. Pouliot,[1¶] R. Roy,[1] M. Samri,[1] and C. St-Pierre[1]

[1]Laboratoire de physique nucléaire, Département de physique, Université Laval,
Québec, Canada G1K 7P4
[2]AECL, Chalk River Laboratories, Ontario, Canada K0J 1J0

ABSTRACT

A study of the reaction mechanisms in central ^{24}Mg + ^{12}C collisions at 25A and 35A MeV has been carried out. Global variables, such as anisotropy ratios and source-velocity ratios, computed for those events in which the total charge of the system has been detected, are compared to simulations based on statistical fragmentation codes. For violent events, a binary mechanism appears to be competing successfully with compound nucleus formation.

1. INTRODUCTION

Central collisions between heavy ions at intermediate energies (*i.e.*, between 10 and 100 MeV/nucleon), are known to result in multifragmentation[1,2] and other interesting phenomena. In an extrapolation from well-understood processes at lower energy, complete or, more likely, incomplete fusion accompanied by pre-equilibrium emission of light particles (for examples, see Refs. 3, 4, 5, 6) have been proposed as the mechanisms responsible for formation of the highly excited nuclear systems, which subsequently decay to the observed final states. Recently, however, a persistent binary nature has been observed in collisions involving a high level of dissipation with the two excited partners retaining most of the kinematic characteristics of the projectile and the target.[7,8] The present paper reports an analysis of the

---

*Experiment performed at TASCC, Chalk River, Ontario, Canada.
†Present address: NSRL, University of Rochester, 271 East River Road, New York 14627.
‡Present address: Institut de Physique Nucléaire d'Orsay, B. P. 91406, Orsay Cedex, France.
§Present address: AECL Research, Chalk River Laboratories, Ontario, Canada, K0J-1J0.
¶Present address: Hôtel-Dieu de Québec, Département de Radio-Oncologie, Québec, Canada.

most central collisions for a light system of 36 nucleons. Since a determination of the source properties for these events is essential, detection of the entire system is a prerequisite for a proper event-shape analysis. Our analysis is limited to collisions in which the total charge of the system (projectile + target) has been detected in an array of charged-particle counters. We compare the anisotropy and velocity distributions of the data with those predicted by two excitation scenarios representing opposite, limiting cases, namely complete fusion and dissipative binary collisions. A general trend, favoring a competition between compound nucleus formation and a dissipative binary mechanism is observed for violent channels.

2. EXPERIMENTAL SET-UP

The experiment was performed at the TASCC facility of Chalk River Laboratories (CRL), with beams of ^{24}Mg ions at $25A$ and $35A$ MeV, incident on a 2.4 mg/cm^2 C target. The inverse-kinematics experiment focussed reaction products into the 80-detector CRL-Laval Array. The array is comprised of three rings of 16 plastic phoswich scintillation detectors each, covering angles from 6° to 24° (sensitive to particles of $Z = 1$ to 12) and two rings of 16 CsI(Tl) scintillators each, covering angles from 24° to 46° (sensitive to particles of $Z = 1$ and 2). Each phoswich detector consists of a thick, slow-plastic E-detector and a 0.7 mm ΔE layer of fast plastic scintillator, heat-pressed to the front of the E detector.[9] Energy calibration data were obtained from elastically scattered ^{24}Mg ions and secondary beams of $Z = 1$ through 11 scattered on ^{197}Au targets located at various distances from the detectors. The phoswich detectors were calibrated according to the relation given in Ref. 10 and the CsI(Tl) detectors by the energy–light relation from Ref. 11. The intrinsic energy resolution of the detectors was better than 5% and the precision of the energy–light relation was close to 5% for both types of detectors. The energy threshold varied from 7.5 to 19.6 MeV/nucleon for $Z = 1$ to $Z = 12$ for the phoswich detectors and was approximately 2 MeV/nucleon for $Z = 1$ and $Z = 2$ in the CsI(Tl) detectors. Isotopic identification for charge $Z = 1$ was possible in the CsI(Tl) detectors in the $25A$ MeV experiment. For the other particles the mass given was that of the most abundant isotope.

The reconstructed velocity of the center of mass (CM) for the complete events provides a good test of whether pileup has been successfully excluded. Fig. 1 shows such a reconstruction for events with total detected charge $\Sigma Z = 18$ and $\Sigma Z = 12$, at $25A$ and $35A$ MeV, along with the calculated CM and beam velocity for both energies. Velocity reconstruction for detected events with $\Sigma Z = 12$ demonstrates the bias that incomplete events would make on the analysis. As indicated by the vertical lines in the figure, complete events lying in the tails toward lower or higher velocities were excluded from the analysis. For exit channels with less ambiguity in the isotopic identification of the reaction products, further improvements in the reconstructed CM velocity may be achieved, as seen in Fig. 8 of Ref. 11. This fact is exploited in our analysis of the 9-He exit channel.

3. DATA ANALYSIS AND SIMULATIONS

The experimental events in which the total system charge was detected ($\Sigma Z = 18$) had charged-particle multiplicities between 5 and 12, and represented a wide variety of exit channels. Of the observed events, 63% combined one intermediate-mass fragment (IMF) of element number 3 to 11 with hydrogen and helium ions, 28% had two or three IMFs plus light ions, and the remaining 9% were completely disassembled into light ions. The completely

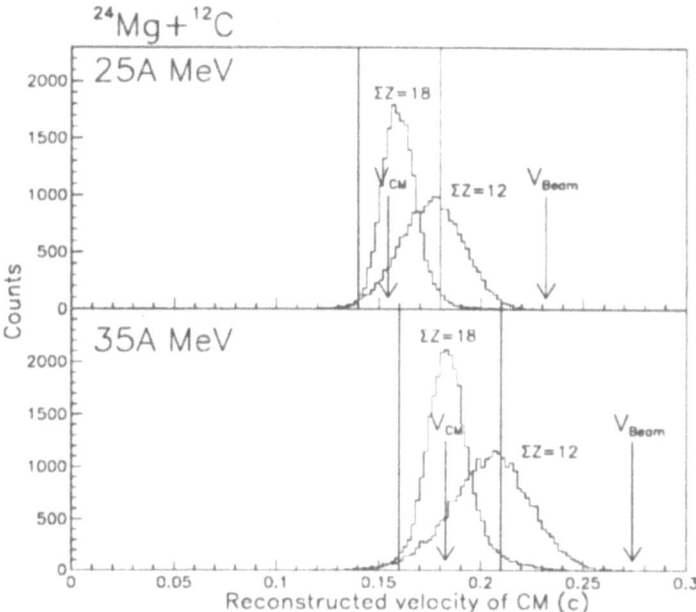

Figure 1. Reconstructed center-of-mass velocity for exit channels with $\Sigma Z = 12$ and $\Sigma Z = 18$ from the ^{24}Mg + ^{12}C reaction at 25A MeV (top) and 35A MeV (bottom). The arrows indicate beam velocity and CM velocity for the complete system of target plus projectile. Vertical lines show the $\Sigma Z = 18$ events selected for further analysis.

detected events represent an absolute cross-section of 8 mb at 25A MeV and 20 mb at 35A MeV, which is a small percentage of the total reaction cross-section, estimated at 1600 mb, following prescription from Ref. 12. The cross-section increase between 25A MeV and 35A MeV may be due to the increased detector acceptance for higher velocities.

A mid-rapidity charge parameter (Z_{mr})[13] was used to evaluate the centrality of the events. At 25A MeV, 76% of the events had Z_{mr} greater than 15 and at 35A MeV the corresponding fraction was 62%, indicating the violent nature of the majority of the $(\Sigma Z = 18)$ events. Another way to probe the violence of a collision is to extract the ratio between the total transverse energy of a given event and the total energy in CM frame of the reaction, for each beam energy. For the completely detected events, that ratio averaged 0.30 at 25A MeV and 0.27 at 35A MeV.

Possible reaction mechanisms for these violent events include dissipative binary collisions, as well as complete and incomplete fusion. Incomplete fusion reactions are largely eliminated from our analysis by the $\Sigma Z = 18$ requirement. It has been shown[3,14] that in incomplete fusion reactions viewed in reverse kinematics, the pre-equilibrium emission of target-like spectators is not forward-peaked in the laboratory frame. Since the probability is very low that all pre-equilibrium, target-like charged particles are emitted forward of 46° and above detector threshold, we would not detect incomplete fusion reactions as complete events. Similarly, pre-equilibrium emission of projectile-like spectators, though rare in reverse kinematics reactions, would be very forward-peaked, and mostly lost in the beam-exit port of the array. Simulations of incomplete fusion reactions were done with the code GENEVE[15] for the system at 35A MeV. Fig. 2 shows the parallel versus perpendicular velocity of pre-equilibrium proton emission for such a mechanism. As seen in the figure, most pre-equilibrium particles

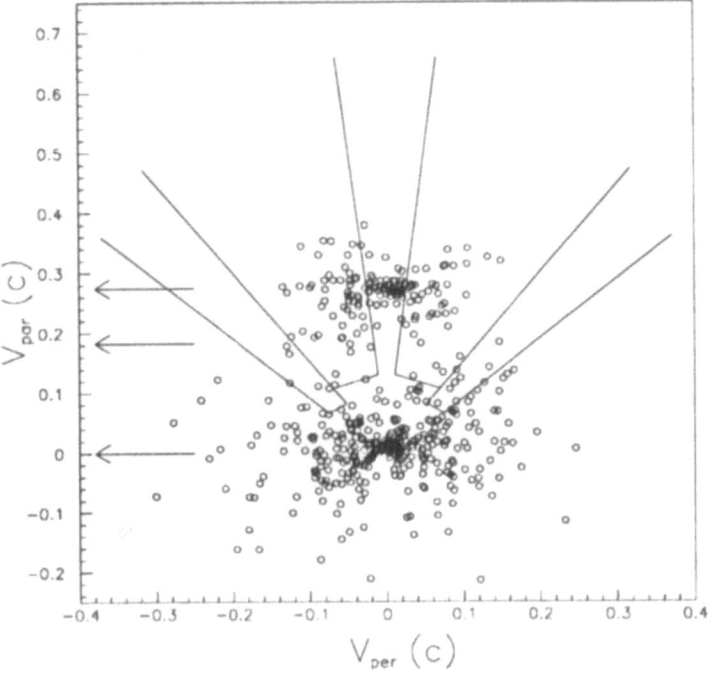

Figure 2. Parallel versus perpendicular velocity plot of pre-equilibrium proton emission in an incomplete fusion scenario of ^{24}Mg + ^{12}C at 35A MeV simulated with the code GENEVE. Lines represent the geometric and energetic thresholds of the array of detectors for protons. Arrows show projectile (0.27c), CM (0.18c) and target (0.0c) velocities.

are eliminated by the geometric and energetic thresholds of the detector arrays. The events with complete charge detection were compared to simulations generated with the statistical code, GEMINI[16] and filtered by the detector acceptance. Two extreme excitation scenarios were considered: complete fusion in which the projectile and the target form a thermalized compound nucleus, and binary dissipative collisions in which a two-source system is produced, composed of a quasi-projectile and a quasi-target with different kinematic and energetic characteristics. The angular momentum input to the GEMINI code was determined from a comparison between experimental data and simulations made for different values of angular momentum. For the complete fusion scenario, we used the maximum angular momentum that can be sustained by the nucleus (25$\hbar$), as determined from the prescription of Ref. 17. For the dissipative binary collision scenario, the value of angular momentum was taken as proportional to the excitation energy of the nucleus, without exceeding the maximum sustainable angular momenta. The simulated disintegrations produced with GEMINI were transformed into the laboratory frame. The complete events were then passed through the experimental filter reproducing the geometry and energy thresholds of the multi-detector array. The filtering also took into account isotopic variations and the position uncertainty due to the large solid angle of each detector. Only the filtered events with $\Sigma Z = 18$ were retained for the comparison. The charged-particle multiplicities of the filtered events in both excitation scenarios were consistent with, but slightly higher than, the experimental data.

The elongation of an event in momentum space is an important observable for distinguishing binary and fusion-like reaction mechanisms. Quantitatively, a global variable can

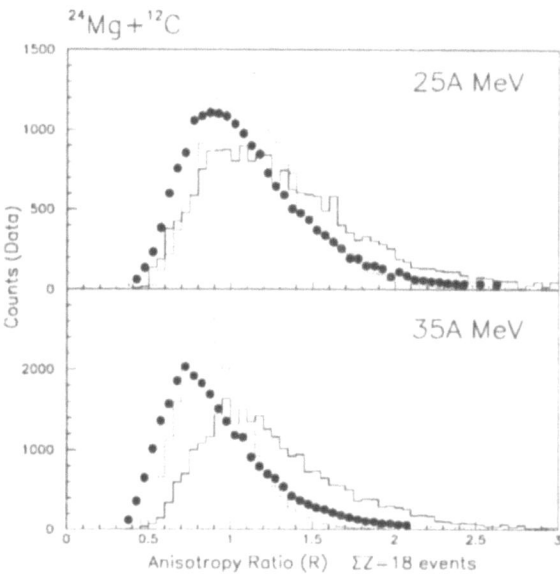

Figure 3. Centroids of the anisotropy ratios versus multiplicity for incompletely detected events (ΣZ = 15, 16, 17) at 25A MeV (top left) and 35A MeV (bottom left). Completely detected experimental events (ΣZ = 18) and the corresponding filtered simulations are shown for 25A MeV (top right) and 35A MeV (bottom right). Filled circles represent experimental data, open squares complete fusion simulations with GEMINI, and open triangles dissipative-binary collision simulations with TORINO and GEMINI. Error bars are the root-mean square divided by the square-root of the number of counts. The horizontal lines represent R centroids for unfiltered simulations for complete fusion (full lines) and dissipative binary collisions (dashed lines).

be constructed from a comparison of the longitudinal and transverse momentum components of the event's constituent particles.[18] This anisotropy ratio, R, is defined as

$$R = \frac{2}{\pi} \frac{\sum_{i=1}^{M} |P_{iCM\perp}|}{\sum_{i=1}^{M} |P_{iCM\parallel}|}, \tag{3.1}$$

where $\frac{2}{\pi}$ is a geometric normalization constant, M is the charged-particle multiplicity, and $P_{iCM\parallel}$ and $P_{iCM\perp}$ are momenta of the ith particle in the CM frame, perpendicular and parallel to the beam axis. Fig. 3 shows anisotropy ratio distributions compared to simulations, for both beam energies. The dissipative binary simulations show a better fit to the data at both beam energies. Fig. 4 shows the anisotropy distribution as a function of charged-particle multiplicity for experimental events with ΣZ = 15, 16 or 17, and for both experimental and simulated events with ΣZ = 18 at both beam energies. The effect of the experimental acceptance on the anisotropy ratio distributions may be deduced from the horizontal lines representing the R distributions for primary or "unfiltered" simulations, averaged over all multiplicities. It is important to note that the R distributions are skewed about their centroids and that the widths of the distributions vary. Consequently, the experimental acceptance may highlight the difference between two distributions having similar "unfiltered" centroids, as is the case in the 25A MeV simulations.

Simulated complete fusion events were obtained from the disintegration of an argon compound nucleus, recoiling at center-of-mass velocity. The excitation-energy was set to

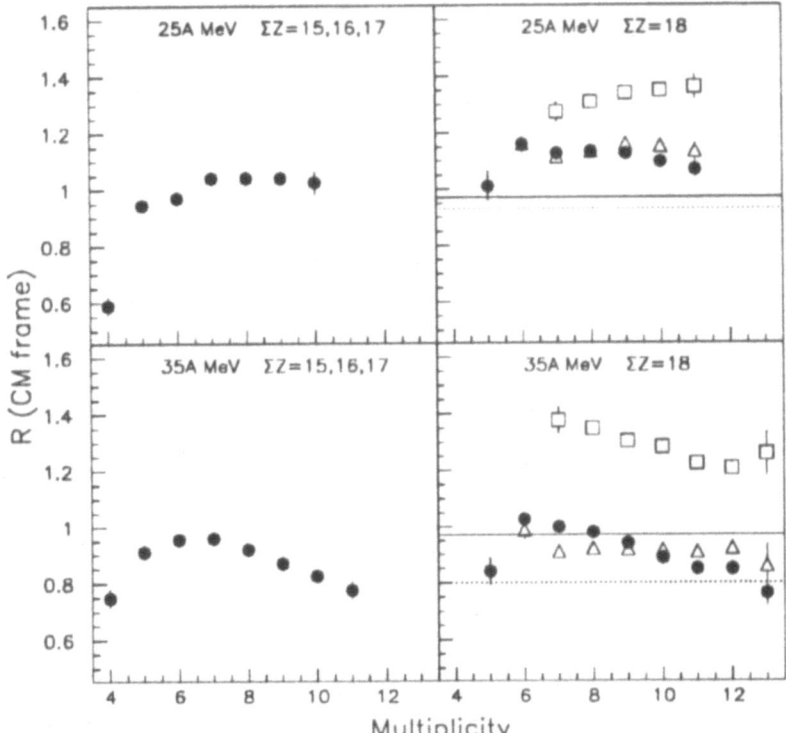

Figure 4. Anisotropy ratio distributions, with R as defined in Eq. (1), for completely detected experimental events ($\Sigma Z = 18$) and the corresponding filtered simulations at 25A MeV (top) and 35A MeV (bottom). Filled circles represent experimental data, full lines complete fusion simulations with GEMINI, and dashed lines dissipative-binary collision simulations with TORINO and GEMINI.

the maximum available in the CM reference frame, *i.e.*, 200 MeV for events from the 25A MeV reaction and 280 MeV for the 35A MeV reaction. In the simulation of dissipative binary collisions, the excitation energy and scattering angle of both the quasi-projectile and the quasi-target were provided by the semi-classical coupled-channels code, TORINO.[19] The code requires the input of an impact parameter, from which it deduces the subsequent course of the reaction. For systems as light and energetic as ours, this should not necessarily be taken as the geometric trajectory of the entrance channel, but rather as a relative scale for the violence of the interation. The impact parameter that best reproduces the anisotropy ratios at 25A MeV is a nominal 5.1 fm, which gives excitations of 95 and 81 MeV and velocities of 74% and 57% of the projectile velocity for the projectile-like and target-like sources, respectively. At 35A MeV, the best agreement is obtained with 4.1 fm, giving the corresponding excitation energies of 145 and 98 MeV, and velocities of 76% and 51% of the projectile velocity.

At both energies, the anisotropies, measured as a function of multiplicity, are found to be similar for the incompletely detected ($\Sigma Z = 15$, 16, or 17) and completely detected ($\Sigma Z = 18$) events. For the complete events, at both energies, the anisotropy ratios lie close to the values predicted by the dissipative binary simulation at all multiplicities. Of particular interest is the dependence of R upon beam energy. Clearly, at 35A MeV, the anisotropy values deviate more from the fusion predictions than at 25A MeV. This suggests that, as beam energy increases, the two sources become increasingly separated in velocity space.

The $\Sigma Z = 18$, 9-He exit channel of this reaction was studied in somewhat greater

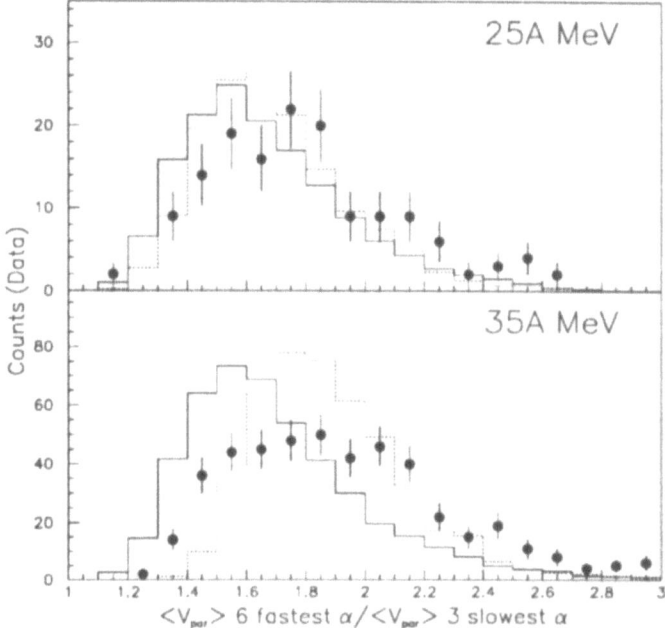

Figure 5. Ratio between the average parallel velocities of the six fastest alpha-particles and the three slowest alpha-particles in the ^{24}Mg + ^{12}C → 9-He channel at 25A MeV (top) and 35A MeV (bottom). Experimental events are shown as filled circles. The full line represents filtered complete fusion events simulated with SOS and the dotted line indicates the filtered TORINO-SOS simulation of dissipative binary collisions. Error bars show the square root of the number of counts.

detail, permitting an analysis with less isotopic ambiguity. We were able to treat this as a 9-α channel. A similar competition between fusion and binary mechanisms was observed. Here the simulations were done with the sequential part of the statistical code SOS,[20] modified by its author to produce exit channels containing only α-particles. The initial conditions for the complete fusion simulation were the same as those for the GEMINI calculations, but with no angular momentum consideration. As before, TORINO was used to produce the conditions for the binary simulations, with a small range of impact parameters, rather than a single value, for each energy. In order to get the best possible fit to the experimental data, impact parameters from 5.0 to 6.0 fm and from 4.0 to 5.0 fm were selected for the 25A MeV and 35A MeV reactions, respectively. Fig. 5 shows the ratio between the average parallel velocity of the 6 fastest alpha-particles and the average velocity of the 3 slowest alpha-particles. This choice of variable is based on the products expected from the decay of a ^{24}Mg* quasi-projectile and a ^{12}C* quasi-target. Both simulated distributions resemble the data at 25A MeV, though the data may be somewhat closer to the dissipative binary collision case. At 35A MeV, the binary character of the 9-α channel is evident.

4. CONCLUSIONS

In conclusion, the reconstructed CM velocity distributions have demonstrated both the validity of the energy calibration and the value of total charge detection in the correct determination of source properties. We have compared the anisotropies and source-velocity ratios observed experimentally for complete events with the results of simulations filtered by detector response. The comparisons indicate that as the energy of the collision increases, binary mechanisms may compete successfully with simple compound nucleus formation.

The fact that the binary simulations, which assume two very different sources in size, velocity and excitation energy, give a better fit to part of the data may be explained by various reaction mechanisms. To properly assess the phenomena, detailed data, such as those presented here, should be compared with a model that treats the dynamics of source formation as well as pre-equilibrium emission and the subsequent statistical decay.

We would like to thank R. J. Charity, J. A. Lopez and J.-P. Wieleczko for the use of their statistical codes. This work was supported in part by the Natural Sciences and Engineering Research Council of Canada.

REFERENCES

1. D. H. E. Gross, Rep. Prog. Phys. 53(1990)605, and references therein.
2. L. G. Moretto and G. J. Wozniak, Ann. Rev. Nucl. Phys. 1993, and references therein.
3. H. Morgenstern et al., Phys. Rev. Lett. 52(1984)1104.
4. D. R. Bowman et al., Phys. Lett. B189(1987)282.
5. Y. Blumenfeld et al., Phys. Rev. Lett. 66(1991)576.
6. K. Hanold et al., Phys. Rev. C48(1993)723.
7. B. Lott et al., Phys. Rev. Lett. 68(1992)3141.
8. J. F. Lecolley et al., Phys. Lett. B325(1994)317.
9. C. A. Pruneau et al., Nucl. Instr. and Meth. in Phys. Res. A297(1990)404.
10. J. Pouliot et al., Nucl. Instr. and Meth. in Phys. Res. A270(1988)69.
11. Y. Larochelle et al., Nucl. Instr. and Meth. in Phys. Res. A348(1994)167.
12. S. Kox et al., Nucl. Phys. A420(1984)162.
13. C. A. Ogilvie et al., Phys. Rev. C40(1989)654.
14. A. Malki et al., Z. Phys. A - Atoms and Nuclei 339(1991)283.
15. J. P. Wieleczko, E. Plagnol and P. Ecomard, Proc. of the 2nd TAPS Workshop, Editors Diaz, Martinez and Shutz, Guardemar 1993, World Scientific, p. 145.
16. R. J. Charity et al., Nucl. Phys. A483(1988)371.
17. R. Schmidt and H. O. Lutz, Phys. Rev. A45(1992)7981.
18. H. Ströbele, Phys. Rev. C27(1983)1349.
19. C. H. Dasso and G. Pollarolo, Comp. Phys. Comm. 50(1988)341.
20. J. A. Lopez and J. Randrup, Comp. Phys. Comm. 70(1990)92.

COHERENT PIONS IN NUCLEUS–NUCLEUS COLLISIONS AT INTERMEDIATE ENERGIES

A. Badalà,[1] R. Barbera,[1,2] A. Palmeri,[1] G. S. Pappalardo,[1] F. Riggi,[1,2]
A. C. Russo,[1] G. Russo,[1,3] and R. Turrisi[1]

[1]Istituto Nazionale di Fisica Nucleare, Sez. di Catania
Corso Italia, 57 — I 95129 Catania, Italy
[2]Dipartimento di Fisica dell'Università di Catania
Corso Italia, 57 — I 95129 Catania, Italy
[3]Istituto Nazionale di Fisica Nucleare, Laboratorio Nazionale del Sud
Via S. Sofia, 44 — I 95123 Catania, Italy

ABSTRACT

Charged pion production has been observed in coincidence with the 4-He projectile breakup channel in ^{16}O + ^{27}Al reactions at 94 MeV/nucleon. Charged pions have been detected using a *range* telescope of plastic scintillators placed at 70°, 90° and 120° with respect to the beam direction. Helium nuclei have been detected using a multielement array of plastic scintillators covering the zenithal angular domain between 3° and 30° and the total azimuthal one. The velocity of the primary projectile-like nucleus (v_{PLN}) has been reconstructed. The v_{PLN} distribution of coincidence events is shifted toward lower velocities with respect to that of inclusive events indicating a sharp coherent deceleration of the projectile when a pion is created. This could be an indication of collective mechanisms for pion production.

1. INTRODUCTION

Subthreshold pion production is still a matter of debate.[1] Many phenomenological features, both in inclusive and exclusive experiments, can be described using statistical models as well as microscopic models based on incoherent production in nucleon–nucleon (N–N) scatterings with an appropriate description of pion reabsorption.[1,2,3,4,5] However, single N–N collisions may not be the unique source for pion production when the total available energy is very close to that of the produced particles. At very low bombarding energy,[6] for example, it has been observed that up to 75% of the total energy available in the center of mass frame is concentrated on the production and emission of an energetic pion. This is only conceivable

Advances in Nuclear Dynamics
Edited by Wolfgang Bauer and Alice Mignerey, Plenum Press, New York, 1996

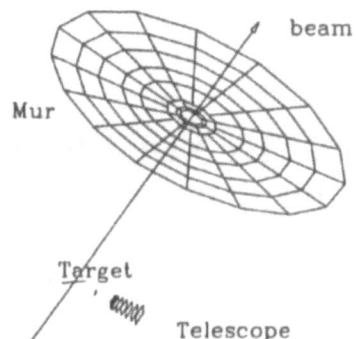

Figure 1. Global view of the experimental setup.

if pion production mechanisms going beyond single nucleon–nucleon collisions play a role. Apart statistical models, in the last ten years several collective/coherent theoretical models, based on various approaches like the isobar production,[7,8,9,10,11,12] the pion condensation,[9] the microscopic many-body mechanism,[13] and the so-called pionic bremmstrahlung,[14,15,16] have been proposed. However, a clear experimental proof capable to distinguish coherent pion production from other mechanisms in heavy ion collisions is still lacking. In fact, at these incident energies both projectile and target generally break into many pieces and loose their identity. Thus, the usual observation of only a part of them in coincidence with a pion do not prove unambiguously the original mechanism, since the final state of the reaction cannot be easily attributed to a well defined process. Recently, some attention has been devoted to the study of projectile breakup in peripheral collisions induced by light nuclei on different targets, spanning the whole range of stable masses, at various bombarding energies.[17,18,19,20,21] Projectile disassembly channels, involving charged particles, have been observed in those events where the total detected charge was equal to the projectile one. Some physical quantities concerning the whole primary projectile-like nucleus, like its velocity and excitation energy distributions, have been reconstructed from the measured kinetic energies of all detected fragments in the event center-of-mass frame. Obviously, the observation of a pion in these breakup reactions could give a clue of the existence of coherent production mechanisms.

2. THE EXPERIMENT

The experiment described in this contribution was performed at GANIL using a ^{16}O beam impinging on an ^{27}Al target at 94 MeV/nucleon .[22]

The overall detection apparatus is sketched in Fig. 1. Charged pions of kinetic energy ranging from 23.4 MeV up to 74.1 MeV have been detected, at laboratory angles 70°, 90° and 120°, using a *range* telescope of ten NE110 plastic scintillators covering a 58.8 msr solid angle. Other particles impinging on the telescope were on-line rejected by adjusting the discriminator thresholds in order to eliminate high energy losses due to particles heavier than pions (mainly protons). The residual contaminants (electrons, not rejected protons, etc.) were eliminated in the off-line data reduction performing a multiple ΔE-E and ΔE-ΔE analysis of the relative energy loss in each scintillator element of the telescope. For a detailed description of a typical *range* telescope see for example Ref. 3, 23, 24. The system for the detection of light charged particles, the *Mur*,[25] is already described in Ref. 20. A charge identification of the particles, from hydrogen to oxygen isotopes, was obtained by means of a standard ΔE–TOF technique. However, a very clean separation of the different charges was possible only for particles crossing the scintillators, while those stopped in the detectors could not be easily identified.[26] Only charge identified particles have been taken into account and a velocity threshold of about 5 cm/ns, due to the thickness of the detectors, has been imposed on the data.

The analysis has been restricted only to $\pi^{\pm} \cap$ 4-He coincidence events where all four $Z = 2$ particles impinging on the *Mur* satisfied the "projectile breakup" conditions discussed in Ref. 20. Essentially these conditions consist in taking into account all Helium ions having $v_{He} > 8$ cm/ns and detected at angles $\theta_{He} \leq 16.5°$. These severe conditions have been chosen to select those events resulting from the breakup of the primary projectile-like nucleus,[20] in which there is the minimum contamination due to fragments coming from more relaxed sources relative to different mechanisms at smaller impact parameters.[2] It is in this framework that we speak about peripheral pion production in projectile breakup reactions. In order to increase the statistics as much as possible, both pion charge states and detection angles are summed up. The contribution of random coincidences has been statistically evaluated by counting the number of $\pi^{\pm} \cap$ 4-He events contained in the spurious coincidence peaks of the time-of-flight spectrum of each detector of the *Mur*[26] but none of these events has been found.

3. RESULTS

Using the standard relativistic kinematics, we have reconstructed, on an event-by-event basis, the projectile-like nucleus velocity (v_{PLN}) defined as that of the center of mass of the four He ions detected in each event. Figure 2 shows the distribution of this quantity for inclusive (upper part)[20] and coincidence events (lower part). Right and left arrows indicate projectile and compound nucleus velocities, respectively. The inclusive velocity distribution has the usual shape observed in projectile fragmentation at the intermediate energies. The centroid of the PLN velocity spectrum is slightly lower than the beam velocity in agreement with the existing phenomenology. The net effect of a charged pion detected in coincidence with the four He ions is to shift the velocity spectrum to lower values. The peaks of the PLN velocity spectra of Fig. 2 are shifted by about 1.5 cm/ns which is, at these velocities, more than five times the velocity indetermination due to the experimental uncertainties. This corresponds to a projectile energy loss of 200–250 MeV, large enough to produce a charged pion in the measured kinetic energy range. So, by detecting all projectile breakup fragments inclusively and in coincidence with the pion and measuring their velocities, we are able in principle to select for the first time a well defined class of events in which the total pion energy (which is an appreciable fraction of the energy available) is provided by the coherent slowing down of the projectile.

Notwithstanding the coherent process discussed above is energetically possible, the number of true coincidences (18 events) is very small as compared to the total number of detected pions ($\sim 1.4 \cdot 10^4$) and, therefore, we had to perform several tests in order to verify that the particles observed in the $\pi^{\pm} \cap$ 4-He events were really pions and to check whether light charged particles hitting the *Mur* in coincidence with a pion effectively come from the slowed down projectile or from a more relaxed source (*fireball*)[2] (i.e. from more central collisions). We have reconstructed the v_{PLN} distribution for those events where the four He ions have been detected in coincidence with protons of the same *range* of the observed pions (i.e. stopping in the same elements of the telescope). The mean value of this distribution is equal, within the error bars, to that of the inclusive projectile-like nucleus velocity distribution.

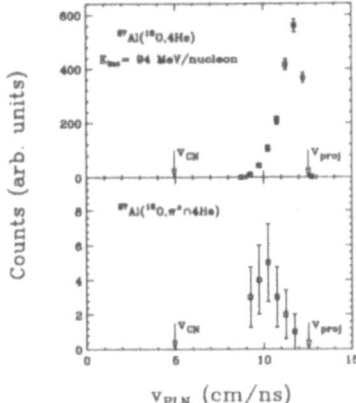

Figure 2. Projectile-like nucleus velocity distributions for 4-He inclusive breakup events (upper part) and for coincidence ones (lower part). In each plot right and left arrows indicate the projectile velocity and the compound nucleus velocity respectively.

Moreover, we simulated the simultaneous emission of pions and light charged particles from an excited source using a Montecarlo event-generator code based on the Modified Fireball Model (MFM),[27] which has been successfully used in the last few years to reproduce several features of heavy-ion dynamics at intermediate energies.[2,4,28] The experimental cuts used to obtain the results plotted in Fig. 2 were also imposed on the simulated data. The aim of the calculation was to estimate the probability of the occurrence of a $\pi^{\pm} \cap 4$-He coincidence event coming from the *fireball*. The number of computer generated events was equal to that of measured events. None of these events was found for the studied reaction. In Fig. 3 the total (left part) and single fragment (right part) experimental parallel momentum distributions for inclusive 4-He breakup events (upper part) and for $\pi^{\pm} \cap 4$-He coincidence ones (lower part) are reported. All momenta are normalized to the projectile momentum (p_b). In both cases coincidence distributions are shifted toward lower parallel momentum values with respect to the inclusive distributions.

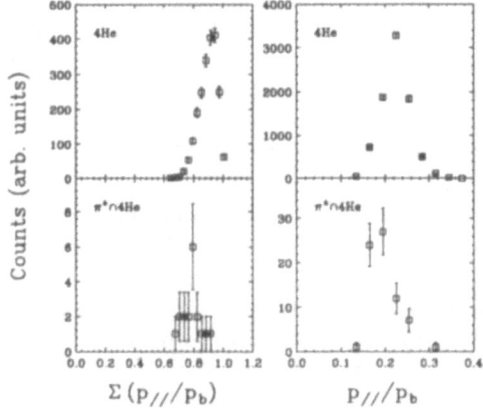

Figure 3. Total (left part) and single fragment (right part) parallel momentum distributions for 4-He inclusive breakup events (upper part) and for $\pi^{\pm} \cap 4$-He coincidence ones (lower part). All momenta are divided by that of the projectile (p_b).

Table 1. Weighted Mean Values of Total and Single Fragment Parallel Momenta for Inclusive Breakup Events and for Coincidence Ones. All momenta are divided by that of the projectile (p_b). The errors are statistical

Event	$\sum (p_{\parallel}/p_b)$	$p_{\parallel}/p_b$
^{27}Al(^{16}O,4-He)	0.898 ± 0.001	0.224 ± 0.001
^{27}Al(^{16}O,$\pi^{\pm} \cap$ 4-He)	0.787 ± 0.014	0.197 ± 0.004

Weighted mean values of the distributions plotted in Fig. 3 are also reported in Table 1. One can easily calculate $[\langle \sum (p_{\parallel}/p_b)\rangle_{4He} - \langle \sum (p_{\parallel}/p_b)\rangle_{\pi^{\pm}\cap 4He}] \simeq 4 \cdot [\langle p_{\parallel}/p_b\rangle_{4He} - \langle p_{\parallel}/p_b\rangle_{\pi^{\pm}\cap 4He}]$ which indicates that the slowing down of the center of mass of the four He ions detected in coincidence events is due to coherent slowing down of each of these particles. Following Ref. 20, we have also reconstructed the projectile-like nucleus excitation energy distribution (E^*_{PLN}) for $\pi^{\pm} \cap$ 4-He events. The E^*_{PLN} mean value is 52 ± 5 MeV which is equal, within the statistical errors, to the value relative to inclusive events,[20] 54 ± 1 MeV. Thus, the four He ions detected in coincidence with a charged pion reasonably come from the same kind of projectile disassembly process as that inclusively observed in Ref. 20. Finally, we have applied the statistical Student t-test to the two distributions plotted in Fig. 2 in order to quantitatively evaluate the probability that they come from different populations. The result of the test allows to conclude that the two distributions are different within a confidence level larger than 99.5%.

4. CONCLUSIONS

In conclusion, charged pion production has been observed in coincidence with the 4-He breakup channel of the projectile in reactions induced by a ^{16}O beam at 94 MeV/nucleon on a ^{27}Al target. An event-by-event analysis has allowed to reconstruct the projectile-like nucleus velocity distributions both for inclusive and coincidence data. The v_{PLN} distribution of $\pi^{\pm} \cap$ 4-He events is shifted toward lower velocities with respect to that of inclusive events and a coherent slowing down of each He ion is evidenced. Notwithstanding random coincidence contamination has been evaluated and taken into account many tests have been performed on coincidence data in order to verify their reliability. Even though the statistics is poor we are confident that the mechanism for pion production is in our case associated to the coherent deceleration of the projectile nucleus. A detailed study of the shape of both pion energy spectra and angular distribution, in coincidence with the 4-He projectile breakup, may be a powerful tool to definitively disentangle a coherent production.

REFERENCES

1. G. Bertsch, S. Das Gupta, Phys. Rep. **160**, 189 (1988); W. Cassing, V. Metag, U. Mosel, K. Niita, Phys. Rep. **188**, 235 (1990); A. Bonasera, F. Gulminelli, J. J. Molitoris, Phys. Rep. **243**, 1 (1994).
2. R. Barbera et al., Nucl. Phys. A **518**, 767 (1990).
3. A. Badalà et al., Phys. Rev. C **43**, 190 (1991).
4. A. Badalà, R. Barbera, A. Palmeri, G. S. Pappalardo, F. Riggi, G. Bizard, D. Durand, J. L. Laville, Phys. Rev. C **45**, 1730 (1992); J. L. Laville, A. Badalà, R. Barbera, G. Bizard, R. Bougault, A. Palmeri, G. S. Pappalardo, F. Riggi, Nucl. Phys. A **564**, 564 (1993).
5. A. Badalà et al., Phys. Rev. C **47**, 231 (1991).

6. G. R. Young, F. E. Obenshain, F. Plasil, P. Braun-Munzinger, R. Freifelder, P. Paul, J. Stachel, Phys. Rev. C **33**, 742 (1986).
7. G. E. Brown, P. A. Deutchman, *Proceedings of the Workshop on High Resolution Heavy Ion Physics at 20–100 MeV/A*, Saclay (France), (1978).
8. P. A. Deutchman, L. W. Townsend, Phys. Rev. Lett. **45**, 1622 (1980).
9. H. J. Pirner, Phys. Rev. C **22**, 1962 (1980).
10. B. Hiller, H. J. Pirner, Phys. Lett. B **109**, 338 (1982).
11. L. W. Townsend, P. A. Deutchman, R. L. Madigan, J. W. Norbury, Nucl. Phys. A **415**, 520 (1984).
12. M. Prakash, C. Guet, G. E. Brown, Nucl. Phys. A **447**, 625c (1985).
13. K. Klingenbeck, M. Dillig, M. G. Huber, Phys. Rev. Lett. **47**, 1655 (1981).
14. D. Vasak *et al.*, Phys. Scr. **22**, 25 (1980).
15. D. Vasak *et al.*, Nucl. Phys. A **428**, 291c (1984).
16. T. Stahl *et al.*, Z. Phys. A **327**, 311 (1987).
17. J. Pouliot, Y. Chan, D. E. Di Gregorio, B. A. Harmon, R. Knop, C. Moisan, R. Roy, R. G. Stokstad, Phys. Rev. C **43**, 735 (1991).
18. J. Pouliot *et al.*, Phys. Lett. B **263**, 18 (1991).
19. R. J. Charity *et al.*, Phys. Rev. C **46**, 1951 (1992).
20. A. Badalà, R. Barbera, A. Palmeri, G. S. Pappalardo, F. Riggi, G. Pollarolo, C. H. Dasso, Phys. Lett. B **299** (1993) 11; A. Badalà, R. Barbera, A. Palmeri, G. S. Pappalardo, F. Riggi, Phys. Rev. C **48**, 633 (1993).
21. J. Pouliot, D. Doré, R. Laforest, R. Roy, C. St-Pierre, J. A. lopez, Phys. Lett. B **299**, 210 (1993).
22. A. Badalà, R. Barbera, A. Palmeri, G. S. Pappalardo, F. Riggi, A. C. Russo, G. Russo, R. Turrisi, G. Bizard, and J. L. laville, Phys. Lett. B **316**, 240 (1993).
23. V. Bernard *et al.*, Nucl. Phys. A **423**, 511 (1984).
24. J. Julien, *Proceedings of the 3rd Interational Conference on Nuclear Reaction Mechanism*, Varenna (Italy), (1982).
25. G. Bizard, A. Drouet, F. Lefebvres, J. P. Patry, B. Tamain, F. Guilbault, C. Lebrun, Nucl. Instr. Meth. Phys. Res. A **244**, 483 (1986).
26. D. Durand *et al.*, Nucl. Phys. A **511**, 442 (1990).
27. A. Bonasera, M. Di Toro, C. Gregoire, Nucl. Phys. A **483**, 738 (1988).
28. A. Badalà, R. Barbera, A. Palmeri, G. S. Pappalardo, F. Riggi, A. C. Russo, Z. Phys. A **344** (1993) 455; A. Badalà *et al.*, *Proceedings of the 9th Winter Workshop on Nuclear Dynamics*, Key West (USA), (1993); A. Badalà *et al.*, Phys. Rev. C **48**, 2350 (1993).

17

^{129}Xe-INDUCED PERIPHERAL REACTIONS AT E/A = 50 MeV

H. Madani,[1] A. C. Mignerey,[1] D. E. Russ,[1] J. Y. Shea,[1] G. Westfall,[2]
W. J. Llope,[2] D. Craig,[2] E. Gualtieri,[2] S. Hannuschke,[2] R. Pak,[2] T. Li,[2]
A. Vandermolen,[2] J. Yee,[2] E. Norbeck,[3] and R. Pedroni[3]

[1]Chemistry Department
University of Maryland
College Park, MD
[2]National Superconducting Cyclotron Laboratory
and Department of Physics and Astronomy
Michigan State University, East Lansing, MI
[3]Department of Physics and Astronomy
University of Iowa
Iowa City, IA

ABSTRACT

The charge distributions of the projectile-like fragments (PLF's) produced in the reactions of ^{129}Xe on ^{27}Al, $^{nat.}$Cu, ^{139}La, and ^{165}Ho at 50 MeV/u are compared to those obtained with ^{129}Xe on ^{209}Bi at 28.2 MeV/u by Baldwin et al.[1] The predictions of Tassan-Got's nucleon exchange model[2] are compared to the data to investigate the persistence of the deep-inelastic mechanism in the intermediate energy regime.

Studies of heavy-ion-induced reactions at intermediate energies are useful tools to probe the transition from mean-field dominated mechanisms at low bombarding energies to reactions governed by nucleon–nucleon interations at higher energies. It is generally thought that this transition does not occur at one particular energy, but rather at a continuum of bombarding energies as a function of system size.[3] The nature and size of the interacting system are important factors in governing the formation of the hot nuclear system produced in the reaction. A number of studies have focused on the role of incomplete fusion in this energy regime.[4,5] One of the goals of this current study is to explore the role of the deep-inelastic mechanism in the formation of hot systems, in particular the projectile-like products. Therefore, the first set of systems studied at the Michigan State University National Superconducting Cyclotron Laboratory (MSU-NSCL) by our group was the 50-MeV/u ^{129}Xe on ^{27}Al, $^{nat.}$Cu, ^{139}La, and ^{165}Ho reactions. The projectile-like fragments produced in the peripheral collisions of these systems were detected with the Maryland Forward Array (MFA) in coincidence with target-like fragments, and light charged particles in the MSU-NSCL 4π array. More details about the experiment and the detection system, namely the MFA, can be found in reference 6. The deep-inelastic mechanism, which is prevalent at lower bombarding energies, has been shown to be present in reactions at intermediate energies up to 35 MeV/u[7-].[10] To explore up to what extent this mechanism persists, the PLF distributions obtained in the 50-MeV/u ^{129}Xe on ^{27}Al, $^{nat.}$Cu, ^{139}La, and ^{165}Ho reactions were parameterized in terms of the characteristics of a deep-inelastic process, and some of the results have been previously reported in reference 11.

It is a well known fact that in a deep-inelastic process the fragment mass and charge distributions broaden with increasing energy damping. This broadening, which is indicative of nucleon exchanges between the two interacting heavy ions, was not observed for the systems studied here.[11] In contrast, a broadening of the charge variances with increasing energy dissipation has been reported for the 28.2-MeV ^{129}Xe on ^{209}Bi system by Baldwin et al.[8] It is interesting to compare the present data to those results obtained with reactions at lower bombarding energies to explore the changes that would occur when the bombarding energy is increased.

A comparison between the charge distribution centroids and variances obtained with the 50-MeV/u ^{129}Xe-induced reactions on ^{27}Al, $^{nat.}$Cu, ^{139}La, and ^{165}Ho with those obtained with the 28.2-MeV/u ^{129}Xe on ^{209}Bi by Baldwin et al.[8] is shown in Figure 1. The obvious difference between the lower bombarding energy, and the higher bombarding energy ones is the behavior of the charge variance. For the 28.2-MeV/u ^{129}Xe on ^{209}Bi the value of σ_Z^2 increases with decreasing PLF laboratory energy and reaches a maximum at 1000 MeV. The increase of σ_Z^2 is achieved with 2 different slopes, the steepest one being for PLF energies between 1000 and 2500 MeV. The charge centroids show a change of slope at the same PLF laboratory energy. On the other hand, the values of σ_Z^2 obtained with the 50-MeV/u systems seem to be nearly constant, and the $\langle Z \rangle$'s decrease nearly linearly with decreasing PLF laboratory energy.

A further exploration of this point is possible by comparing distributions predicted by a nucleon exchange model to the present data and to the data obtained with the 28.2-MeV ^{129}Xe on ^{209}Bi. The model of choice in this study was the stochastic nucleon exchange model of Tassan-Got[2], which has shown a good description of low bombarding energy data[12]. Due to the model limitations, centroids and variances of the charge distributions are available only for the lower range of PLF laboratory energy, which approximately corresponds to the range of total kinetic energy loss (TKEL) determined by the entrance channel Coulomb barrier.

The primary distributions (solid lines) predicted by the nucleon exchange model are corrected for light particle evaporation using the decay code GEMINI,[4] and the results are indicated by the dashed lines. It can be seen that, even with evaporation corrections, the data are not reproduced by the model for the 50-MeV/u ^{129}Xe-induced reactions, especially in the case of the two heavier targets. However, it is important to point out that preequilibrium emission of light particles, which can be very sizeable at these intermediate energies, has not been taken into account. On the other hand, a qualitative agreement between the data and the model predictions can be observed for the 28.2-MeV ^{129}Xe on ^{209}Bi system. A good quantitative agreement may be possible if preequilibrium emission is taken into account.

Another mechanism that is explored with the present data is incomplete fusion. In the scenario considered here, a piece of the projectile is sheared off and fuses with the target. The remainder proceeds forward with little change in its momentum. It has been previously shown that beam velocity components are present at high light charged particle multiplicities for the present systems.[11] However, the V_{PLF}/V_{beam} centroids of the velocity distributions obtained

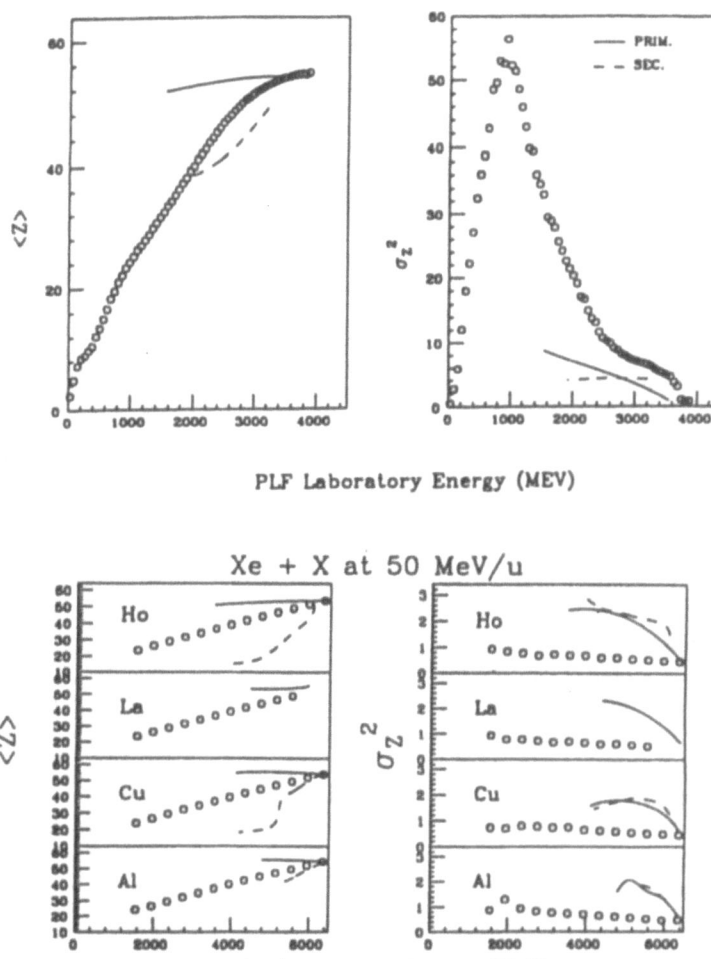

Figure 1. Experimental charge centroids and variances as a function of the PLF laboratory energy for the 28.2-MeV/u ^{129}Xe on ^{209}Bi system (top panel), and for the 50-MeV/u ^{129}Xe-induced reactions (bottom panel). The predictions of the nucleon exchange model before and after evaporation corrections are indicated by the solid and dashed lines, respectively. The data for the Xe on Bi system is courtesy of Baldwin et al.[1]

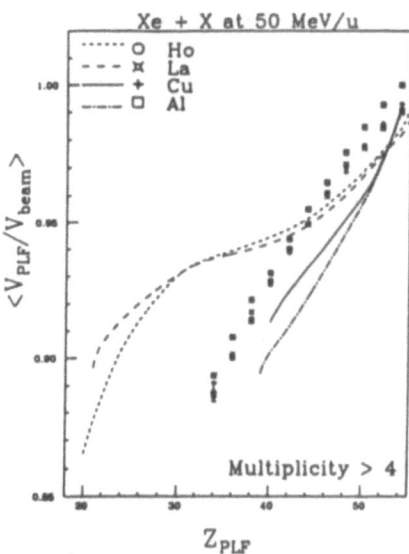

Figure 2. Experimental $\langle V_{PLF}/V_{beam}\rangle$ as a function of PLF secondary charge for the 50-MeV ^{129}Xe-induced reactions. Tassan-Got's model predictions of $\langle V_{PLF}/V_{beam}\rangle$ are indicated.

for the four systems decrease with decreasing PLF charge, as can be seen in Figure 2. As no mass parameter was available from the present data, a mass parameterization was used to determine the experimental PLF velocity. The velocity predicted by the nucleon exchange model for the four systems is also displayed in Figure 2. It is interesting to note the generally good qualitative agreement between the data and the model calculations for the cases of the two lighter targets (Al and Cu), although the predictions are consistently lower than the experimental values. In contrast, for the two heavier targets, the behavior of the experimental velocity is reproduced only when the PLF charges are larger than 42. For lower values of $\langle Z\rangle$, the dependence of the experimental velocity on $\langle Z\rangle$ seems to be steeper than that of the predicted velocity, and the experimental values in this case are consistently larger than the calculated ones, maybe indicating some incomplete fusion mechanism.

      The present parameterization of the PLF's obtained in the 50-MeV/u ^{129}Xe on ^{27}Al, $^{nat.}$Cu, ^{139}La, and ^{165}Ho reactions in terms of a deep-inelastic mechanism has yielded ambiguous results about the presence of this mechanism in this energy regime. Although the experimental centroids and variances of the charge distributions have not been reproduced, the general behavior of the V_{PLF}/V_{beam} centroids with the PLF charge is described for the cases of the Al, and Cu targets. The implementation of a preequilibrium emission calculation is underway to better estimate the available excitation energy in these reactions. In addition, calculations based on the Boltzmann–Uehling–Uhlenbeck (BUU) transport equation are being performed. In these calculations it is possible to explore the influence of non-binary decay channels. An example is the formation of a projectile-like fragment, a target-like fragment and a smaller fragment from the neck region formed by the overlap of the two nuclei.[13]

REFERENCES

1. S. Baldwin *et al.*, Private Communications.

2. L. Tassan-Got and C. Stephan, Internal Report INPO-DRE-89-46, 1989. Nucl. Phys. **A524**, 121 (1991)
3. J. P. Bandorf *et al.*, Phys. Lett. B **62** 30 (1985)
4. R. J. Charity *et al.*, Nucl. Phys. **A476** 516 (1988)
5. D. R. Bowman *et al.*, Nucl. Phys. **A523** 386 (1991)
6. A. Mignerey *et al.*, Progress Report (1992), DOE/ER/40321-9, Progress Report (1993), DOE/ER/40321-12, Progress Report (1993), DOE/ER/40802-1.
7. B. Lott *et al.*, Phys. Rev. Lett. **68**, 3141 (1994)
8. S. Baldwin *et al.*, Proceedings of the 9th Winter Workshop on Nuclear Dynamics, Key West, Florida January 30-February 6, 1993, Ed. B. Back, W. Bauer, and J. Harris World Scientific, Singapore, 1993, P.36
9. B. Quednau, *et al.*, Phys. Lett. B **309**, 10 (1993).
10. V. Skulski *et al.*, Proceedings of the 10th Winter Workshop on Nuclear Dynamics, Snowbird, Utah January 16–22, 1994, Ed. J. Harris, A. Mignerey, and W. Bauer, World Scientific, Singapore, 1994, P.30
11. H. Madani *et al.*, Proceedings of the 10th Winter Workshop on Nuclear Dynamics, Snowbird, Utah January 16–22, 1994, Ed. J. Harris, A. Mignerey, and W. Bauer, World Scientific, Singapore, 1994, P.151
12. H. Madani *et al.*, Phys. Rev. C. (accepted for publication)
13. Montoya *et al.*, Proceedings of the 10th Winter Workshop on Nuclear Dynamics, Snowbird, Utah January 16–22, 1994, Ed. J. Harris, A. Mignerey, and W. Bauer, World Scientific, Singapore, 1994, P.168

18

POSSIBLE SYNTHESIS OF ELEMENT 110 AND THE FUTURE PROSPECTS FOR SUPERHEAVY ELEMENTS

W. Loveland

Deptartment of Chemistry
Oregon State University
Corvallis, OR

ABSTRACT

An experiment to synthesize element 110 by the ^{59}Co + ^{209}Bi reaction has been performed. One event with many of the characteristics of a successful synthesis of ^{267}110 was observed. The best reactions for the synthesis of the superheavy elements are discussed.

1. INTRODUCTION

The past year was a spectacular year for heavy element nuclear science. Beginning with the long-awaited confirmation of the discovery of element 106,[1] we heard a few months later of the synthesis of new n-rich isotopes of seaborgium(106) and hassium(108).[2,3] The significance of this work was that when one compared the measured halflives of these new isotopes with theoretical predictions[4] (Figure 1), remarkable agreement between theory and experiment was observed. Similar agreement was found for all the new very heavy nuclides made during this year. These developments were followed by reports of the synthesis of four different isotopes of element 110, by three different experimental groups.[5,6,7,8] Three of these nuclei were made using cold fusion reactions, ^{209}Bi (^{59}Co, n) ^{267}110 ($t_{1/2}$ = 4μs, σ = 0.8pb^5); ^{208}Pb (^{62}Ni, n) ^{269}110 ($t_{1/2}$ = 170 μs, σ = 3.5 pb^6); and ^{208}Pb (^{64}Ni, n) ^{271}110 ($t_{1/2}$ = 1.4 ms, σ = 15 pb^7) while the fourth isotope was made using a hot fusion reaction, ^{244}Pu (^{34}S, 5n) ^{273}110 ($t_{1/2}$ = 400 μs, σ = 0.3 pb^8). All of these nuclei have halflives in agreement with theoretical predictions[4] based on the special stability associated with N = 162 configurations. The remarkable experiments at GSI which involved the synthesis of several atoms of two different isotopes of element 110 concluded with the first successful synthesis of element 111 ($t_{1/2}$ = 1.5 ms, σ = 3.5 pb^9). It can be argued that element 111 is the first superheavy element.

In this contribution, I will focus on two points: (a) an expanded, informal description of the report of a group at LBL,[5] of which I was a part, about the synthesis of element 110 and

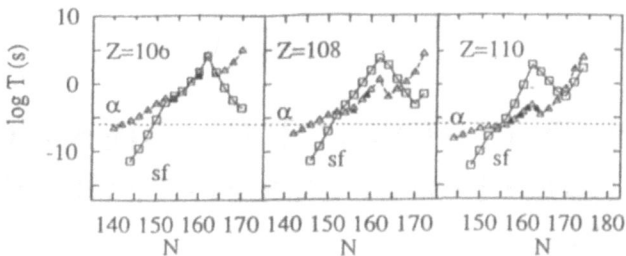

Figure 1. Comparison of measured (solid symbols) and predicted (open symbols) decay properties of the isotopes of elements 106, 108, and 110.

(b) examination of the prospects for future advances in this field. The expanded description of the work of Ref. [5] is offered to help gain perspective on the difficulty of experiments with cross sections at the pb level while I hope to reflect the genuine optimism regarding future work in the second part of this report.

2. POSSIBLE SYNTHESIS OF ELEMENT 110[5]

The Berkeley experiment to synthesize element 110 was one of the last experiments carried out at the LBL SuperHILAC accelerator in the late summer of 1991. The experimental team can be characterized as "Al Ghiorso and friends." The experiment involved the use of a gas-filled magnetic separator, SASSY2,[10] specially constructed for this experiment. The notable feature of this separator was its high transmission of evaporation residues (measured to be 96–98 percent), and its rejection of beam particles (10^{15}) and target-like fragments (10^{3}). The irradiations took place over a period of 41 days (24 days of this time was spent taking data for the ^{59}Co + ^{209}Bi reaction) with a total of 1.5×10^{18} ions passing through the 0.5 mg/cm^2 Bi targets. (This disappointing low average current of $0.1 p\mu a$ was due to repeated failures of the separator/target system and did not reflect the available SuperHILAC beam current of $1 p\mu a$.) The average beam energy was about 5.0 MeV/nucleon, producing an expected completely fused system excitation energy of 15 MeV. During this entire irradiation, one event was observed, with multiple correlated signals, which corresponded to a possible synthesis of ^{267}110. If all beam particles were considered to be equally effective, this event corresponds to a production cross section of 0.8 pb.

To understand the association of the observed event with the synthesis of element 110, one needs to understand the predicted decay properties of ^{267}110 and its daughters (Figure 2).

^{267}110 is predicted to decay by the emission of an 11.5 MeV α-particle with $t_{1/2}$ 20 μs. Its daughter, ^{263}Hs, is predicted to decay by the emission of a 10.8 MeV α-particle with a halflife of 170 μs. The resulting nucleus, ^{259}Sg, is known[11] to decay by α-emission with $t_{1/2}$ = 0.5s. We think it is probable that ^{259}Sg$_{153}$ also decays, in part, by electron capture, forming ^{259}Ha since significant (10–35 percent) EC branches are predicted by EC systematics[12] and these EC branches have been observed for all of the other heavy N = 153 nuclei, i.e., ^{252}Es, ^{253}Fm, ^{254}Md, ^{255}No, ^{256}Lr, ^{257}Rf, and ^{258}Ha. The EC decay of ^{261}Sg has also been inferred[11] with an EC branching ratio of 25 percent. The nucleus ^{259}Ha is unknown and is predicted to decay by the emission of a 9.0 MeV α-particle with a halflife of a few seconds. The daughter of ^{259}Ha (^{255}Lr) is known to decay by the emission of 8.43 and 8.37 MeV α-particles with $t_{1/2}$ = 22s. The α-decay of ^{255}Lr leads to unobservable products decaying by EC to long-lived ^{251}Cf.

Expected   Observed

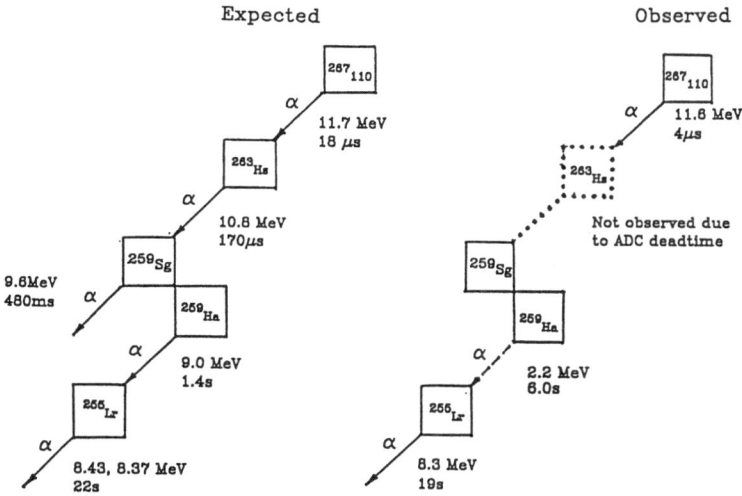

Figure 2. Expected and observed decay properties of ^{267}110.

The observed decay sequence is shown in Table 1. A recoil was implanted in the focal plane with a $B\rho$ and dE/dx characteristic of ^{267}110. The energy deposited by the recoil was 34 MeV; a calculation[5] of the expected energy losses in the spectrometer would lead one to expect an energy of 38 MeV for the complete fusion recoil. Four microseconds after implantation of the recoil, an 11.6 MeV α-particle was observed, followed 6s later by a 2.2 MeV α-particle and 19s after the implantation by an 8.35 MeV α-particle. There were no full energy α-signals observed in this region of the focal plane for the rest of this run (900s).

Our interpretation of this event is shown in Figure 2. The recoil implantation was that of ^{267}110 followed 4 μs later by its decay by the emission of an 11.6 MeV α-particle. The next decay in the chain, that of ^{263}Hs, was missed because at the time of this event, a transient recorder that was supposed to record the daughter decay was not working creating effective electronic deadtime of 280 μs. We suggest that the next member of the chain, ^{259}Sg, undergoes undetected EC decay to ^{259}Ha. ^{259}Ha decays leading to the detection of a 2.2 MeV α-particle, representing a partial escape of the ^{259}Ha α-particle. Finally at 22s ^{255}Lr decays with the emission of an 8.35 MeV α-particle. The period of 900s with no signals being recorded is consistent with the occurrence of EC decays to long-lived ^{251}Cf.

Table 1. Observed Decay Sequence[a]

Event	Time(s)
Recoil implantation	0.0
11.6 MeV alpha	0.000004
2.2 MeV alpha	6.0
8.35 MeV alpha	19
"no events"	900

[a]In addition to this sequence of spatially correlated events, there occurred a "background sequence" in which a 17.4 MeV recoil implanted 1.4 mm from the site of the observations 26s before the 34 MeV recoil. The 17.4 MeV recoil decayed 26.150 s later by the emission of an 8.1 MeV α-particle.

Table 2. Alternative Scenario for the Event

$t(s)$	Observation	Explanation	Comment
0	Recoil Implantation E_{recoil} = 17.4 MeV	$^{212}Po'''$	
26	Recoil Implantation E_{recoil} = 34 MeV	^{213}Rn implant	E_{recoil} too high
26.000004	11.6 MeV α	Decay of $^{212}Po'''$	
26.150	8.1 MeV α	Decay of ^{213}Rn	prob. = 0.02
32	2.2 MeV α	"chance correlation"	prob. = 10^{-4}
46	8.35 MeV α	"chance correlation"	
900	no signals	???	

What can we say positively about this scenario? The $B\rho$ and dE/dx of the implanted recoil are consistent with element 110. The recoil energy indicates a CN recoil rather than a transfer product. The 11.6 MeV α-particle indicates either Z = 110 or $^{212}Po'''$. The correlation between the initial recoil and the 11.6 MeV α-particle can not be due to chance as the probability of such an accidental coincidence can be calculated[13] to be 7×10^{-8} for the run in question. A partial escape of the ^{259}Ha α-particle is expected 45 percent of the time with simulations showing the most probable escape energy to be the observed value of 2.2 MeV. The overall probability of seeing a spatial correlation between the implanted recoil, the 11.6 MeV α-particle, the 2.2 MeV α-particle and the 8.31 MeV α-particle can be calculated[13] to be $\leq 10^{-12}$. Only four nuclei (^{255}Lr, ^{256}Lr, ^{257}No and $^{211}Po'''$) show decay sequences that can be interpreted as having an 8.3 MeV α-particle followed by a prolonged period of no observable decays.

In an experiment lasting many days and searching for phenomena occurring at the picobarn level, one must deal with "background events." There was an important background event with a close spatial and temporal association with the event of interest (Table 1). A recoil corresponding to a TLF (E_{recoil} = 17.4 MeV) implanted 26s before the event of interest at a location that was 1.4 mm away from the event of interest. An α-particle of energy 8.1 MeV from this implant was detected 26.150s later. A survey of all the data taken in this experiment revealed 16 such rogue events with a distribution of times between implantation and decay consistent with a parent halflife of 35s.

In trying to assign an origin to a single event such as this, one must consider alternate explanations. The "background event" plays an important role in the most important alternate explanation that is shown in Table 2.

In this alternate explanation of the data, the 17.4 MeV recoil is taken to correspond to the implantation of $^{212}Po'''$ which decays 26s later by emitting an 11.6 MeV α-particle. The 34 MeV recoil is taken to correspond to the implantation of ^{213}Rn which decays 150 ms later by the emission of an 8.1 MeV α-particle. The 2.2 MeV α-particle and the 8.35 MeV α-particle are assigned to chance correlations. This scenario, we believe, has certain serious flaws. The recoil energy of 34 MeV is too high for a TLF such as ^{213}Rn. The probability of having ^{213}Rn decay 6 halflives after implantation is 0.02. The probability of getting the correlations with the 2.2 and 8.35 MeV α-particles by chance is 10^{-4}.

In summary, we believe we have evidence for the possible synthesis of $^{267}110$ in the LBL experiment. The evidence is clearly not of the same high quality as that presented in the GSI studies.[6,7] Interestingly the measured properties of $^{267}110$ agree reasonably well with theoretical predictions[4] (Figure 1).

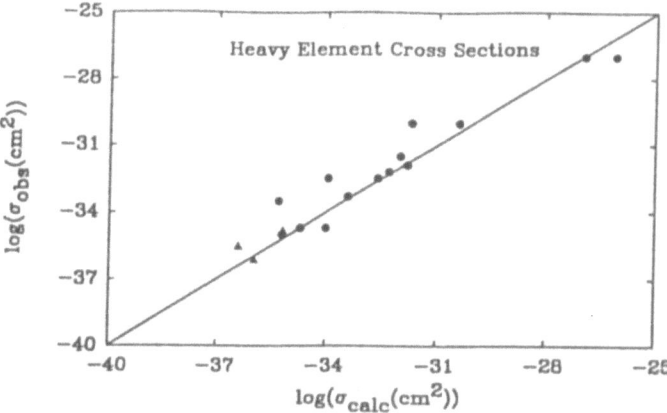

Figure 3. Comparison of observed and predicted cross sections for heavy element production.

3. FUTURE PROSPECTS

Because of the agreement of measured and predicted values of the halflives of the isotopes of elements 106, 108, and 110, there is little doubt about the existence of the superheavy elements. It turns out, for reasons discussed below, that there is little doubt about our ability to synthesize these elements as well. In doing so, we shall fulfill one of the long range goals of nuclear physics and solve one of the outstanding problems in physics.[14] In addition, the well known increases in the halflives of the translawrencium nuclei with increasing neutron number offer us an avenue to explore the atomic physics and chemistry of the heaviest elements where relativistic effects are expected to change, in a fundamental manner, our view of the periodic table and associated phenomena.[15] These opportunities only become available if we are able to make these very n-rich nuclei.

The essential ideas of now to reliably predict the production rates of heavy nuclei in charged particle-induced reactions have been known for some time. It is clear that the production and survival of high nuclei (against fission) involve processes that utilize the poorly characterized tails of distributions in excitation energy, spin, etc. Because of this, one prefers semi-empirical approaches to these predictions rather than more fundamental attacks such as those involving coupled channel calculations of sub-barrier fusion probabilities, etc. For several years, I have been a proponent of a simple semi- empirical formalism[16] for estimating heavy element production rates in reactions with stable or radioactive beams. s-wave fusion cross sections, at the Bass barrier, V_B, are calculated using a semi-empirical formalism of Armbruster[17] that tries to incorporate dynamical fusion hindrance. Excitation energies are evaluated using the masses of Liran and Zeldes[18] or the more recent masses of Möller and Nix,[19] for $V_B - B_n \leq E_{proj} \leq V_B$. The survival probabilities of the fused systems are calculated using a mean of Γ_n/Γ_f values from Sikkeland[20] and Cherepanov.[21] A similar formalism, due to Magda,[22] exists for multinucleon transfer reactions.

The success of formalism is shown in Figure 3 in which we compare the measured and calculated production cross sections for the reactions used to synthesize elements 101–111. (The data measured during the past year is indicated by the solid triangles.) Using this formalism,[16] assuming all stable nuclei as projectiles with beam currents of $1p\mu a$, and considering all stable or long-lived nuclei as target materials (thicknesses 1.0 mg/cm² or less), I have evaluated the best reactions for synthesizing elements 112, 113, and 114 and their

Table 3. Best Superheavy Element Synthesis
Reactions

Reaction	Yield(atoms/day)
$^{66}Zn + {}^{207}Pb \rightarrow {}^{272}112 + n$	0.5
$^{65}Cu + {}^{209}Bi \rightarrow {}^{273}112 + n$	0.09
$^{66}Zn + {}^{209}Bi \rightarrow {}^{274}113 + n$	0.02
$^{48}Ca + {}^{244}Pu \rightarrow {}^{290}114 + 2n$	0.01

yields. The best reactions are cold fusion reactions and are shown in Table 3.

Clearly the synthesis of element 112 is within our grasp using current technology while the synthesis of elements 113 and 114 is possible with improved technologies that offer the use of increased target thicknesses (2×) and beam currents (5×).

Previously I have performed a similar evaluation[16] of the prospects of synthesizing new heavy nuclei using radioactive beams. The principal conclusions of that work are that the use of radioactive beams is not the best way to synthesize elements 110 and higher, but that significant opportunities do exist to make new n-rich isotopes of elements 104–108. The former conclusion stems mostly from the projected low intensities of radioactive beams (which are factors of 10^3–10^6 less than stable beam intensities).

This work was supported in part by the USDOE, Office of Nuclear and High Energy Physics.

REFERENCES

1. K. E. Gregorich, et al., *Phys. Rev. Lett.* **72** (1994) 1423.
2. Yu. A. Lazarev, et al., *Phys. Rev. Lett.* **73** (1994) 624.
3. Yu. A. Lazarev and K. J. Moody, private communication.
4. See, for example, R. Smolanczuk, J. Skalski and A. Sobiczewski, GSI-94-77, November, 1994.
5. A. Ghiorso et al., *Nucl. Phys.* **A** (to be published).
6. S. Hofmann, et al., *Z. Phys.* **A** (to be published).
7. P. Armbruster, private communication.
8. Yu. A. Lazarev, private communication.
9. S. Hofmann, et al., *Z. Phys.* **A** (to be published).
10. A. Ghiorso, *J. Radioanal. Nucl. Chem.* **124** (1988) 407.
11. G. Munzenberg et al., *Z. Phys.* **A322**(1985) 227.
12. K. J. Moody and R. B. Firestone, private communication.
13. A. Ghiorso et al., *Phys. Rev.* **C** (submitted for publication).
14. V. L. Ginzburg, *Phys. Today*, May, 1990, p. 9.
15. O. L. Keller and G. T. Seaborg, *Ann. Rev. Nucl. Sci.* **27** (1977) 139.
16. W. Loveland, in *Radioactive Nuclear Beams*, D. J. Morrissey, Ed. (Editions Frontieres, Gif-sur-Yvette, 1993) pp. 527–531.
17. P. Armbruster, *Ann. Rev. Nucl. Sci.* **35** (1985) 135.
18. S. Liran and N. Zeldes, *At. Data and Nucl. Data Tables* **17** (1976) 431.
19. P. Möller and J. R. Nix, *At. Data and Nucl. Data Tables* (to be published).
20. T. Sikkeland, A. Ghiorso and M. Nurmia, *Phys. Rev.* **172** (1968) 1232.
21. E. A. Cherepanov, A. S. Iljinov, and M. V. Mebel, *J. Phys.* **G9** (1983) 931.
22. M. Magda, in *Frontier Topics in Physics*, W. Schied and A. Sandulescu, Ed. (Plenum, New York, 1994) pp. 169–180.

ENERGY CALIBRATION OF INTERMEDIATE MASS FRAGMENTS FROM THE 930-MeV ^{79}Br ON ^{27}Al REACTION

H. Madani,[1] E. R. Chávez,[1] M. E. Ortiz,[1] J. Suro,[1] A. Dacal,[1] J. Gómez Del Campo,[2] and D. Shapira[2]

[1]Instituto de Física, UNAM, AP. 20-364
Del. A. Obregón, México City, México
[2]Oak Ridge National Laboratory, Physics Division
Oak Ridge, TN

ABSTRACT

The intermediate mass fragments produced in the reaction ^{79}Br on ^{27}Al at 930 MeV have been characterized using the Heavy Ion Light Ion (HILI) detector.[1] Their laboratory energy has been determined using two different approaches.

1. INTRODUCTION

The intermediate mass fragments (IMF's) emitted in heavy-ion collisions at low bombarding energies ($\leq$ 12 MeV/u) are a useful tool for probing the properties of the hot nuclear matter produced in these reactions. The productions of IMF's in the reactions ^{86}Kr on ^{63}Cu at 486, 550, 640, and 730 MeV has been attributed to the asymmetric binary breakup of the equilibrated compound nucleus obtained in these collisions.[2] One of the goals of our present study of the 930-MeV ^{79}Br on ^{27}Al reaction is to investigate the sources and mechanism of the IMF production. It is therefore essential to have a good characterization of these IMF's as well as an accurate determination of their energies.

2. THE EXPERIMENT

In the present study IMF's produced in the reaction ^{79}Br on ^{27}Al at 930 MeV have been measured using the HILI detection system. The HILI consists of three main components. A multi-wire parallel plate avalanche counter (PPAC), followed by an ionization chamber (IC), and finally a hodoscope of 192ΔE-E plastic phoswich scintillators (HODO). This arrangement

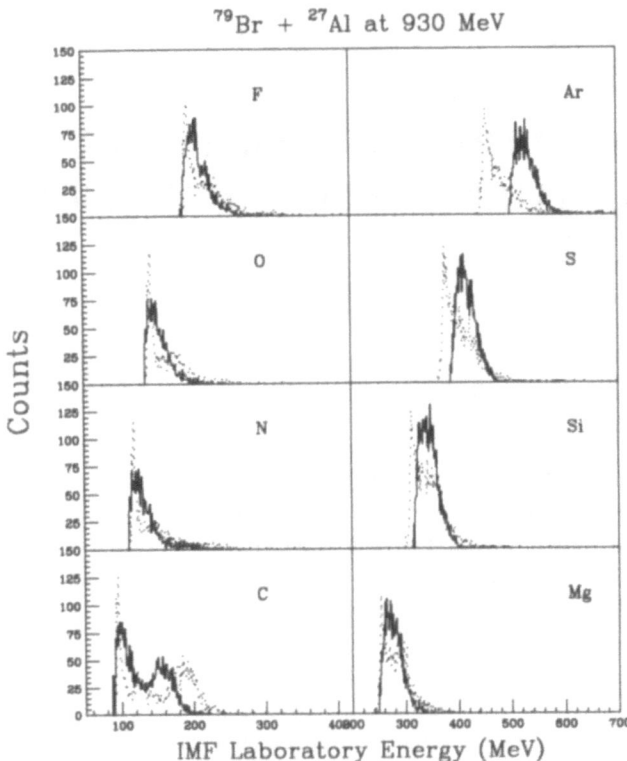

Figure 1. IMF laboratory energy spectra obtained with the ionization chamber calibration (solid lines), and the HODO calibration (dotted lines).

allows the detection of the reaction fragments over a wide dynamic range. A more detailed description of the HILI and its various components can be found in Ref. [1].

3. IMF ENERGY CALIBRATION

Since the hodoscope has been precisely calibrated only for protons, indirect approaches have been used to determine the laboratory energies of any other particle that punches through the ionization chamber into the hodoscope. Here we describe two prescriptions. In both, one essential ingredient is the correction for the energy lost in all material layers before the active detection region. Energy losses in each layer have been calculated for the elements from Li (Z = 3) to Ca (Z = 23) and for a range of kinetic energies from 1 to 1050 MeV. The results were stored in tables which were then used to look up the incident energy of the IMF based on the energy it lost in any layer of the HILI.

In the first approach, referred to as HODO calibration, the calibrations of the plastic scintillators of the hodoscope were performed for protons as described in reference 1. An extrapolation to heavier particles was then made following Ref. [3], and the energy lost by the IMF in the plastic scintillators was thus determined. Finally, the initial energy of the IMF before it entered the HILI detector was extracted from the lookup tables mentioned above.

In the second approach, the same procedure of determining the IMF energy from the lookup tables has been used. However, in this case, the calibration of the ionization chamber was used to determine the energy left by the IMF's in the IC before they punch through to the hodoscope.

Energy spectra obtained for selected isotopes from carbon to argon are displayed in Figure 1. The dotted lines refer to the approach based on the HODO calibration, and the solid lines refer to the second approach. A difference between the energies obtained from the two methods can be observed. The fact that the threshold energy obtained for elements of charge less than 12 is found to be the same for the two methods, shows that the low energy end of the calibration is reliable. A difference between the threshold energies is observed for Z's higher than 12. This discrepancy, which increases with increasing Z, proves the need for a more accurate calibration of the plastic phoswiches.

Such an alternate calibration technique based on the specific luminescence model and developed by Michaelian et al.,[4] and Belmont et al.,[5] is currently being implemented for the hodoscope calibration.[6] In this technique, the light response function $L(E)$ of the detector can be calculated for one or more element for which calibration data exist. The results can then can be interpolated or extrapolated to other elements with no calibration points. It is necessary to point out that protons and alphas have a distinct behavior from heavier element, and therefore one calibration point for particles of $Z > 2$ is needed for optimum results. As no calibration beams with particles heavier than alphas were available during the experiment, those cases in Figure 1, for which the threshold energy was found to be the same for both calibration techniques, provide calibration points for particles heavier than alphas.

REFERENCES

1. D. Shapira et al., Nucl. Inst. and Meth. A **301** (1991) 76
2. J. Boger et al., Phys. Rev. C **49** (1994) 1597
3. Pouliot et al., Nucl. Inst. and Meth. A **270** (1988) 69
4. K. Michaelian et al., Nucl. Inst. and Meth. A **334** (1993) 457, *Phys. Rev. B* **49** (1994) 15550
5. E. Belmont et al., Nucl. Inst. and Meth. A **332** (1993) 202
6. M. E. Ortiz et al., submitted to Revista Mexicana de Física.

20

PROGRESS IN COLLECTIVE FLOW STUDIES FROM THE ONSET TO BEVALAC/SIS

M. A. Lisa

Nuclear Science Division
Lawrence Berkeley Laboratory
Berkeley, CA

ABSTRACT

Collective flow in heavy ion collisions was first observed experimentally more than a decade ago at the Bevalac by the Plastic Ball collaboration. Although early calculations had suggested that measurement of the flow would place tight constraints on the nuclear equation of state, uncertainties in other input parameters of microscopic models, which also affect the flow, led to large ambiguities in the equation of state. This talk will discuss recent flow studies that attempt to overcome these difficulties. The EOS and FOPI experiments at the Bevalac and SIS accelerators have measured flow in the 200–2000A·MeV bombarding energy range with better acceptance, particle identification, and systematics than was previously available. Meanwhile, programs at MSU and GANIL are studying the disappearance of flow around 50A·MeV. Systematic comparison of these data with predictions of microscopic models is beginning to reduce the ambiguities in the extraction of physics quantities. Also, new directions in flow studies, such as the flow of produced particles and radial flow, offer the possibility of further information from flow studies. Recent accomplishments and new directions in flow studies are discussed, and areas where further study is needed are pointed out.

1. INTRODUCTION

Early hydrodynamical models of heavy ion collisions predicted the existence of a strong collective component, called flow, in the emission pattern of reaction products (see 1 and references therein). These models suggested that the strength of nuclear flow is strongly dependent on the nuclear compressibility, raising the hope that the nuclear equation of state (EoS) could be mapped out through flow measurements 1,2. Reaction models, such as the internuclear cascade 3, which lacked an explicit nuclear EoS, predicted little or no flow 4.

Advances in Nuclear Dynamics
Edited by Wolfgang Bauer and Alice Mignerey, Plenum Press, New York, 1996

Directed flow was first identified experimentally 5 in heavy ion collisions at the Bevalac more than 10 years ago. Comparative studies (see, e.g., 6) of early sidewards flow measurements with theoretical predictions proved unable to extract definitively the nuclear EoS. Ambiguities arose from both experimental and theoretical uncertainty. Experimentally, limited acceptance and particle identification (PID) capabilities hampered systematic comparison with model predictions 7. On the theoretical side, parameters in the models, other than the EoS, were found to have a strong effect on flow 8. The effect on the flow of varying "physics" inputs to models, such as momentum-dependent terms in the mean field 9,10, the in-medium nucleon–nucleon cross section σ_{nn} 11, and fragment formation mechanisms 12 has been studied. Also, the effects of more "technical" details of the models have been studied, such as energy and angular momentum conservation 10,13, realistic treatment of the nuclear surface 14 and binding energy 4, and effects of the nucleon "uncertainty principle" 15. It became increasingly clear that flow measurements with better acceptance, accuracy, and systematics were needed, if the subtle effect of the EoS were to be teased out of the data through comprehensive comparison to theory.

In the past decade, a large subfield has developed around flow studies, and much has been learned. New aspects of flow, such as the squeeze-out and radial flow, have been identified, and the systematics of flow are being mapped out over a large energy range. New 4π heavy ion detectors 16,17,18, designed to overcome the shortcomings of the Plastic Ball and Streamer Chamber, are providing a much clearer and complete view of the flow phenomenon. Furthermore, more realistic dynamical models are now in place that allow more direct comparison with experiment. Some recent developments in flow studies at bombarding energies in the range $E \approx 10–2000A \cdot \mathrm{MeV}$ are discussed here.

2. REPULSIVE SIDEWARD FLOW

Repulsive sidewards flow is characterized by a non-isotropic emission pattern, in which particles emitted forward and backward in the c.m. are preferentially found in the reaction plane on opposite sides of the beam direction. Sidewards flow is commonly measured with Transverse Momentum Analysis 19, in which the average momentum per nucleon in the reaction plane is plotted as a function of c.m. rapidity. The slope of the resulting "S"-shaped curve at $y_{c.m.} = 0$ is taken as a measure of the collective motion, and is often simply called "flow." Figure 1 shows flow curves recently measured by the EOS collaboration for Au + Au at $E/A = 0.25–1.15$ GeV 20.

It had been suggested that the flow of composite fragments might be more sensitive to the EoS than proton flow or inclusive flow measurements 12. With good PID, the EOS TPC can study flow for the first time separately for different particle species. These detailed data have been used to construct more stringent tests of dynamical models than had previously been possible 20.

An attempt to understand the flow mass systematics themselves — independent of a dynamical model — has recently been undertaken 21; some results are shown in Figure 2. Here, the azimuthal distribution with respect to the reaction plane is shown for protons, deuterons, and $A = 3$ particles in 5 rapidity windows in the c.m. frame. Directed flow manifests itself in a tighter focussing in the reaction plane for particles at forward rapidity, and the mass dependence of flow 20 is shown by a tighter focussing of the heavier particles. When a low p_t is applied to the data, one observes that a simple coalescence prescription is able to reproduce the mass dependence of flow remarkably well — squaring (cubing) the proton azimuthal distribution reproduces the deuteron ($A = 3$) distribution 21. Thus, a clearer

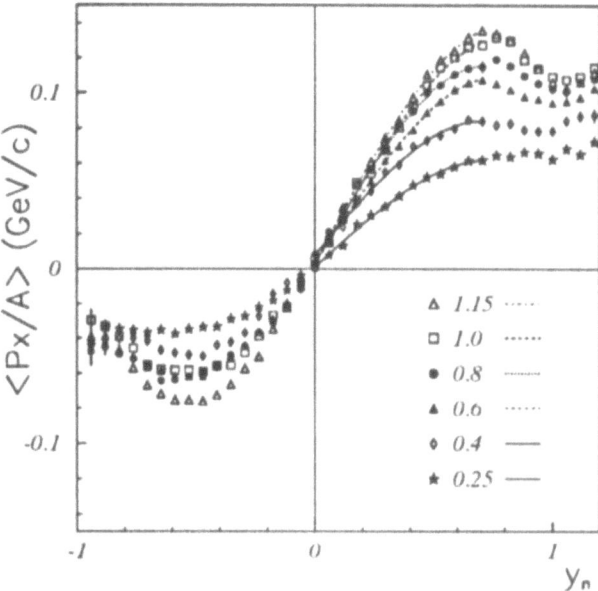

Figure 1. "S"-shaped flow curves measured for semi-central Au + Au collisions at the Bevalac by the EOS collaboration. From 20.

and more complete understanding of some flow systematics is emerging from the new flow studies. Further, this new finding validates the use of one-body models that lack composite cluster formation, such as the BUU, in the study of flow physics 6.

Recently, predictions of a BUU model have been compared to two data sets in an attempt to reduce the ambiguity in determining the EoS from flow measurements 24. Figure 3 shows measured flow as a function of event centrality by the Plastic Ball collaboration for the symmetric system Nb + Nb 22, and by the Diogene collaboration for the asymmetric system Ar + Pb 23, both at $E/A = 400A\cdot$MeV. BUU flow predictions using four parametrizations of the EoS are compared to the measurements. It is seen that only a soft EoS with momentum dependence interations (MDI) in the mean field reproduces both data sets, whereas comparison

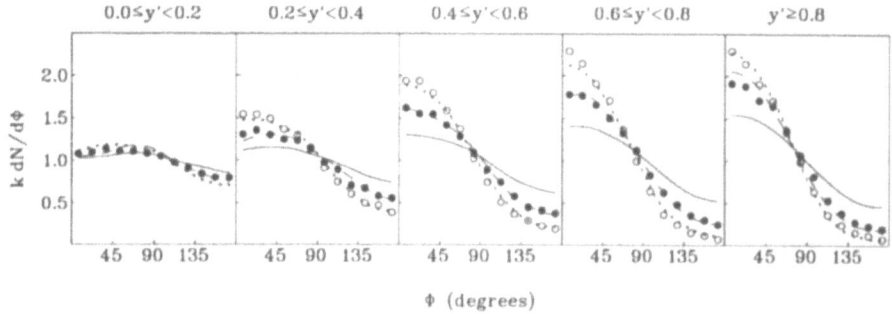

Figure 2. Azimuthal distributions with respect to the reaction plane for protons (solid line), deuterons (filled points), and $A = 3$ fragments (open points) are shown, as well as the square of the proton distribution (dashed line), and the deuteron distribution taken to the power of 3/2. From 21.

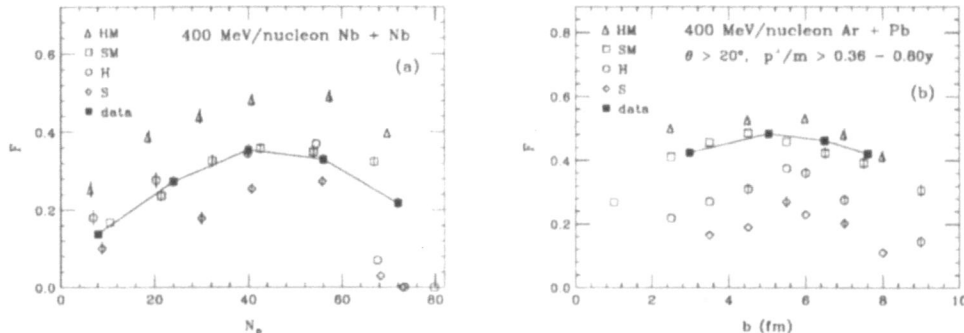

Figure 3. Flow as a function of event centrality is shown for Nb + Nb 22 and Ar + Pb 23 at 400A·MeV. Also shown are predictions of a BUU model with soft and hard EoS, with and without momentum dependence in the mean field. From 24.

with either one of the measurements alone could not uniquely identify the correct EoS. This would leave the largest uncertainty in the model associated with the in-medium cross section σ_{nn}. It has been suggested that flow sytematics as a function of system mass may reduce this ambiguity as well. Thus, there is hope that, upon comparison with the more complete data sets now coming from the new flow measurements, dynamical models may yet provide reliable information on the nuclear EoS.

A new direction in the study of collective decay modes looks at the flow of produced particles. With VUU and QMD model calculations Bass and collaborators 25,26 have explored π flow in heavy ion collisions. For all but the most central collisions, pions are predicted to "anti-flow" — that is, flow in the reaction plane in the direction opposite the direction of nucleon flow. Preliminary confirmation of this prediction has been seen in Au + Au collisions at 1A·GeV by the FOPI collaboration 27. The magnitude and sign of the pion flow for varying event centrality may depend strongly on the nuclear EoS 26, providing another sensitive aspect of the flow phenomenon. Analysis of π flow is in progress with the new data sets at LBL and GSI 28.

3. THE DISAPPEARANCE OF FLOW

The nuclear force is repulsive at Bevalac/SIS bombarding energy, leading to projectile fragment scattering to positive angles in the reaction plane, which we define as "positive flow." At lower bombarding energy, around 50A·MeV, the nuclear interation is attractive, and one expects the projectile-target pair to partially rotate around each other before scattering to negative angles, leading to "negative flow." Flow disappears at the so-called balance energy, E_{bal}, where these two effects cancel each other. The disappearance of flow was first observed in the La + La system 29, and has since been measured for a large number of target-projectile combinations from C + C 30 to (a lower limit for) Au + Au 31.

The mass dependence of the balance energy is shown in Figure 4. The $A^{-1/3}$ dependence of E_{bal} can qualitatively be understood in terms of the competition between the repulsive effect of hard collisions between projectile and target nucleons (which scales as A), and the attractive interation of the nuclear surfaces (which scales as $A^{2/3}$). Comparisons of measurements with predictions of a BUU model 30 show that E_{bal} is relatively insensitive to the EoS used in the BUU, but that there is a sensitivity to the in-medium σ_{nn}. In particular, as shown in

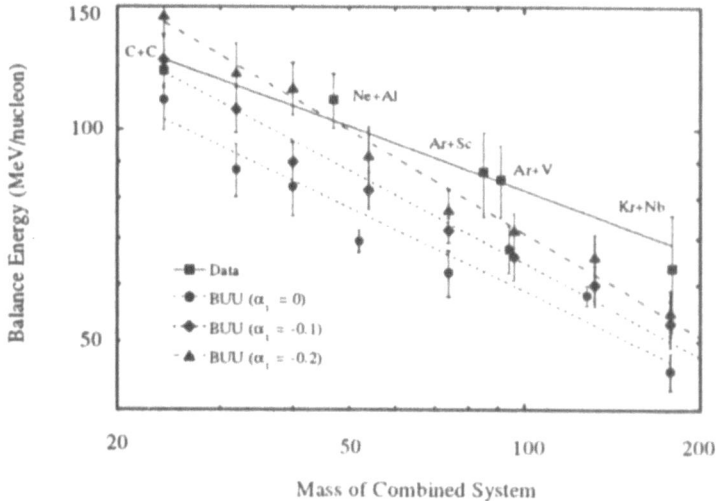

Figure 4. The observed mass dependence of the disappearance of flow is compared to BUU predictions using a soft EoS, and σ_{nn} at nuclear matter density reduced by 0%, 10%, and 20% from its free value. Lines guide the eye. From 30.

Figure 4, the model reproduces the data best if σ_{nn} at nuclear matter density is reduced by 20% compared to its free value. Thus, flow studies at these low energies may provide a handle on one of the main physics inputs to models.

A potential source of ambiguity in extracting physics from comparisons with the BUU model, however, arises from the fact that the treatment of the nuclear surface is not well defined in BUU models. The details of the treatment of the nuclear surface in the model can strongly affect the predicted value of E_{bal} 14. One way to probe the details of the nuclear surface on flow may be to to measure the impact parameter dependence of the disappearance of flow 32. So far, this measurement has been done in few systems only 33, but the systematics are currently being extended 34.

4. SQUEEZE-OUT

Although I have no time to discuss this mode in detail, another collective mode observed in particle emission patterns is the so-called "squeeze-out" effect 35, corresponding to a preference for particles to be emitted perpendicular to the reaction plane. Dynamical models indicate that the squeeze-out magnitude in heavy ion collisions may be quite sensitive to the nuclear EoS 36. However, for the early measurements, the effects of detector acceptance are of similar magnitude as the change in squeeze-out using different equations of state 36.

The squeeze-out effect is being measured by new detector systems with better and simpler acceptance for protons and neutrons 37, as well as for charged and neutral pions 38,39. It will be exciting to see what insights are gained by comparison with model predictions.

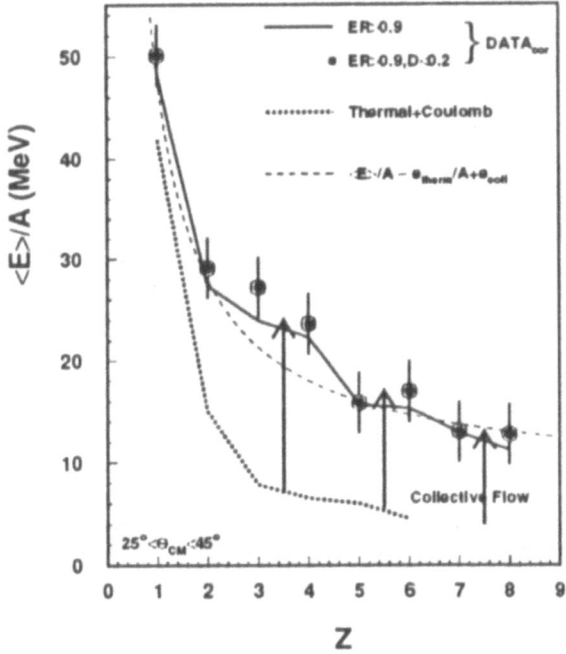

Figure 5. The average energy per nucleon of fragments emitted from central Au + Au collisions at $E = 150A \cdot$MeV is plotted as a function of the fragment charge. Average energy values expected from thermal and Coulomb effects alone, indicated by the dotted line, fall below the observed values. The effects of radial flow are indicated by the arrows, and bring the average energies to the dashed line. From 42.

5. RADIAL FLOW

Besides the directed modes of collective motion — squeeze-out and attractive and repulsive flow — there is another possible collective mode in the decay of an excited nuclear system — isotropic radial flow. In the simple thermodynamic "fireball" model 40, one assumes a non-isentropic expansion resulting in transverse momentum spectra that are, in the classical limit, of Maxwell–Boltzmann type, the inverse exponential slope representing the source temperature T. If, on the other hand, the decay resembles more hydrodynamic expansion, due to a short mean free path of escaping nuclear matter, one would expect a "shoulder arm" on the spectra resulting from the ordered radial motion, or "blast wave." The shape of the spectra then would be dictated by the radial flow velocity β and the intrinsic temperature (that is, in the flow frame, not the overall source frame) of the source. Such a scenario was considered to explain proton and pion spectra at 90° in the Ne + NaF reaction at 800$A \cdot$MeV 41. However, only with the availability of the recent high quality data sets have systematic studies of the radial flow in heavy ion collisions been possible.

A systematic study of Au + Au reactions at 150$A \cdot$MeV 42 reveals a strong isotropic collective component to the energy of fragments emitted from central collisions. If there were no radial flow, the average energy of a fragment would be proportional to the source temperature and independent of its mass (a Coulomb barrier adds a small additional component proportional to the fragment charge). As is shown in Figure 5, especially for the heavier fragments, the average fragment energy in central collisions is too high to be explained by

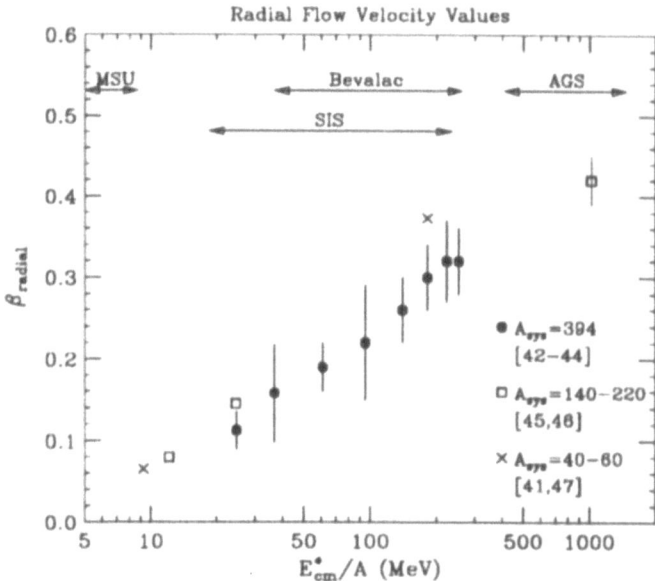

Figure 6. Radial flow velocities extracted from heavy ion measurements as a function of the available energy in the center of mass. Flow values for various size systems (projectile target combinations) are indicated by different marker types. For each measurement reported, the average flow velocity is plotted. References indicated in figure.

thermal and Coulomb effects alone. The energies are better reproduced if one assumes the presence of a radial flow which takes 30–50% of the available energy. Studies of energy spectra shapes for Au + Au collisions at other bombarding energies 43,44 also indicate the presence of radial flow that accounts for about half of the available energy in the c.m. system. Surprisingly (unfortunately?), model calculations suggest that the strength of the radial flow is quite insensitive to the EoS used in the model 44. Figure 6 summarizes most radial flow measurements published to date. The radial flow phenomenon appears to be a general feature of heavy ion collisions, not restricted to a given energy range.

The cause of the radial flow in heavy ion collisions deserves theoretical investigation, in light of the apparent insensitivity of predicted radial flow values to model parameters. However, the *effects* of having 50% of the available energy tied up in radial flow are of equal interest. These effects may include reduced entropy 48 and temperature of the participant source, and significant effects on fragment yields 49.

6. CONCLUSIONS

More than ten years after its discovery, the study of collective flow in heavy ion collisions continues as an active and productive area of research. Today, more sensitive and complete flow measurements challenge more sophisticated theoretical models. Here, a summary of some recent achievements and new directions in flow studies is given, followed by suggestions where future work should be devoted.

In the Bevalac/SIS energy range, a clearer understanding of the systematics of repulsive flow is emerging. These systematics are being used to stringently test microscopic reaction models. Ambiguities that had plagued early attempts to extract the nuclear EoS from flow measurements are being reduced through systematic model comparisons. In particular, varying the system asymmetry may disentangle the effects of the momentum and density dependence of the EoS, while the system size dependence of flow may hold independent information about the contribution from hard nucleon–nucleon collisions. Current investigations into the flow of produced particles holds great potential to yield physical insights into the EoS and particle transport properties.

At lower bombarding energies, the balance energy has been measured over a large mass range for symmetric systems, and evidence for a reduced in-medium σ_{nn} is obtained through comparisons with BUU predictions.

Clear observations have now been made over a large range of bombarding energies of a new mode of collective flow: isotropic radial flow, which accounts for about 50% of the available energy for participant nucleons in central collisions. We now see in the literature initial investigations into the effects of radial flow on other physical quantities. Such studies may cause us to review in a new light old questions about participant matter decay.

All of these studies have led to greater refinement and sophistication of dynamical reaction models. Initially, this had the frustrating result of revealing ambiguities in EoS studies, as it became clear that other effects could modify the flow signal significantly. However, as the interplay between these effects becomes better understood, we find that flow studies can provide information about a range of physical processes, and hope again arises that the original goal of constraining the nuclear EoS may be in sight.

Further work needs to be done in flow studies. Directed flow effects have been observed at the AGS already 50. The flow excitation function should be mapped out in the energy range spanning Bevalac/SIS energies, up to the maximum AGS energy, since the increasing importance of resonances in hot nuclear matter may affect the EoS. A measure of the evolution of the EoS with bombarding energy (temperature) may be important when considering the effect of collective motion at higher energy — even at CERN and RHIC. Another high priority should be measuring the flow excitation function in the energy range from the balance energy up to Bevalac/SIS energies. It is in this range that flow changes most rapidly as a function of energy, and here that hydrodynamic scaling is most strongly violated 51. In the spirit of the study by Pan and Danielewicz 24, it may be worthwhile to study balance energy systematics for asymmetric systems, to investigate possible effects of momentum dependent interations at low energy. As mentioned above, the impact parameter dependence of E_{bal} should be measured, to reduce ambiguities associated with model treatment of the nuclear surface. Finally, much of our justification for investigating the nuclear EoS is derived from astrophysics, where one would like to understand *neutron* stars. Therefore, as radioactive beams become available, the collective aspects of nuclear reactions should be surveyed as a function of isospin.

Instructive conversations with Drs. P. Danielewicz, W. G. Lynch, G. Odyniec, A. M. Poskanzer, G. Rai, H. G. Ritter, and G. D. Westfall are gratefully appreciated.

REFERENCES

1. H. Stöcker and W. Greiner, Phys. Rep. 137, 227 (1986).
2. *The Nuclear Equation of State*, W. Greiner and H. Stöcker, ed., NATO ASI Series B, Vol 216A (1989).
3. J. Cugnon, T. Mizutani, and J. Vandermeulen, Nucl. Phys. A352, 505 (1981).
4. J. Molitoris, *et al.*, Phys. Rev. C33, 867 (1986).
5. H. A. Gustafsson, *et al.*, Phys. Rev. Lett. 52, 1590 (1984).
6. B. Blättel, V. Koch, and U. Mosel, Rep. Prog. Phys. 56, 1 (1993).

7. V. Koch, *et al.*, Nucl. Phys. **A532**, 715 (1991).
8. B. Blättel, *et al.*, Phys. Rev. **C43**, 2728 (1991).
9. J. Aichelin, *et al.*, Phys. Rev. Lett. **58**, 1926 (1987).
10. J. Zhang, S. Das Gupta, and C. Gale, Phys. Rev. **C50**, 1617 (1994).
11. G. F. Bertsch, *et al.*, Phys. Lett. **B189**, 384 (1987).
12. G. Peilert, *et al.*, Phys. Rev. **C38**, 1402 (1989).
13. C. Gale and S. Das Gupta, Phys. Rev. **C42**, 1577 (1990).
14. D. Klakow, G. Welke, W. Bauer, Phys. Rev. **C48**, 1982 (1993).
15. J. Jänicke and J. Aichelin, Nucl. Phys. **A547**, 542 (1992).
16. G. Rai, *et al.*, IEEE Trans. Nucl. Sci. **37**, 56 (1990).
17. A. Gobbi *et al.*, Nucl. Instrum. Meth. Phys. Res., Sect. A **324**, 156 (1993).
18. G. D. Westfall *et al.*, Nucl. Instrum. Meth. Phys. Res., Sect. A **238**, 347 (1985).
19. P. Danielewicz and G. Odyniec, Phys. Lett. **157B**, 146 (1985).
20. M. D. Partlan, *et al.*, preprint LBL-36280, submitted to Phys. Rev. Lett.
21. S. Wang, *et al.*, Phys. Rev. Lett., in press (1995).
22. H. A. Gustafsson, *et al.*, Mod. Phys. Lett. **A3**, 1323 (1988).
23. J. Gosset, *et al.*, in Ref. 2.
24. Q. Pan and P. Danielewicz, Phys. Rev. Lett. **70**, 2062 (1993).
25. S. A. Bass, *et al.*, Phys. Lett. **B302**, 381 (1993).
26. S. A. Bass, *et al.*, Phys. Rev. bf C51, R12 (1995).
27. D. Pelte, W. Reisdorf, T. Wienold, and the FOPI collaboration, GSI 09-93.
28. J. Kintner and T. Wienold, private communications.
29. D. Krofcheck, *et al.*, Phys. Rev. Lett. **63**, 2028 (1989).
30. G. D. Westfall, *et al.*, Phys. Rev. Lett. **71**, 1986 (1993).
31. W. M. Zhang, *et al.*, Phys. Rev. **C42**, R491 (1990).
32. J. P. Sullivan, *et al.*, Phys. Lett. **B249**, 8 (1990).
33. J. C. Angelique, *et al.*, contribution to XXXI Interational Winter Meeting on Nuclear Physics, Bormio, ed I. Iori
34. R. Pak, contribution to this conference.
35. H. H. Gutbrod, *et al.*, Phys. Lett. **B216**, 267 (1989).
36. C. Hartnack, *et al.*, Mod. Phys. Lett. **A9**, 1151 (1994).
37. D. Lambrecht, *et al.*, Z. Phys. **A350**, 115 (1994).
38. L. B. Venema, *et al.*, Phys. Rev. Lett. **71**, 835 (1993).
39. D. Brill, *et al.*, Phys. Rev. Lett. **71**, 336 (1993).
40. G. D. Westfall, *et al.*, Phys. Rev. Lett. **37**, 1202 (1976).
41. P. J. Siemens and J. O. Rasmussen, Phys. Rev. Lett. **42**, 880 (1979).
42. S. C. Jeong, *et al.*, Phys. Rev. Lett. **72**, 3468 (1994).
43. W. C. Hsi, *et al.*, Phys. Rev. Lett. **73**, 3367 (1994).
44. M. A. Lisa, *et al.*, preprint LBL-35504
45. H. W. Barz, *et al.*, Nucl. Phys. **A531**, 453 (1991).
46. W. Bauer, *et al.*, Phys. Rev. **C47**, R1838 (1993).
47. D. Heuer, *et al.*, Phys. Rev. **C50**, 1943 (1994).
48. C. Kuhn, *et al.*, Phys. Rev. **C48**, 1232 (1993).
49. G. J. Kunde, *et al.*, Phys. Rev. Lett. **74**, 38 (1995), and contribution to this conference.
50. J. Barrette, *et al.*, Phys. Rev. Lett. **73**, 2532 (1994).
51. A. Bonasera and L. P. Csernai, Phys. Rev. Lett. **59**, 630 (1987), and contribution to this conference.

21

THE IMPACT PARAMETER DEPENDENCE OF THE DISAPPEARANCE OF FLOW

R. Pak,[1] W. J. Llope,[1]* D. Craig,[1]† E. E. Gualtieri,[1] S. A. Hannuschke,[1]
N. T. B. Stone,[1] A. M. Vander Molen,[1] G. D. Westfall,[1] J. Yee,[1] R. A. Lacey,[2]
J. Lauret,[2] A. C. Mignerey,[3] and D. E. Russ[3]

[1]National Superconducting Cyclotron Laboratory
and Department of Physics and Astronomy
Michigan State University
East Lansing, MI
[2]Department of Chemistry
State University of New York at Stony Brook
Stony Brook, NY
[3]Department of Chemistry and Biochemistry
University of Maryland
College Park, MD

ABSTRACT

The Michigan State University 4π Array has been recently upgraded to include the High Rate Array (HRA), a close-packed 45 element phoswich array which covers laboratory polar angles from $\approx 3°$ to $\approx 18°$. The HRA subtends all solid angle between the Maryland Forward Array and the detectors of the main Ball resulting in $\approx 90\%$ geometric efficiency for the entire array. With this improved detector configuration, we studied the reaction ^{40}Ar + ^{45}Sc at beam energies from 35 to 115 MeV/nucleon. We present preliminary results on the impact parameter dependence of the disappearance of flow in nuclear collisions, extending the systematics of our previous work.

1. INTRODUCTION

The study of collective flow in nucleus–nucleus collisions can provide information about the nuclear equation of state (EOS).[1,2] The disappearance of collective flow occurs at an incident energy termed the balance energy, E_{bal}.[3] The balance energy corresponds to the

*Present address: T. W. Bonner Nuclear Laboratory, Rice University, Houston, TX 77251-1892
†Present address: Dept. of Physics, Univ. of Wisconsin at Madison, Madison, WI 53706

Advances in Nuclear Dynamics
Edited by Wolfgang Bauer and Alice Mignerey, Plenum Press, New York, 1996

point where the attractive scattering, dominant at incident energies around 10 MeV/nucleon, balances the repulsive interations observed at energies around 400 MeV/nucleon. Several groups have performed experiments relating to the disappearance of flow.[3,4,5,6,7,8,9,10] We have recently completed a systematic study of the disappearance of flow for central collisions in the systems ^{12}C + ^{12}C, ^{20}Ne + ^{27}Al, ^{40}Ar + ^{45}Sc, and ^{84}Kr + ^{93}Nb.[11] We find that E_{bal} scales as $A^{-1/3}$ where A is the mass of the combined projectile-target system. Physically this scaling corresponds to the competition between the attractive mean field interation which should scale as the surface area, $A^{2/3}$, and the repulsive nucleon–nucleon scattering which should scale as the volume of the two nuclei, A. The general trend of this result is reproduced by Boltzmann–Uehling–Uhlenbeck (BUU) model calculations.[11]

However, the impact parameter dependence of the disappearance of flow has not been established. We expect E_{bal} to vary as much as 50% as the impact parameter is increased.[6] The extraction of flow from peripheral collisions is made more difficult by the presence of excited spectator matter that can emit fragments sequentially.[12] By carefully selecting participant fragments, flow can be extracted from peripheral collisions and comparison with dynamical model calculations can be used to determine information about the nuclear EOS.

2. EXPERIMENT AND METHOD

The present measurements were carried out with the Michigan State University 4π Array[13] at the National Superconducting Cyclotron Laboratory (NSCL) using beams from the K1200 cyclotron. A target of 1.0 mg/cm^2 natural scandium was bombarded with ^{40}Ar projectiles ranging in energy between 35 and 115 MeV/nucleon in 10 MeV/nucleon steps. Beam intensities were approximately 100 electrical pA. Data were taken with a minimum bias trigger requiring at least two hits in the main Ball (charged particle multiplicity $N \geq 2$).

The MSU 4π Array has been upgraded with the High Rate Array (HRA).[14] The HRA is a close-packed configuration of 45 phoswich detectors spanning laboratory polar angles from $\approx 3°$ to $\approx 18°$. The array has good granularity and high data rate capability. The HRA subtends all solid angle between the Maryland Forward Array and the detectors of the main Ball resulting in $\approx 90\%$ geometric efficiency for the entire array. Design and fabrication of the HRA was carried out entirely at the NSCL. With the HRA we obtained Z resolution up to the charge of the ^{40}Ar projectile.

We use the standard technique for flow analysis,[11] in which the impact parameter, b, and the orientation of the reaction plane must be determined. The impact parameter is assigned through cuts on centrality variables measured with the improved acceptance of the upgraded 4π Array. Next, a particle of interest (POI) is selected from the event. Autocorrelation is avoided by omitting this POI in the calculation of the reaction plane.[15] A recoil correction according to the prescription given in Ref. [4] is applied to boost the remaining particles toward the POI. Finally, the fraction of the POI's transverse momentum in the reaction plane is evaluated. This procedure is repeated for each particle in the event.

The average $\langle p_x/p_t \rangle$ over all events is plotted versus rapidity in the center-of-mass (CM) frame, y_{CM}. From this plot the flow is extracted by fitting a straight line to the data over the midrapidity region. The slope of this line is the flow, which is a measure of the amount of collective momentum transfer in the reaction.

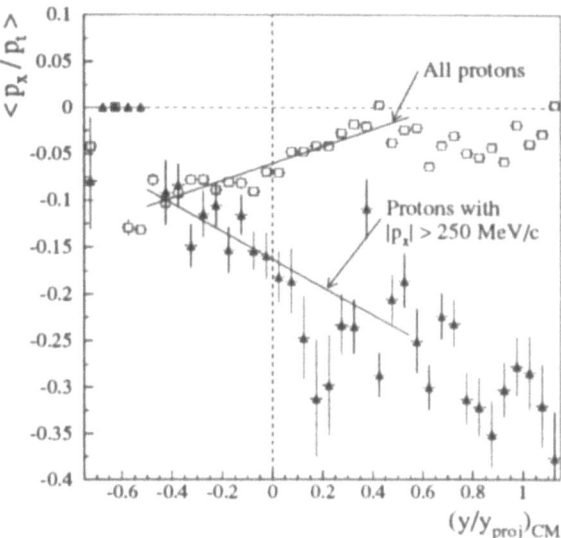

Figure 1. Average fraction of the transverse momentum in the reaction plane for protons versus the normalized rapidity in the center-of-mass frame for ^{40}Ar + ^{45}Sc reactions. The reaction plane is determined using the PLF. The open squares are for an inclusive data set. The solid triangles are the result of applying a cut that allows only protons with transverse momenta in the reaction plane greater than 250 MeV/c. The straight lines are fit in the region $-0.5 \leq (y/y_{proj})_{CM} \leq 0.5$.

3. RESULTS

3.1. Reaction Plane by PLF

A simple method to determine the reaction plane is to choose the projectile-like-fragment (PLF) as the POI. Events are selected in which a PLF (here $Z \geq 8$) was measured in the HRA. The azimuthal angle of the PLF is then taken to be the azimuthal angle of the reaction plane. The open squares in Fig. 1 are the result of plotting $\langle p_x/p_t \rangle$ for protons versus y_{CM} normalized by the projectile rapidity in the CM frame for an inclusive data set. The data, which exhibit the characteristic shape associated with flow, are offset from the origin because no recoil correction was applied in this case.

The solid triangles are the result of applying a momentum cut to suppress the contribution to the flow from excited spectator matter. This cut allows only protons with transverse momenta in the reaction plane greater than 250 MeV/c. Following this cut, the slope of the straight line fit is negative. Thus protons emitted from the hot midrapidity source are preferentially deflected by the mean field away from the PLF. These conclusions are consistent with those for the ^{12}C + ^{12}C data at 50 MeV/nucleon in Ref. [12], which is a similar relatively light system measured below the balance energy (E_{bal} = 122 MeV/nucleon).[11]

3.2. Reaction Plane Correlation

A more refined technique to determine the reaction plane is the azimuthal correlation method,[16] outlined in these proceedings in the talk given by S. Hannuschke. At least four particles are required in the event for this method, because removing the POI leaves three to determine the plane. The azimuthal angles of the reaction planes, Φ_{RP}, calculated with $Z = 3$

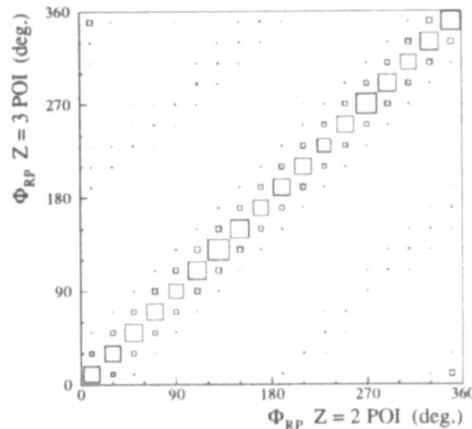

Figure 2. Azimuthal angles of the reaction planes, Φ_{RP}, calculated with $Z = 3$ as the POI versus Φ_{RP} calculated with $Z = 2$ as the POI for ^{40}Ar + ^{45}Sc reactions.

as the POI versus Φ_{RP} calculated with $Z = 2$ as the POI are shown in Fig. 2. Clearly a strong correlation exists between the reaction plane determination for different POI.

To investigate whether there is any multiplicity dependence of this correlation, we extracted the difference of the azimuthal angles of the reaction planes with He and Li fragments as the POIs, $\Delta\Phi_{RP} = \Phi_{RP}(Z = 2) - \Phi_{RP}(Z = 3)$, as a function of the multiplicity, N. We observe that the width of this distribution of $\Delta\Phi_{RP}$ is independent of N, so that multiplicity distortions can be ignored in the present analysis. This quantity, $\Delta\Phi_{RP}$, is also a measure of the dispersion in the calculated reaction planes, which for the case considered here is $\Delta\Phi_{RP} = \pm90°$. Comparison between other particle types yields similar results.

3.3. Impact Parameter Dependence

In Fig. 3 we show $\langle p_x/p_t \rangle$ plotted versus $(y/y_{proj})_{CM}$ for central collisions with He as the POI at four beam energies. Impact parameter selection is based on the top 10% most central events determined through a cut on the reduced total transverse kinetic energy, $KE_t^{red} = (KE_t/KE_{proj})$ where $KE_t = KE \sin^2 \theta$. The reaction plane is found using the azimuthal correlation method with the recoil correction. The data are fit with a straight line over the midrapidity region, $-0.5 \le (y/y_{proj})_{CM} \le 0.5$. The slope of the line is termed the normalized flow. From Fig. 3 it is evident that as the beam energy increases the flow decreases, passes through a minimum, and reappears again at the highest beam energy.

The extracted values of the normalized flow versus the beam energy for central collisions are shown as open squares in Fig. 4. Clearly the points pass through a minimum which corresponds to the balance energy, E_{bal}. The curve drawn is merely to guide the eye. Repeating the entire process for 10% consecutive cuts on KE_t^{red} (i.e., 10% to 20%, 20% to 30%, etc.) results in the other excitation functions displayed in Fig. 4. From the shift in the minima it is evident that E_{bal} increases as the impact parameter increases. This result is in qualitative agreement with Ref. [6] in which the reverse kinematic system ^{40}Ar + ^{27}Al was studied. The balance energy corresponds to the energy where the attractive scattering balances the repulsive interations. For the more peripheral collisions the overlap volume, i.e., the compressed participant zone, is smaller and consequently a larger deposition energy is necessary to overcome the effects of the mean field.

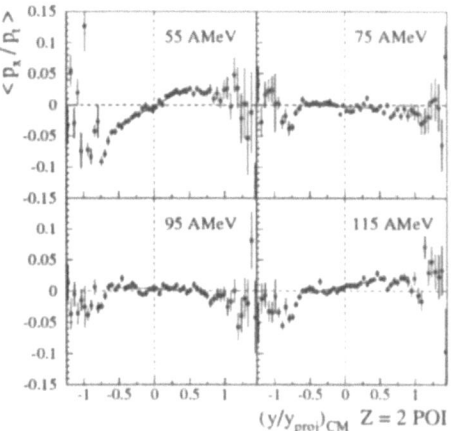

Figure 3. Average fraction of the transverse momentum in the reaction plane for $Z = 2$ as the POI in central collisions at four beam energies versus the normalized rapidity in the center-of-mass frame for ^{40}Ar + ^{45}Sc reactions. The reaction plane is determined using the azimuthal correlation method with the recoil correction. The straight lines are fit in the region $-0.5 \leq (y/y_{proj})_{CM} \leq 0.5$.

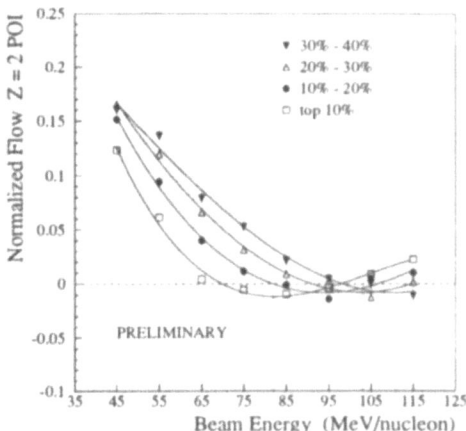

Figure 4. Excitation functions of the measured normalized flow for $Z = 2$ as the POI at four impact parameter bins for ^{40}Ar + ^{45}Sc reactions. Impact parameter bins are assigned by cuts on the reduced total transverse kinetic energy, KE_t^{red}. The solid curves and the dotted lines are meant to guide the eye.

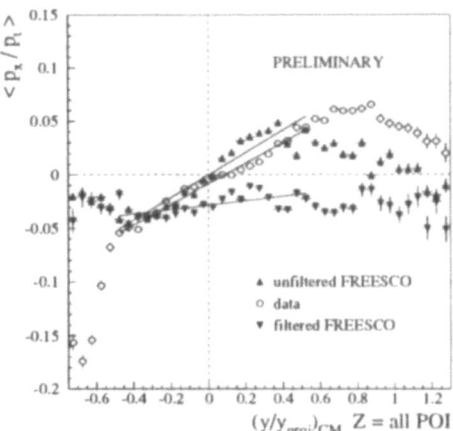

Figure 5. Average fraction of the transverse momentum in the reaction plane for all POI in central collisions at a beam energy of 45 MeV/nucleon versus the normalized rapidity in the center-of-mass frame for ^{40}Ar + ^{45}Sc reactions. Open circles are data, and solid triangles are events generated using FREESCO. The inverted triangles are FREESCO events filtered through our detector acceptance.

We have simulated flow using the model FREESCO.[17] All effects of the experimental acceptance were included in these simulations. The simulated events and the actual data are analyzed with the same flow routine using the azimuthal correlation method with the recoil correction to determine the reaction plane. In Fig. 5 we show a comparison between data and simulation for central collisions at a beam energy of 45 MeV/nucleon. The straight lines are fit in the region $-0.5 \leq (y/y_{proj})_{CM} \leq 0.5$. From a comparison of the unfiltered events to the data, it is evident that these results are consistent with the fact that measured flow is always less than the "actual" flow generated by the code.[18] More importantly, Fig. 5 demonstrates that the measured flow is not an artifact of our detector acceptance or analysis method.

4. CONCLUSIONS

Using the MSU 4π Array upgraded with the High Rate Array, we studied the impact parameter dependence of the disappearance of flow for ^{40}Ar + ^{45}Sc reactions. Our preliminary results indicate that the balance energy increases for events with larger impact parameters. In peripheral collisions the overlap volume is smaller, $i.e.$, the compressed participant zone, therefore a larger deposition energy is necessary to overcome the effects of the mean field.

ACKNOWLEDGMENTS

This work was supported by the National Science Foundation under Grant No. PHY-92-14992. We gratefully acknowledge the assistance of D. Swan and J. Wagner during the design and assembly stages of the HRA.

REFERENCES

1. H. H. Gutbrod, A. M. Poskanzer, and H. G. Ritter, Rep. Prog. Phys. **52**, 1267 (1989).
2. P. Danielewicz et al., Phys. Rev. C **38**, 120 (1988).
3. C. A. Ogilvie et al., Phys. Rev. C **42**, R10 (1990).
4. C. A. Ogilvie et al., Phys. Rev. C **40**, 2592 (1989).
5. D. Krofcheck et al., Phys. Rev. Lett. **63**, 2028 (1989).
6. J. P. Sullivan et al., Phys. Lett. B **249**, 8 (1990).
7. W. M. Zhang et al., Phys. Rev. C **42**, R491 (1990).
8. D. Krofcheck et al., Phys. Rev. C **43**, 350 (1991).
9. J. Péter, Nucl. Phys. **A545**, 173c (1992).
10. J. Lauret et al., Phys. Lett. B **339**, 22 (1994).
11. G. D. Westfall et al., Phys. Rev. Lett. **71**, 1986 (1993).
12. W. K. Wilson et al., Phys. Rev. C **43**, 2696 (1991).
13. G. D. Westfall et al., Nucl. Inst. and Methods **A238**, 347 (1985).
14. R. Pak et al., NSCL/MSU Ann. Rep. 244 (1993).
15. P. Danielewicz and G. Odyniec, Phys. Lett. **157B**, 146 (1985).
16. W. K. Wilson et al., Phys. Rev. C **45**, 738 (1992).
17. G. Fái and J. Randrup, Nucl. Phys. **A404**, 551 (1983).
18. J. P. Sullivan and J. Péter, Nucl. Phys. **A540**, 275 (1992).

22

THE DISAPPEARANCE OF FUSION/FISSION

G. D. Westfall,[1] J. Yee,[1] E. E. Gualtieri,[1] A. M. Vander Molen,[1]
W. J. Llope,[1*] S. Hannuschke,[1] R. Pak,[1] N. T. B. Stone,[1] D. Craig,[1] T. Li,[1]
J. S. Winfield,[1] S. J. Yennello,[1†] R. A. Lacey,[2] A. Nadasen,[3] and E. Norbeck[4]

[1]National Superconducting Cyclotron Laboratory
and Department of Physics and Astronomy
Michigan State University, East Lansing, Michigan
[2]Department of Chemistry
State University of New York at Stony Brook
Stony Brook, New York
[3]Department of Natural Sciences
University of Michigan, Dearborn, Michigan
[4]Department of Physics
University of Iowa
Iowa City, Iowa

ABSTRACT

The disappearance of fusion/fission has been studied for reactions of ^{40}Ar +
^{232}Th at incident energies ranging from 15 to 115 MeV/nucleon using the MSU 4π
Array. We have studied these reactions using a variety of observables including
fission fragment opening angles, fission fragment azimuthal correlations, intermediate mass fragment and light charged particle production, and event shape
analysis. We observe a change in the decay characteristics of high momentum
linear transfer collisions as a function of incident energy from fission-like to
multifragment emission.

1. INTRODUCTION

The disappearance of fusion/fission has been observed[1,2] in inclusive reactions of
^{40}Ar + ^{232}Th at incident energies between 39 and 44 MeV/nucleon using fission fragment
opening angle distributions. This disappearance may signal a change in the reaction
mechanism and decay characteristics of the nuclear systems formed in these reactions.

---

*Present address: T. W. Bonner Nuclear Laboratory, Rice University, Houston, Texas 77251.
†Present address: Cyclotron Institute, Texas A&M University, College Station, Texas 77843.

Advances in Nuclear Dynamics
Edited by Wolfgang Bauer and Alice Mignerey, Plenum Press, New York, 1996

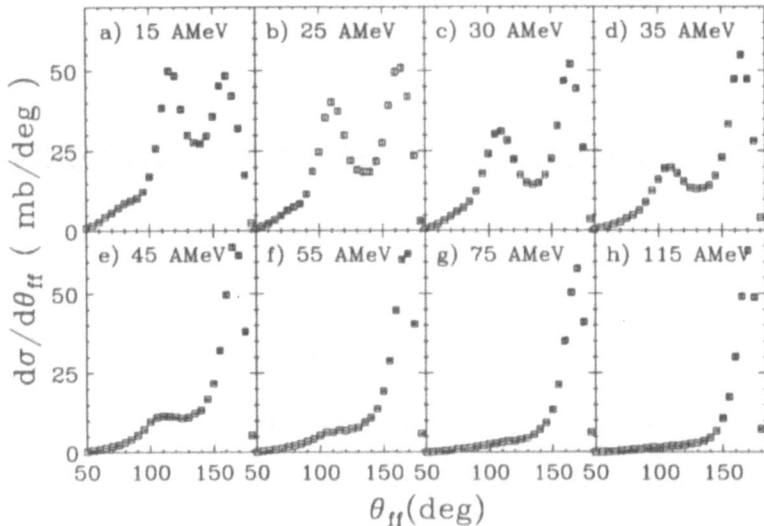

Figure 1. The fission fragment folding angle distributions for Ar + Th reactions at incident energies ranging from 15 to 115 MeV/nucleon.

More inclusive measurements have been done for the Ar+Th system[3,4] but the character of the reaction mechanism remains unclear.

To investigate these changes, we have carried out a systematic study of fusion/fission and particle production[5,6,7,8] using the MSU 4π Array[9] at the National Superconducting Cyclotron Laboratory and beams from the K1200 cyclotron. The system studied was ^{40}Ar + ^{232}Th at 15, 25, 30, 35, 40, 45, 55, 75, and 115 MeV/nucleon. The 4π Array incorporates 30 low pressure multiwire proportional counters which provide a nearly 4π acceptance for fission fragments allowing exclusive studies of fusion/fission reactions. Using the observables of the fission fragment opening angle, θ_{ff}, the fission fragment azimuthal angle, ϕ_{ff}, the azimuthal correlation function for three fragments, $C(\psi_3)$, and the average sphericity, $\langle S \rangle$, we show that there is a distinct transition in the reaction mechanism and decay of the system as the beam energy is raised from 15 to 115 MeV/nucleon. In addition we show that the multiplicity, N_{imf}, of intermediate mass fragments (IMFs), and the total charge, Z_{lcp}, bound in light charged particles (LCPs), continue to increase with incident energy for central collisions in contrast to previous inclusive measurements.[3]

2. FISSION FRAGMENT DISTRIBUTIONS

The measured cross sections for θ_{ff} are shown in Fig. 1 for reactions of ^{40}Ar + ^{232}Th from 15 to 115 MeV/nucleon. These distributions reflect the absolute cross section for observing two fission fragments triggered on events with at least one fission fragment. The absolute values of ϕ_{ff}, defined as the azimuthal angle between the planes formed by each fission fragment and the beam axis, are constrained to be less than 90° where 0° corresponds to coplanar emission.

Clearly visible in this figure are two peaks in θ_{ff}. The peak at angles around 170° corresponds to low linear momentum transfer (LMT) reactions which are associated

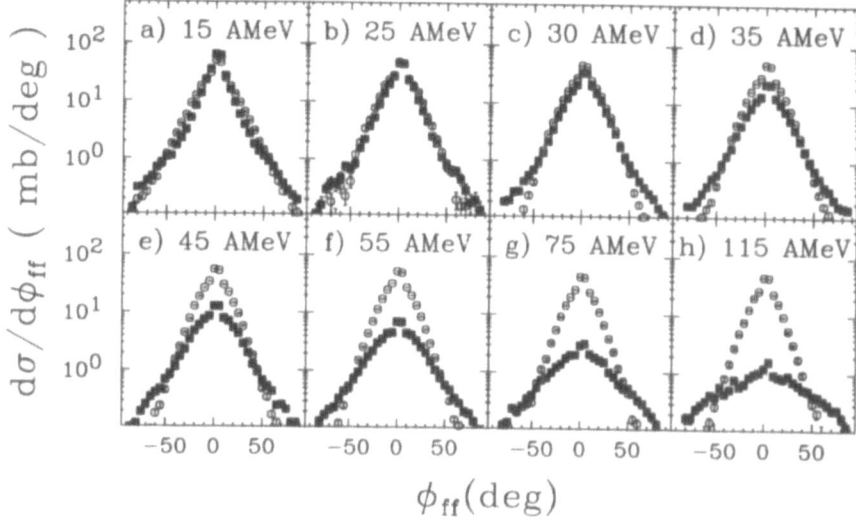

Figure 2. Fission fragment azimuthal angle distributions for Ar + Th reactions gated on high LMT (filled squares) and low LMT (open circles).

with peripheral collisions. The peak around 115° represents high LMT events related to central collisions. As the beam energy is raised from 15 to 115 MeV/nucleon, the high LMT peak disappears while the low LMT peak remains relatively constant.

To gain a better understanding of the fusion/fission channel at high incident energies, we have extracted the distributions of ϕ_{ff} gated on LMT using the θ_{ff} distributions. In Fig. 2, these distributions are shown for Ar+Th reactions from 15 to 115 MeV/nucleon gated on high and low LMT. The distributions are peaked strongly around 0° at the lowest incident energies indicating strongly coplanar decay.

As the energy is increased, the strong peaking remains for the low LMT reactions while the width of the distribution broadens dramatically for high LMT reactions at the highest incident energy. This broadening implies that the two-body character of the decay is diminishing and multifragment decay channels are beginning to dominate.

This change from two-body decay to multifragment decay can also be demonstrated using the three particle azimuthal correlation function, $C(\psi_3)$, between two fission fragments and an IMF. ψ_3 is the reduced azimuthal angle of the three fragments, $\psi_3 = (\phi_{f_1-f_2} * \phi_{f_1-i} * \phi_{f_2-i})^{1/3}$, where $\phi_{f_1-f_2}$ is the azimuthal angle between the two fission fragments, ϕ_{f_1-i} is the angle between the first fission fragment and the IMF, and ϕ_{f_2-i} is the angle between the second fission and the IMF. This correlation function is defined as the correlated yield divided by the uncorrelated yield. The results for Ar + Th reactions from 15 to 115 MeV/nucleon are shown in Fig. 3 gated on high LMT (squares) and low LMT (circles).

At all incident energies, the correlation function for low LMT is flat. At low incident energies, the correlation function for high LMT is peaked at 50° indicating either ϕ_{f_1-i} or ϕ_{f_2-i} is close to 0°. This result implies that the IMF is emitted from one of the fission fragments. As the beam energy is raised, the correlation disappears again indicating the onset of a multifragment decay channel.

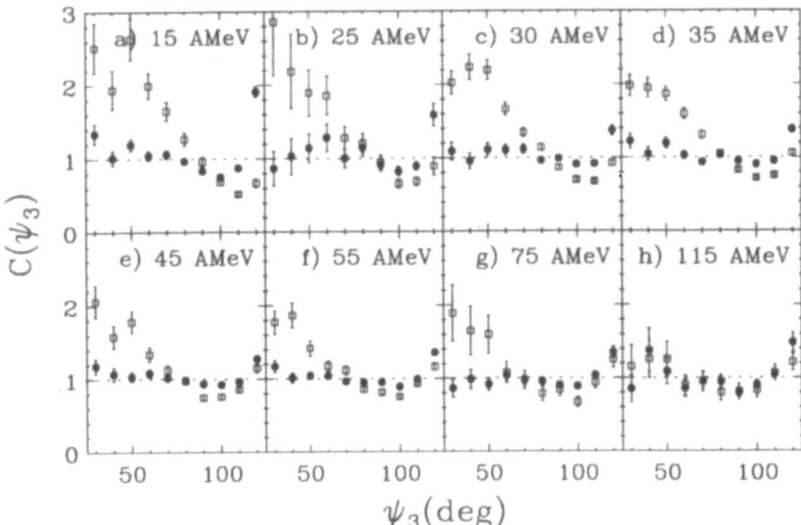

Figure 3. Three particle azimuthal correlation function for two fission fragments and an IMF from Ar + Th reactions gated on high LMT (open squares) and low LMT (filled circles).

3. FRAGMENT PRODUCTION

A recurring theme in fusion/fission work has been the saturation of linear momentum transfer and limiting excitation energy. To study this type of phenomena, we have made a systematic study of fragment production in Ar + Th reactions. We have extracted N_{imf} and Z_{lcp} from Ar + Th reactions from 15 to 115 MeV/nucleon.

These results are shown in Fig. 4 for three different centrality cuts. The first selection is just inclusive data. The second selection is central collisions characterized by the 10% highest values of transverse kinetic energy ($KE_t = KE \sin^2 \theta$). The third set requires large KE_t and θ_{ff} in the high LMT peak. The surprising result is seen that the IMF multiplicities and charge bound in LCPs seem to saturate for inclusive measurements while the values selected on central collisions increase smoothly with the incident energy.

The continued increase of N_{imf} suggests the onset of multifragmentation although the uncorrected average number of IMFs is only 2 even at the highest incident energy. The uncorrected N_{imf} distributions are shown in Fig. 4. At 115 MeV/nucleon, 40% of the central cross section leads to events in which three or more IMFs are emitted. Thus, these results for fragment production suggest that in central collisions a change in the decay mechanism occurs when the beam energy is raised from 15 to 115 MeV/nucleon.

4. SHAPE ANALYSIS

Another observable which can be used to address the question of the onset of multi-fragmentation is the average sphericity of the event in momentum space. The sphericity is defined in terms of unit vectors of the all the particles in each event. In this method, each event is characterized using a momentum tensor defined by[10,11] $F_{ij} = \sum_n \frac{p_i^n p_j^n}{p_n^2}$. Ordered eigenvalues of $\mathbf{F}, f_1 < f_2 < f_3$ are used to define the quantities $q_i = \frac{f_i^2}{\sum_{j=1}^{3} f_j^2}$. In terms of these variables, the

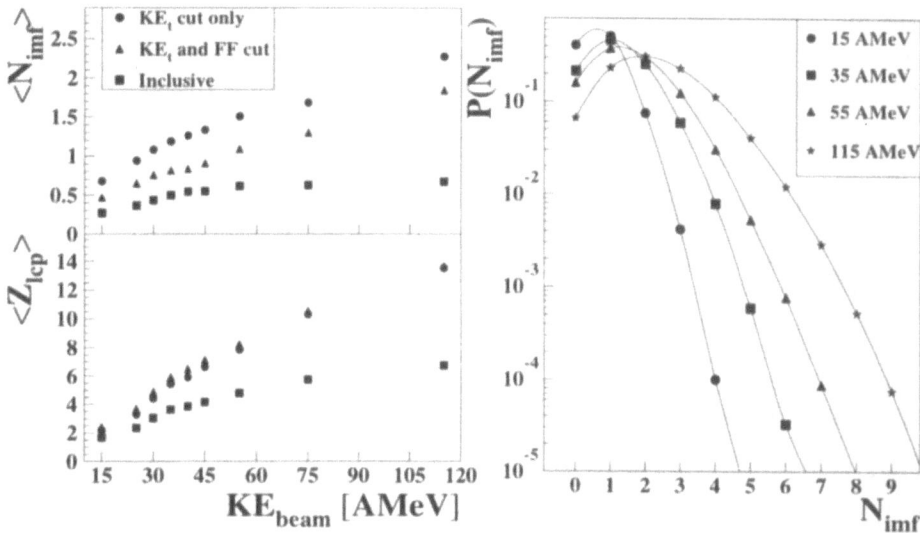

Figure 4. The left panel shows the average charge emitted as light charges particles and the average IMF multiplicity as a function of incident energy for Ar+Th reactions. The squares represent inclusive measurements while the circles and triangles represent different centrality cuts. The right panel shows the IMF multiplicity distributions in central collisions for Ar + Th reactions. These results are not corrected for acceptance.

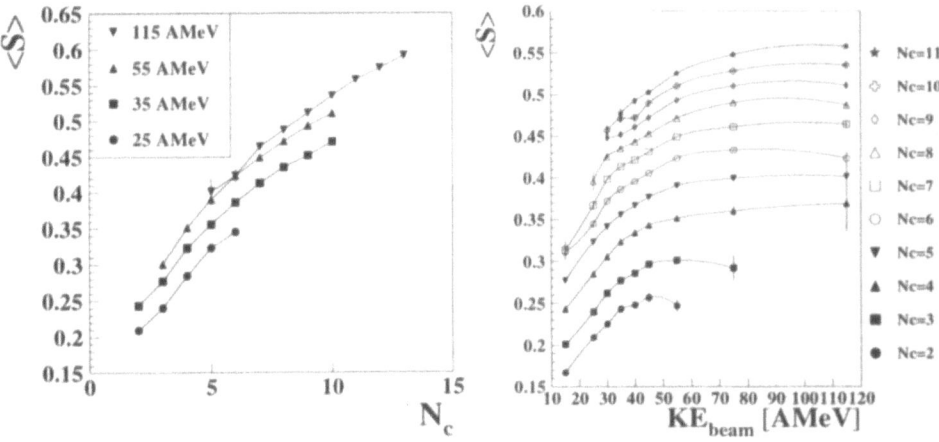

Figure 5. The left panel shows the average sphericity as a function of charged particle multiplicity for four incident energies of Ar+Th reactions selected on central collisions. The right panel shows the average sphericities from Ar + Th reactions for all measured incident energies and charged particle multiplicities.

sphericity is defined as $S = \frac{3}{2}(1 - q_3)$. Events that decay by two-body, sequential emission will have relatively small values of $\langle S \rangle$. Events that decay by simultaneous multifragmentation will have relatively large values of $\langle S \rangle$. Thus, sphericity can be used to distinguish between these two processes. However, $\langle S \rangle$ depends on the number of particles used to determine it. This effect can be treated by plotting the extracted values of $\langle S \rangle$ as a function of multiplicity. In the left panel of Fig. 5, the extracted values of $\langle S \rangle$ are shown as a function of charged particle multiplicity and incident energy. The well-known dependence of $\langle S \rangle$ on multiplicity is seen.

In the right panel of Fig. 5, the values of $\langle S \rangle$ are plotted as a function of incident energy for each observed multiplicity. A definitive increase in $\langle S \rangle$ is observed for each set of multiplicities which is followed by a plateau. This increase signals a change from elongated shapes in momentum space to more spherical values. This change can be associated with the onset of multifragmentation because one expects that multifragment events will be nearly spherical. The plateau value of the sphericities matches the values one obtains from simulated spherical distributions at the same multiplicity. The observed increase in $\langle S \rangle$ with incident energy and the observed plateau at the highest incident energy is not caused by acceptance effects. We have carried out a set of simulations filtered through the acceptance of the 4π Array and demonstrated that a non-spherical distribution is not made spherical by our acceptance.[7] We also demonstrated that sampling particles from a non-spherical distribution produces a non-spherical distribution and sampling from a spherical distribution produces a spherical distribution.

5. CONCLUSIONS

We have carried out a systematic study of fusion/fission in the system Ar + Th. We find that as the beam energy is increased from 15 to 115 MeV/nucleon, the fusion/fission component changes over from two-body fission fragment emission to the simultaneous emission of many fragments. We find that the IMF multiplicities and LCP charge continue to increase with incident energy. We observe a dramatic change in the event shapes which indicates the change from two-body sequential emission toward multifragmentation.

ACKNOWLEDGMENTS

This work was supported by the National Science Foundation under Grant number PHY-92-14992.

REFERENCES

1. E. C. Pollacco, M. Conjeaud, S. Harar, C. Volant, Y. Cassagnou, R. Dayras, R. Legrain, M. S. Nguyen, H. Oeschler, and F. Saint-Laurent, Phys. Lett. **B146**, 29 (1984).
2. M. Conjeaud, S. Harar, M. Mostefai, E. C. Pollacco, C. Volant, Y. Cassgnou, R. Dayras, R. Legrain, H. Oeschler, and F. Saint-Laurent, Phys. Lett. **B159**, 244 (1985).
3. D. X. Jiang, H. Doubre, J. Galin, D. Guerreau, E. Piasecki, J. Pouthas, A. Sokolov, B. Cramer, G. Ingold, U. Jahnke, E. Schwinn, J. L. Charvet, J. Frehaut, B. Lott, C. Magnago, M. Morjean, Y. Patin, Y. Pranal, J. L. Uzureau, B. Gatty, and D. Jacquet, Nucl. Phys. **A503**, 560 (1989).
4. E. Schwinn, U. Jahnke, J. L. Charvet, B. Cramer, H. Doubre, J. Fréhaut, J. Galin, B. Gatty, D. Guerreau, G. Ingold, D. Jaquet, D. X. Jiang, B. Lott, M. Morjean, C. Magnago, Y. Patin, J. Pouthas, E. Piasecki, and A. Sokolow, Nucl. Phys. **A568**, 169 (1994).

5. J. Yee, Ph. D. thesis, 1995.

6. J. Yee, E. E. Gualtieri, D. Craig, S. A. Hannuschke, T. Li, W. J. Llope, R. Pak, N. T. B. Stone, A. M. Vander Molen, G. D. Westfall, J. S. Winfield, S. J. Yennello, R. A. Lacey, A. Nadasen, and E. Norbeck, MSUCL-969, submitted to Phys. Lett. B, 1995.

7. E. E. Gualtieri, Ph. D. thesis, 1995.

8. E. E. Gualtieri, J. Yee, D. Craig, S. Hannuschke, R. A. Lacey, T. Li, W. J. Llope, A. Nadasen, E. Norbeck, R. Pak, N. T. B. Stone, A. M. Vander Molen, J. S. Winfield, G. D. Westfall, and S. J. Yennello, MSUCL-970, submitted to Phys. Lett. B, 1995.

9. G. D. Westfall, J. E. Yurkon, J. van der Plicht, Z. M. Koenig, B. V. Jacak, R. Fox, G. M. Crawley, M. R. Maier, B. E. Hasselquist, R. S. Tickle, and D. Horn, Nucl. Inst. and Methods **238**, 347 (1985).

10. J. A. López and J. Randrup, Nucl. Phys. **A491**, 477 (1989).

11. D. A. Cebra, S. Howden, J. Karn, A. Nadasen , C. A. Ogilvie, A. Vander Molen, G. D. Westfall, W. K. Wilson, and J. S. Winfield, Phys. Rev. Lett. **64**, 2246 (1990).

23

FLOW IN ULTRARELATIVISTIC HEAVY ION COLLISIONS AT THE AGS

Johannes P. Wessels,[1] J. Barrette,[2] R. Bellwied,[3] S. Bennett,[3] R. Bersch,[1]
P. Braun-Munzinger,[1] W. C. Chang,[1] W. E. Cleland,[4] M. Clemen,[4] J. Cole,[5]
T. M. Cormier,[3] Y. Dai,[2] G. David,[6] J. Dee,[1] O. Dietzsch,[7] M. Drigert,[5]
K. Filimonov,[2] A. French,[3] S. Gilbert,[2] J. R. Hall,[3] T. K. Hemmick,[1]
N. Herrmann,[8] B. Hong,[1] K. Jayananda,[4] S. Johnson,[1] Y. Kwon,[1]
R. Lacasse,[2] Q. Li,[3] T. W. Ludlam,[6] A. Lukaszew,[3] S. McCorkle,[6]
S. K. Mark,[2] R. Matheus,[3] D. Miśkowiec,[1] J. Murgatroyd,[3] E. O'Brien,[6]
S. Panitkin,[1] P. Paul,[1] T. Piazza,[1] M. Pollack,[1] C. Pruneau,[3] Y. Qi,[2]
M. N. Rao,[1] E. Reber,[5] M. Rosati,[2] N. C. daSilva,[3] S. Sedykh,[1] J. Sheen,[3]
U. Sonnadara,[4] J. Stachel,[1] N. Starinski,[2] H. Takai,[6] E. M. Takagui,[7]
M. Trzaska,[1] S. Voloshin,[4] T. Vongpaseuth,[1] G. Wang,[2] C. L. Woody,[6]
N. Xu,[7] Y. Zhang,[1] Z. Zhang,[4] and C. Zou[1]

[1]Department of Physics
State University of New York at Stony Brook
Stony Brook, NY
[2]McGill University
Montreal, Canada
[3]Wayne State University
Detroit, MI
[4]University of Pittsburgh
Pittsburgh, PA
[5]Idaho National Engineering Laboratory
Idaho Falls, ID
[6]Brookhaven National Laboratory
Upton, NY
[7]University of São Paulo
Brazil
[8]Gesellschaft für Schwerionenforschung
Darmstadt, Germany

ABSTRACT

The azimuthal distributions of produced transverse energy from the reaction
Au + Au at 10.8 AGeV/c have been studied with the nearly 4π calorimetric cov-

Advances in Nuclear Dynamics
Edited by Wolfgang Bauer and Alice Mignerey, Plenum Press, New York, 1996

erage of the E877-setup at the AGS. A Fourier analysis of these distributions has been performed in different regions of pseudorapidity as a function of centrality. The extracted sideward flow is found to be maximal in semi-central collisions. This analysis has also been employed to extract the orientation of the reaction plane. Spectra of particles identified in the forward spectrometer are then analyzed with respect to the reaction plane and their triple differential cross sections are determined. The nucleons are found to be the carriers of the sideward flow identified in the calorimeters. While the pions are also preferentially emitted into the reaction plane they show a slight tendency for emission to the opposite side of where the nucleons are found.

1. INTRODUCTION

In the quest for the nuclear equation-of-state systematic studies have been carried out to achieve a complete picture of the so-called sideward flow as a function of projectile target combination as well as bombarding energy. Experimentally, the phenomenon was first established at the Bevalac 1. The experimental studies were accompanied by the development of hydrodynamical and cascade models 2 in an attempt to relate the experimental observables to the ingredients of these codes such as the incompressibility modulus of the equation-of-state or in-medium cross sections. With the availability of heavy, symmetric projectile target combinations at the AGS, these studies have recently focused on the possibility of identifying the possible phase transition to the quark–gluon plasma (QGP). Hydrodynamic calculations involving such a phase transition 3, 4 have shown that the functional dependence of the mean transverse momentum in the reaction plane is altered in the presence of even small volumes of QGP as compared to predictions by cascade codes in this energy regime (RQMD 5, ARC 6).

In the following we will show how we use calorimeters to establish the sideward flow signal and to determine the reaction plane. This will then allow us to extract triple differential cross sections for particles measured in the spectrometer. Finally, we will extract average transverse momenta into the reaction plane as a function of rapidity for different particle species.

2. EXPERIMENTAL SETUP

A layout of the whole setup can be found elsewhere 7. Here, details of the calorimeters that are used to determine the reaction plane and the transverse energy flow are shown. In Fig. 1 front and side views of the calorimeters surrounding the target are depicted. Both detectors are highly segmented in polar and azimuthal angles. In this setup the target calorimeter (TCal) uses 832 NaI-crystals to cover the interval $-0.5 \leq \eta \leq 0.8$ in pseudorapidity (η) in a nearly projective geometry. The participant calorimeter (PCal) covers the region $0.83 \leq \eta \leq 4.5$. This lead–iron-scintillator sampling calorimeter is subdivided into 4 depth sections, 8 radial, and 16 azimuthal sections for a total of 512 channels.

3. FOURIER ANALYSIS OF AZIMUTHAL TRANSVERSE ENERGY DISTRIBUTIONS

The main purpose of the calorimeters is to provide information about the impact parameter through the measurement of the total amount of transverse energy (E_t) produced

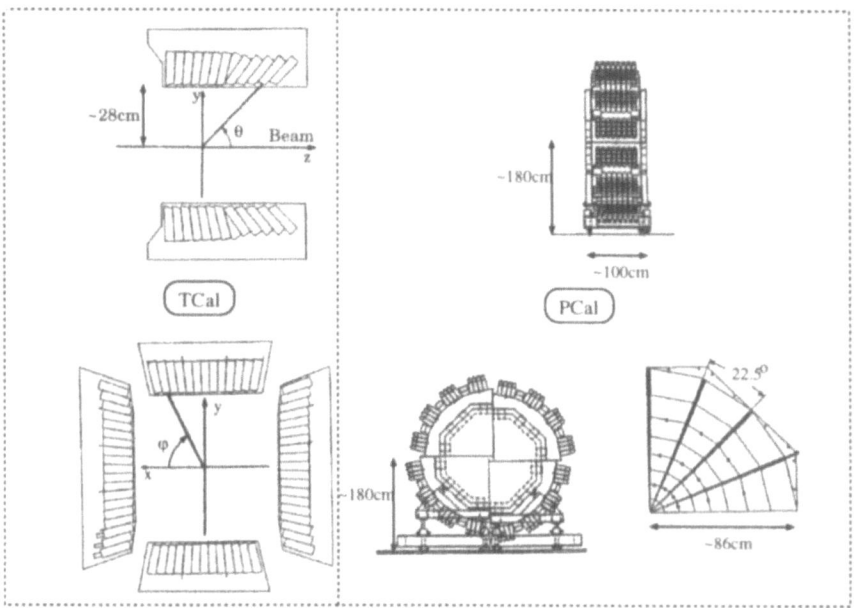

Figure 1. Cross sectional views of the target calorimeter (TCal) and the participant calorimeter (PCal). Top half and bottom half are cross sections in the yz-plane and in the xy-plane. The beam goes along the z-axis.

in the reaction. Each calorimeter cell i is located at a specific polar (ϑ_i) and azimuthal angle (ϕ_i). It detects a fraction ε_i^t of the total transverse energy $E_t = \sum_i \varepsilon_i^t$ at a specific pseudorapidity $\eta_i = -\ln(\tan(\vartheta_i/2))$. Due to the segmentation of the calorimeters in the azimuth, it is possible to construct the transverse energy vector $\vec{E}_t$ in different pseudorapidity intervals. It has been shown previously [7, 9] that this allows for the reconstruction of the reaction plane angle (ψ_r') on an event-by-event basis (the prime is to denote that the quantity has not yet been unfolded with the resolution). In addition to having a good determination of the impact parameter b through the measurement of E_t, it is now possible to determine the impact parameter vector $\vec{b}$ by measuring $\vec{E}_t$ at forward and/or backward pseudorapidity.

In the following the reaction plane angle has been determined using the TCal which determines $\vec{E}_t$ in the backward hemisphere. Then all individual calorimeter cells of the PCal have been subdivided into 12 bins in pseudorapidity from $\langle\eta\rangle = 1.2$ to $\langle\eta\rangle = 4.3$. For each cell of those bins the distribution

$$\frac{d\varepsilon_t^i}{d\psi_i'}$$

was constructed, with $\psi_i' = \phi_i - \psi_r'$. The distributions were added for cells belonging to the same pseudorapidity interval. The resulting distributions were Fourier analyzed in the following fashion:

$$\frac{dE_t}{d\psi' d\eta} = \frac{\langle E_t\rangle_\eta}{2\pi} + \frac{1}{\pi}\sum_{n=1}^{\infty}(a_n\cos(n\psi') + b_n\sin(n\psi'))$$

where $\langle E_t\rangle_\eta$ is the mean transverse energy in the η-bin under consideration. From this analysis

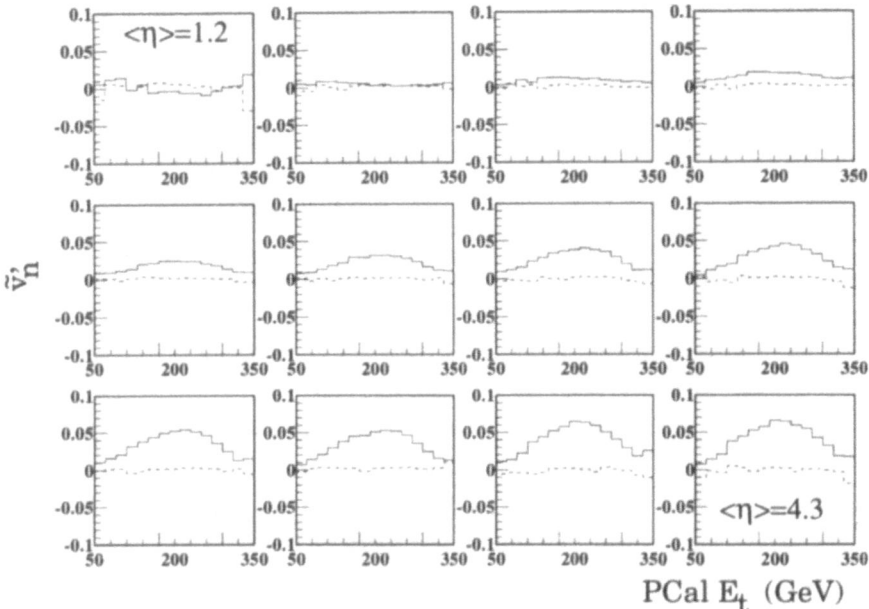

Figure 2. Multipole moments $\tilde{v}_n$ ($n = 1$:solid; $n = 2$:dashed) for different pseudorapidity intervals from $\langle \eta \rangle = 1.2$ to $\langle \eta \rangle = 4.3$ as a function of transverse energy E_t.

one obtains multipole moments:

$$|\tilde{v}_n| = \frac{\sqrt{a_n^2 + b_n^2}}{\langle E_t \rangle_\eta} \quad \text{with} \quad \text{sign}(\tilde{v}_n) = \text{sign}(a_n)$$

The dimensionless quantity $\tilde{v}_1$ indicates the relative fraction of the common shift of all measured transverse energies compared to the total detected E_t. This is a measure of the collective sideward motion of the participants. The quantity $\tilde{v}_2$ would be non-zero if there were more complex deformations of the event-shape than just a global offset due to the sideward flow. These moments are plotted as a function of the total E_t (i.e. the centrality of the collision) in Fig. 2. Here, the reaction plane was determined in backward direction using the TCal. One can see that $\tilde{v}_1$ changes its sign as one goes from below midrapidity forward. Also one clearly sees how in going from peripheral (low E_t) to central (high E_t) collisions $\tilde{v}_1$ first rises and then falls again with a peak around $E_t \sim 230 GeV$, in accordance with the observation at lower energies, where the sideward flow also peaks at mid-impact parameters. In all pseudorapidity bins $\tilde{v}_2$ is compatible with zero.

Determining the reaction plane with the PCal in forward direction allows to do the same analysis for the other calorimeter. A compilation of all the extracted $\tilde{v}_1$ as a function of the total accessible η-space is shown in Fig. 3 for all measured centralities. Since the detectors cannot discriminate between different particles the extracted sideward flow is plotted versus pseudorapidity rather than rapidity, still, the familiar "S"-shape is clearly visible. The data have not been unfolded with the reaction plane resolution (i.e. the true anisotropies are larger). This explains also the sudden jump around $\eta = 0.8$ where data have been merged from two detectors whose resolution for the reaction plane angle is very different. Preliminary evaluations of model calculations 2 show that the measured sideward flow of transverse energy

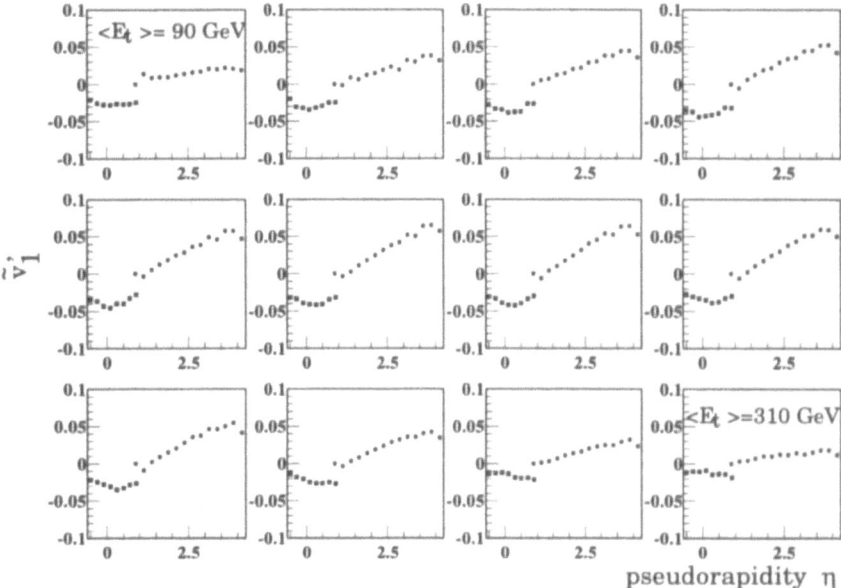

Figure 3. Measurement of $\tilde{v}_1$ as a function of η for different centralities characterized by the total E_t showing a gradual increase of sideward transverse energy flow and eventually the decrease as one approaches the most central collisions.

is larger than predicted by, e.g., RQMD in cascade-mode 8.

4. PARTICLE SPECTRA WITH RESPECT TO THE REACTION PLANE

As was already pointed out, the calorimeters do not allow for particle identification. However, the collimator inside the PCal had an opening of $-134 < \theta_x < 16$ mrad horizontally and $-11 < \theta_y < 11$ mrad vertically through which particles could enter the spectrometer, where their mass and momentum were determined. In the following the transverse momentum p_t of protons and positive pions was decomposed into components $p_{x'}$ and $p_{y'}$ with

$$p_t = \sqrt{p_{x'}^2 + p_{y'}^2}.$$

The component $p_{x'}$ is pointing into the direction of the impact parameter vector, $p_{y'}$ perpendicular to it, and both perpendicular to the beam. Again, the prime denoting that the quantities have not been unfolded with the resolution in determining $\vec{b}$. This allows then to construct triple differential cross sections $d^3N/dp_{x'}dp_{y'}dy$ as they are shown in Fig. 4 for protons and pions. A clear difference is apparent in these spectra. The protons have a distinct asymmetry in $p_{x'}$ while the pions are more or less symmetric about the origin. The observed common shift in the x'-direction is in the same direction as the observed sideward flow of transverse energy in the forward calorimeter. This identifies the nucleons as the carriers of the observed sideward flow in the transverse energy spectra.

If indeed the source of these protons and pions has a common sideward motion, then one can extract the common velocity shift by measuring the mean transverse momentum into

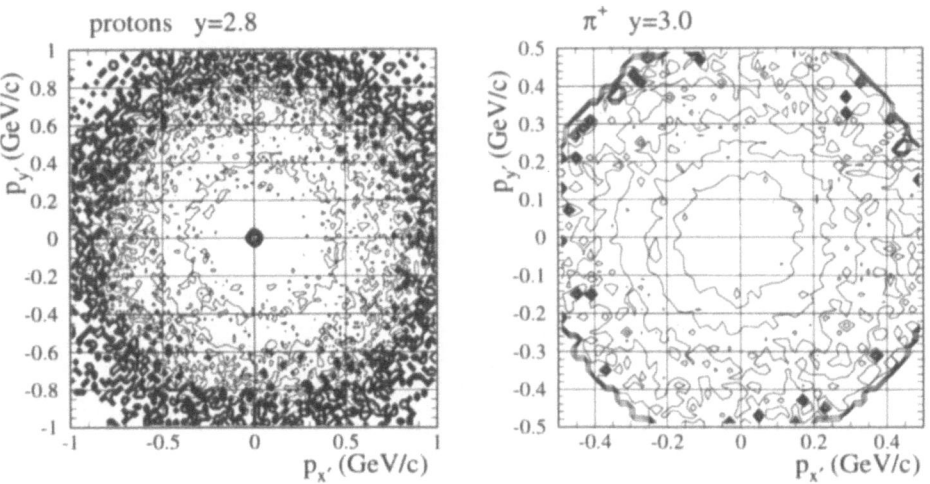

Figure 4. Triple differential cross sections $d^3N/dp_{x'}dp_{y'}dy$ at rapidity $y = 2.8 \pm 0.05$ for protons (left) and $y = 3.0 \pm 0.05$ for π^+(right). The reaction plane angle ψ'_r was determined by the TCal.

the reaction plane divided by the mass of the particle. This has been done in Fig. 5. Protons show the well-known sideward flow phenomenon, while the π^+ appear to flow in the opposite direction. The relative errors are of the order of the fluctuations. While these spectra have been corrected for the reaction plane resolution, the data are limited by the acceptance as one gets close to midrapidity.

Another way of quantifying the azimuthal anisotropy observed, e.g., in the proton spectrum in Fig. 4 is by determining the Boltzmann temperature (T_B) from the transverse mass spectrum ($m_t = \sqrt{p_t^2 + m_p^2}$) in azimuthal slices $\Delta\psi' = 45°$ with respect to the measured reaction plane. In Fig. 6 the relative change of the Boltzmann temperature is plotted as a function of the azimuthal angle ψ' for protons of rapidity $y = 2.8 \pm 0.05$. The curve has roughly a cosine shape with the maximum T_B in the reaction plane to the same side where transverse energy flow is observed.

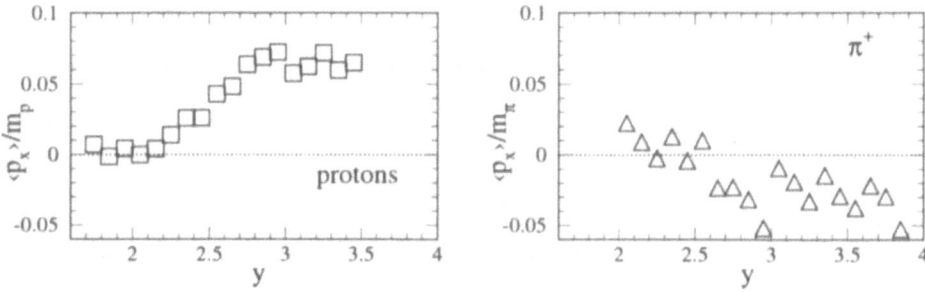

Figure 5. Mean transverse momentum of protons (left) and pions (right) divided by their respective mass as a function of rapidity. Data are obtained from the 12% most central events which includes the maximum of the sideward transverse energy flow observed in the calorimeters.

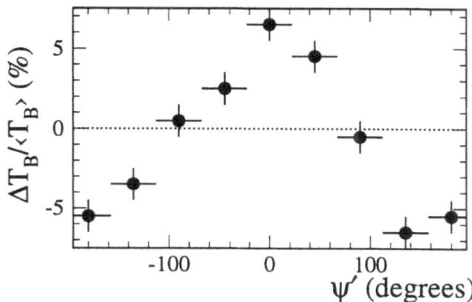

Figure 6. Relative variation of the Boltzmann temperature parameter (T_B) as a function of azimuthal angles ψ' for protons with respect to the reaction plane ($y = 2.8 \pm 0.05$).

5. CONCLUSION AND OUTLOOK

In conclusion, the phenomenon of collective sideward flow as determined by the anisotropy of E_t has been explored over the accessible pseudorapidity region. As compared to model predictions using RQMD in cascade mode the anisotropy seems to be larger. No indications for a squeeze-out signature can be deduced from the mentioned anisotropies. In combining the data of the calorimeters and the spectrometer it is for the first time possible to measure triple differential cross sections of particles at the AGS. From the analysis of these it has been found that the nucleons are the carriers of the observed sideward flow. The magnitude of the sideward flow for the measured π^+ is smaller than that of the nucleons and it is opposite in direction. Whether this is entirely due to shadowing or is due to some true collective motion of the pion-source with respect to the nucleons remains to be clarified. Also, the particular shape of the mean transverse momentum distributions as a function of rapidity around midrapidity needs to be studied with better statistics and reduced systematical errors in order to distinguish between the various models.

ACKNOWLEDGMENTS

This work was financially supported by grants from the U. S. DoE, the U. S. NSF, the Canadian NSERC, and CNPq Brazil. The author would like to acknowledge his support by the Alexander-von-Humboldt Foundation.

REFERENCES

1. H. A. Gustafsson *et al.*, Phys. Rev. Lett. **52** (1984) 1590.
2. H. Stöcker and W. Greiner, Phys. Rep **137**, 277 (1986).
3. N. S. Amelin, E. F. Staubo, L. P. Csernai, V. D. Toneev, K. K. Gudima, D. Strottman, Phys. Rev. Lett. **67**, 1523 (1991); L. Bravina, L. P. Csernai, P. Lévai, D. Strottman, Phys. Rev. **C50**, 2162 (1994).
4. Dirk H. Rischke, Yarış Pürsün, Joachim A. Maruhn, Horst Stöcker, Walter Greiner, Columbia preprint CU-TP-695
5. H. Sorge, A.v. Keitz, R. Mattiello, H. Stöcker, W. Greiner, Phys. Lett. **B243**, 7 (1990).
6. D. E. Kahana, D. Keane, Y. Pang, T. Schlagel, S. Wang, preprint, nucl-th/9405017
7. J. P. Wessels, Y. Zhang, Advances in Nuclear Dynamics, Proc. 10th Winter Workshop on Nuclear Dynamics, Snowbird, Utah, Jan. 1994, J. Harris, A. Mignerey, and W. Bauer. edt., World Scientific, p. 228.
8. Y. Zhang, J. P. Wessels, Proc. of 'Quark Matter '95', in print.
9. J. Barrette et al., Phys. Rev. Lett. **73**, 2532 (1994).

24

MEAN FIELD EFFECTS IN HEAVY-ION COLLISIONS AT AGS ENERGIES

Bao-An Li, C. M. Ko, and G. Q. Li

Department of Physics and Cyclotron Institute
Texas A&M University, College Station, TX

ABSTRACT

In the framework of a newly developed relativistic transport model for high energy heavy-ion collisions we study mean field effects in heavy-ion collisions at AGS energies. It is found that in central collisions of Au + Au at $P_{beam}/A =$ 11.6 GeV/c a simple, Skyrme-type nuclear mean field satisfying the causality requirement reduces the maximum baryon and energy densities reached in the cascade model by about 30% and 40%, respectively. Moreover, the mean field causes a factor of 2.5 increase in the strength of the baryon transverse collective flow.

1. MOTIVATION

The purpose of relativistic heavy-ion collisions at AGS energies is to study the properties of hot and dense hadronic matter and the possible phase transition to the quark–gluon plasma. A vast body of data have been collected and analyzed during the past years.[1] Comparisons of these data with models, such as RQMD,[2] ARC[3] and QGSM,[4] have revealed much interesting physics. In particular, a picture of nearly complete stopping of baryons in central heavy-ion collisions at AGS energies has emerged from these studies.

Here we study effects of the nuclear mean field at AGS energies within the framework of a relativistic transport (ART) model for high energy heavy-ion collisions. Our motivations of studying effects of the nuclear mean field which has been ignored in ARC are mainly the following. First of all, we believe that the mean-field potential is not negligible in heavy-ion collisions at AGS energies. Although the forward scattering amplitudes of hadron–hadron collisions in the high energy limit have been found approximately purely imaginary,[5] the AGS energies may not be high enough for the real part of the scattering amplitude to completely vanish. Of course, the form and strength of the corresponding mean field in the hot and dense hadronic matter is highly uncertain and has been a subject of much discussions. Secondly, although the kinetic energy is much higher than the potential energy in the early stage of the reaction, particles are gradually slowed down and the mean field plays an increasingly important role as the reaction proceeds. In particular, the repulsive mean field in the high density region tends to keep particles from coming too close to each other and therefore reduce the maximum energy and baryon densities reached in the reaction should there be no mean field. Moreover, in the expansion phase of the reaction mean-field effects are expected to be even stronger. It is therefore necessary to study how the reaction dynamics are affected by the nuclear mean field and search for experimentally observable consequences of the nuclear mean field.

2. THE MODEL ART

Our relativistic transport model (ART) is developed from the well known Boltzmann–Uehling–Uhlenbeck (BUU) model (e.g., Refs. 6, 7) for intermediate energy heavy-ion collisions by including more baryon and meson resonances and interations among them.[8] More specifically, we have included in the model the following baryons: N, $\Delta(1232)$, $N^*(1440)$, $N^*(1535)$, Λ, Σ, and mesons: π, ρ, ω, η, K, with their explicit isospin degrees of freedom. Both elastic and inelastic collisions among most of these particles are simulated as best as we can using as much input from the experimental hadron–hadron data as possible.[9] Most of the inelastic hadron–hadron collisions are modeled through the formation of various resonances. The advantage of this approximation is that the finite lifetimes of these resonances can partially take into account effects of the formation time of newly produced secondaries. More specifically, we have included in the model

$$NN \leftrightarrow N\Delta, \ NN^*(1440), \ NN^*(1535), \tag{2.1}$$
$$NN \leftrightarrow \Delta\Delta, \ \Delta N^*(1440), \tag{2.2}$$
$$NN \rightarrow NN\rho, \ NN\omega, \ \Delta\Delta\pi, \tag{2.3}$$
$$NN \rightarrow \Delta\Delta\rho, \tag{2.4}$$
$$N\Delta \leftrightarrow NN^*(1440), \ NN^*(1535), \tag{2.5}$$
$$\Delta\Delta \leftrightarrow NN^*(1440), \ NN^*(1535), \tag{2.6}$$
$$\Delta N^*(1440) \leftrightarrow NN^*(1535), \tag{2.7}$$

and those producing kaons

$$NN \rightarrow N\Lambda(\Sigma)K, \ \Delta\Lambda(\Sigma)K, \tag{2.8}$$
$$NR \rightarrow N\Lambda(\Sigma)K, \ \Delta\Lambda(\Sigma)K, \tag{2.9}$$
$$RR \rightarrow N\Lambda(\Sigma)K, \ \Delta\Lambda(\Sigma)K, \tag{2.10}$$

where R is Δ, $N^*(1440)$ or $N^*(1535)$ resonance. Cross sections for the production of single Δ, $N^*(1440)$ and $N^*(1535)$ are estimated from that for the production of single π and η in nucleon–nucleon collisions. Cross sections for the double resonances production in processes (2) are estimated by subtracting from the inclusive 2π production cross section[9] the contribution of $NN \rightarrow NN\rho$ and the 2π decay of the $N^*(1440)$ in the $NN \rightarrow NN^*(1440)$ process. The cross sections for ρ and ω production in channels (3) are taken directly from the experimental data.[9] The cross section for the process $NN \rightarrow \Delta\Delta\pi$ is taken as the difference between the inclusive 3π and the ω production cross section. We then attribute the difference between the experimental total nucleon–nucleon inelastic cross section and the sum of cross sections for channels (1) to (3) as well as the kaon production cross sections of channels (8) to the process $NN \rightarrow \Delta\Delta\rho$. This is done to ensure the total inelasticity of nucleon–nucleon collisions as only a limited, though large, number of reaction channels have been incorporated so far. The bias introduced by this approximation towards the cross section of the quasi-4π production process $NN \rightarrow \Delta\Delta\rho$ is, however, very small. Cross sections for the inverse processes are calculated by using the detailed balance taking into account the finite widths of resonances. Masses of baryon and meson resonances are generated according to the single or joint Breit–Wigner distributions with momentum dependent widths using the rejection method.

One can also separate meson–baryon collisions into elastic and inelastic parts. We model elastic collisions through both the formation of baryon resonances, i.e., $\pi N \leftrightarrow \Delta(N^*(1440), N^*(1535))$ and $\eta N \leftrightarrow N^*(1535)$, as well as direct processes, such as, $\pi(\rho) + N(\Delta, N^*) \rightarrow \pi(\rho) + N(\Delta, N^*)$, and $K^+ + N \rightarrow K^+ + N$. The formation of the three baryon resonances accounts almost completely the $\pi + N$ elastic cross sections at low energies. At higher energies, the elastic cross section is about 7 mb and is mainly due to the formation of higher resonances which are not included in the present model. We therefore attribute the difference between the experimental elastic cross section and the contribution from the three baryon resonances to the direct process $\pi + N \rightarrow \pi + N$. For experimentally unknown cross sections, such as, $\pi^0 + N$, $\pi + \Delta(N^*)$ and $\rho + N$, we calculate them using the resonance model through the implicit formation of heavier baryon resonances with masses up to 2 GeV.

The $\pi + N$ inelastic collision mainly goes through the production of pions and kaons. Similar to the treatment of baryon–baryon collisions, we model the inelastic $\pi + N$ collisions through the production of resonances, namely, $\pi + N \leftrightarrow \Delta + \pi(\rho, \omega)$ and $\pi(\rho, \omega) + N(\Delta, N^*) \rightarrow \Lambda(\Sigma) + K$. All experimental cross sections for $\pi + N$ collisions with final states having two and three pions are attributed to the production of $\Delta\pi$ and $\Delta\rho$, respectively. The difference between the experimental total $\pi + N$ inelastic cross section and cross sections for the production of two and three pions as well as kaons are then attributed to the production of $\Delta\omega$. Cross sections for the production of kaons in $\pi(\rho, \omega) + N(\Delta, N^*)$ collisions are taken to be the same as that in the $\pi + N$ collision.

We model meson–meson elastic collisions through the formation of the ρ meson and the direct process $\pi\pi \leftrightarrow \pi\pi$. The latter takes into account the case when the quantum numbers of colliding pions forbid the formation of a P_{11} meson ρ. The inelastic meson–meson collisions are modeled through the annihilation $MM \rightarrow K\bar{K}$, while the cross sections are calculated using the K^* exchange model.[11]

Main features of the multiparticle production systematics[10] in energetic nucleon–nucleon collisions are incorporated in the modeling. Such as, the "leading" particle behavior, is ensured by requiring the outgoing baryons to have the same or similar quantum numbers as the incident ones so that their longitudinal directions are retained when performing the momentum transformation from the baryon–baryon c.m. frame to the nucleus–nucleus c.m. frame. Moreover, final state baryons have the typical forward–backward peaked angular distribution or the soft transverse momentum distribution.

3. MEAN FIELD EFFECTS AT AGS ENERGIES

First, we show in Fig. 1 and Fig. 2 the evolution of the local baryon density and energy density in the reaction plane using the cascade mode of ART for the reaction of Au + Au at $P_{beam}/A = 11.6$ GeV/c and $b = 0$. The two nuclei are set in touch at $t = 0$, they reach the maximum compression of about $9\rho_0$ and the maximum energy density of about 3.6 GeV/fm^3 at about 4 fm/c. It is seen that the high density region with $\rho_l \geq 5\rho_0$ and $e_l \geq 2.5$ GeV/fm^3

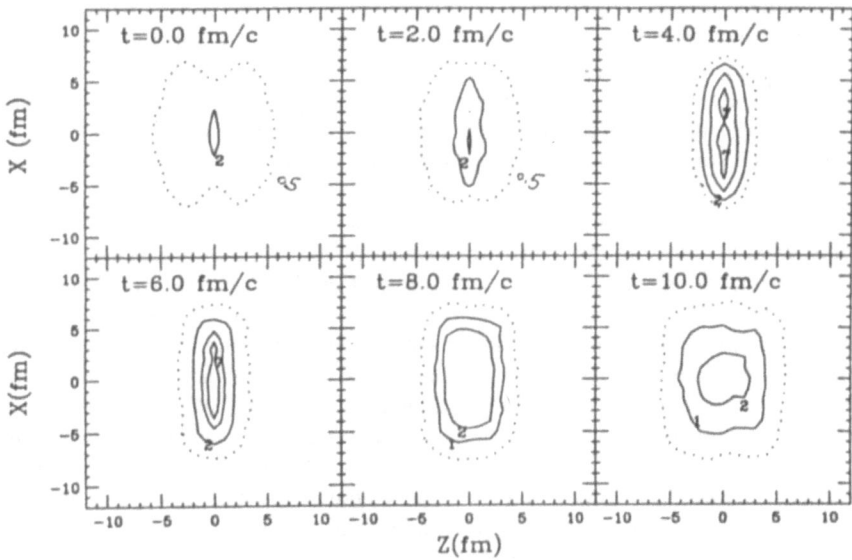

Figure 1. Local baryon density in the reaction plane for the head-on collision of Au + Au at 11.6 Gev/c calculated in the cascade mode of ART

lasts for about 4 fm/c. The maximum volume of about 200 fm³ of the high density region is reached at about 4 fm/c. The currently estimated critical baryon and energy densities for forming the quark gluon plasma is about $5\rho_0$ and 2.5 GeV/fm³, respectively.[5] According to the above cascade model prediction, it is highly possible to form the quark–gluon plasma at AGS energies. However, as we will see in the following the lifetime and volume of the high density region are significantly reduced by the repulsive mean field.

We now turn to the effects of the nuclear mean field. Without much reliable knowledge about the nuclear equation of state in hot and dense medium, we use here the simple, Skyrme-type parameterization widely used at Bevalac energies and below,

$$U(\rho) = -358.1 \left(\frac{\rho}{\rho_0}\right) + 304.8 \left(\frac{\rho}{\rho_0}\right)^{1.167}. \tag{3.1}$$

This is the so-called soft equation of state which satisfies the causality requirement upto about $\rho \approx 7\rho_0$. To see quantitatively effects of the mean filed on the creation of high baryon and energy densities, we show in Fig. 3 the evolution of the local baryon and energy densities in the central cell during the reaction of Au + Au at $P_{beam}/A = 11.6$ GeV/c and $b = 0$ for the three different cases. It is seen that the maximum baryon and energy densities are reduced to about $7\rho_0$ and 2.6 GeV/fm³, respectively, by using the soft nuclear equation of state although the mean field has almost no effect in the early stage of the reaction when the kinetic energy is much higher than the potential energy. With a stiff equation of state corresponding to the compressibility of $K = 380$ MeV the reduction is even larger. Since the stiff equation of state violates causality already at about $3\rho_0$, we will only use the soft equation of state in the following. The reduction of the maximum baryon and energy densities due to the mean field may be large enough to affect significantly the kind of physical processes that can happen during the reaction.

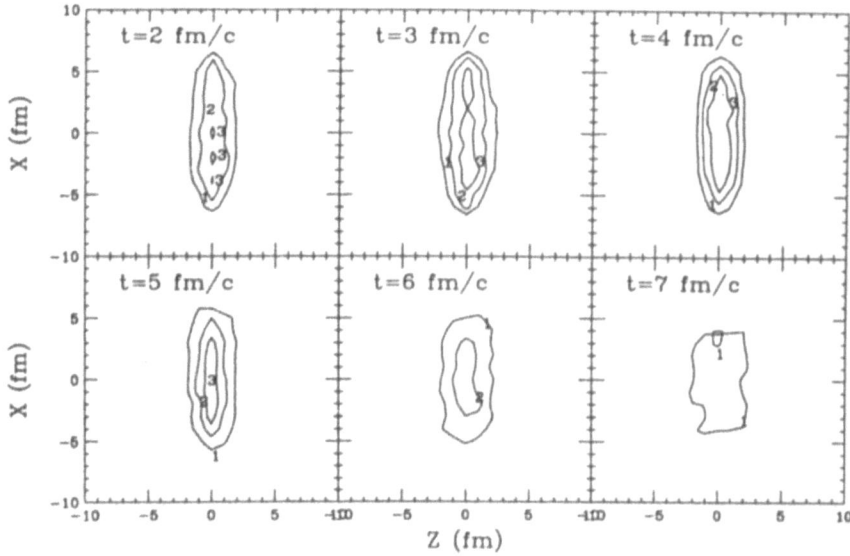

Figure 2. Local energy density in the reaction plane for the head-on collision of Au + Au at 11.6 Gev/c calculated in the cascade mode of ART

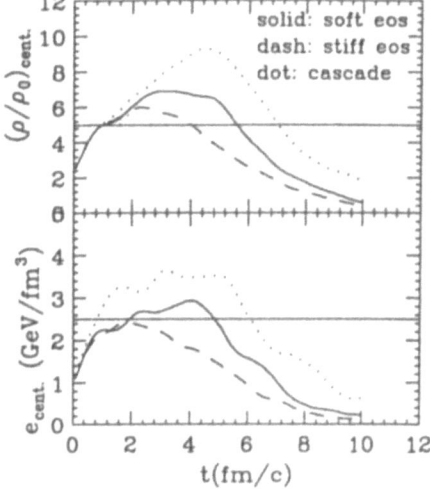

Figure 3. Central, local baryon and energy densities in the head-on collision of Au + Au at 11.6 Gev/c.

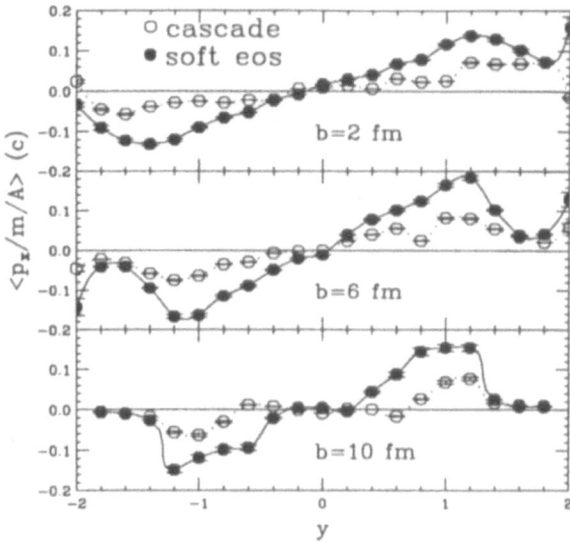

Figure 4. Baryon transverse momentum distribution in the reaction plane of Au + Au at 11.6 Gev/c and impact parameter of 2, 6 and 10 fm.

The validity of any hadronic model for heavy ion collisions at AGS energies is naturally limited by the possible phase transition to the quark–gluon plasma. With this precaution in mind, one can use predictions of hadronic models as a baseline in searching for new phenomena. The formation of the quark–gluon plasma or a mixed phase of hadrons, quarks and gluons is expected to reduce the pressure of the system and leads to a softened nuclear equation of state. It is therefore interesting to search for experimental observables that are sensitive to the equation of state. One such variable at AGS energies is the baryon transverse flow. The three windows in Fig. 4 shows the average transverse momentum of nucleons in the reaction plane as a function of rapidity for the reaction of Au + Au at $P_{beam}/A = 11.6$ GeV/c and impact parameters of 2, 6 and 10 fm respectively. Significant differences exist between calculations with and without the mean field for the reaction at all of the three impact parameters. In particular, the flow parameter defined as the slope of the transverse momentum distribution at midrapidity is about a factor of 2.5 larger in the case with the mean field. The strength of the so-called "bounce-off" effect at target or projectile rapidities is also much stronger in calculations with the mean field. It is also seen that the collective flow is the strongest in the midcentral collisions. More detailed analysis shows that the flow is mainly generated in the high density region and saturates in the expansion phase. The strong mean field effect on the collective flow is therefore very useful for disentangle theoretical models. Moreover, an experimentally measured, much weaker collective flow than the cascade model prediction may be a strong indication of the quark–gluon plasma formation at AGS energies.

4. SUMMARY

In summary, we have developed a new relativistic transport model for heavy-ion collisions at AGS energies. Within the framework of this model, we have found that the mean field can significantly affect the maximum energy and baryon densities reached in the reaction. The transverse collective flow analysis is suggested as a useful tool to disentangle theoretical models and a possible indicator of the formation of quark–gluon plasma at AGS energies.

The study was supported in part by the NSF Grant No. PHY-9212209 and the Welch Foundation Grant No. A-1110.

REFERENCES

1. Proceedings of Heavy Ion Physics at the AGS, HIPAGS'93, 13–15, Jan., 1993, Eds. G. S. F. Stephans, S. G. Steadman, and W. L. Kehoe.
2. H. Sorge, H. Stöcker, and W. Greiner, Ann. of Phys. (NY) **192**, 266 (1989); R. Mattiellò, H. Sorge, H. Stöcker, and W. Greiner, Phys. Rev. Lett. **63**, 1459 (1989); H. Sorge, A. V. Keitz, L. Winckelmann, A. Jahns, H. Sorge, H. Stöcker, and W. Greiner, Phys. Lett. **B263**, 353 (1991).
3. Y. Pang, T. J. Schlagel, and S. H. Kahana, Phys. Rev. Lett. **68**, 2743 (1992); T. J. Schlagel, Y. Pang and S. H. Kahana, Phys. Rev. Lett. **69**, 3290 (1992); S. H. Kahana, Y. Pang, T. J. Schlagel and C. Dover, Phys. Rev. C **47**, R1356 (1993).
4. L. Bravina, L. P. Csernai, P. Levai and D. Strottman, Phys. Rev. C **50**, 2161 (1994); and references therein.
5. C. Y. Wong, Introduction to High Energy Heavy-Ion Collisions, (World Scientific, Singapore), 1994.
6. G. F. Bertsch and S. Das Gupta, Phys. Rep., **160**, 189 (1988).
7. B. A. Li and W. Bauer, Phys. Rev. C **44**, 450 (1991); B. A. Li, W. Bauer and G. F. Bertsch, *ibid.*, C **44**, 2095 (1991).
8. B. A. Li and C. M. Ko, Phys. Rev. C **52**, 2037 (1995).
9. Total cross-sections for the reactions of high energy particles, A. Baldini, V. Flaminio, W. G. Moorhead, D. R. O. Morrison, (Springer-Verlag, Berlin), 1988.
10. Properties and production spectra of elementary particles, A. N. Diddens, H. Pilkuhn and K. Schlüpmann, (Springer-Verlag, Berlin), 1972.
11. C. M. Ko, Z. G. Wu, L. H. Xia and G. E. Brown, Phys. Rev. Lett. **66**, 2577 (1991); Phys. Rev. C **43**, 1881 (1991).

25

PRODUCTION OF HEAVY FRAGMENTS IN THE REACTION ^{40}Ar + ^{232}Th

E. Berthoumieux,[1] E. C. Pollacco,[1] C. Volant,[1] E. De Filippo,[1,2] R. Barth,[3]
B. Berthier,[1] Y. Cassagnou,[1] Sl. Cavallaro,[4] J. L. Charvet,[1] M. Colonna,[5]
A. Cunsolo,[2,4] R. Dayras,[1] D. Durand,[8] A. Foti,[2,4] S. Harar,[5] G. Lanzanó,[2]
R. Legrain,[1] V. Lips,[6] C. Mazur,[1] E. Norbeck,[7] H. Oeschler,[6] A. Pagano,[2]
S. Urso[2]

[1]CEA, DAPNIA/SPhN CEN Saclay, 91191 Gif-sur-Yvette Cedex, France
[2]Istituto Nazionale di Fisica Nucleare and Dipartimento di Fisica, Corso Italia 57, 95129 Catania, Italy
[3]GSI Darmstadt, D-6100 Darmstadt, Germany
[4]Dipartimento di Fisica and INFN-Laboratorio Nazionale del Sud, Viale Andrea Doria, Catania, Italy
[5]Ganil, BP 5027, 14021 Caen, France
[6]Institut für Kernphysik, Technische Hochschule D-64289 Darmstadt, Germany.
[7]Department of Physics, University of Iowa, Iowa City, Iowa 5242, USA
[8]LPC, ISMRA, 14050 Caen Cedex, France

ABSTRACT

Heavy fragments of mass approximately $150u$ are observed in the reaction ^{40}Ar(44 and 77 A.MeV) + ^{232}Th. At 27 A.MeV the yield of heavy fragments is small. The data indicate that the heavy products are formed at high excitation and are accompanied by two intermediate mass fragments, IMF. Boltzmann–Nordheim–Vlasov calculations give a good description and interpret the IMF emission as arising from the overlapping nuclear zone.

The experiment described herein was motivated by a series of measurements which show that in reactions with heavy projectiles, like ^{40}Ar 1, 2, ^{58}Ni3 and ^{84}Kr4 on fissile targets, the probability to transfer high linear momentum diminishes with increasing incident energy. The above measurements were performed using a fission correlation technique which requires that two fission fragments are detected. A possible explanation is that the incident channel leads to a nuclear complex that effectively decays via multi-fragment (see G. Westfall's contribution). Whether the process is simultaneous break-up or a series of binary decays, it will lead to privation of in-plane fission correlation. Coming to our motivation, it was

BaF$_2$ back

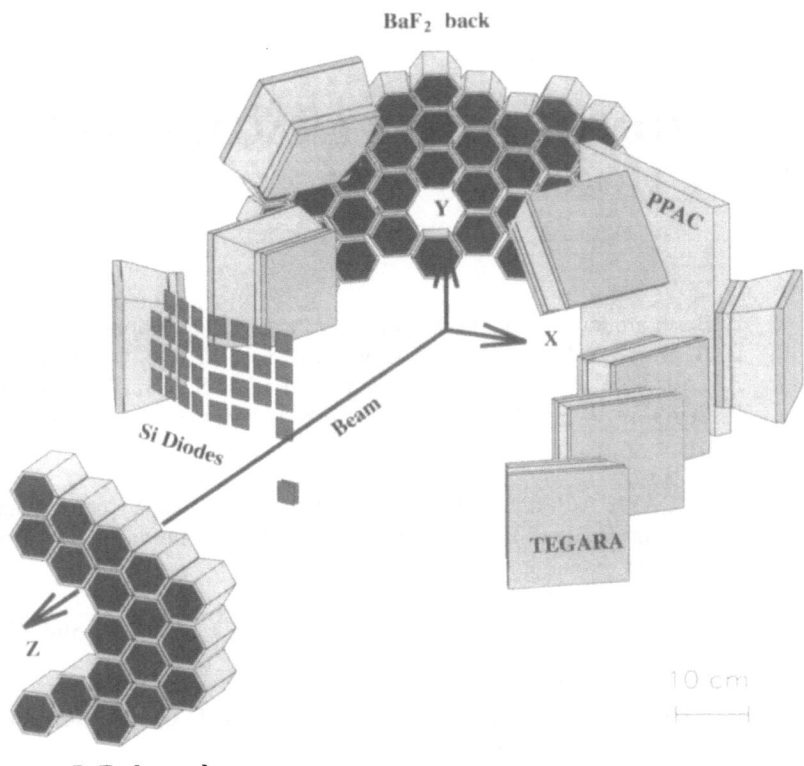

Figure 1. Experimental set-up.

considered that given appropriate entrance channel dynamics a large fraction of the incident kinetic energy is thermalized. This results in a copious loss of mass and leads to a composite whose angular momentum, mass and energy will constrain the fission channel, hence, a mean to scrutinize a heavy system at high excitation. Thus we can then refer to the study of the decay of hot nuclei at the limits of excitation, the delicate balance between evaporation residue and fission, limiting temperatures and so. Thus, our study of the reaction ^{40}Ar + ^{232}Th at 27, 44 and 77 A.MeV concentrated on isolating events leading to a very heavy fragment, HF. Selecting the HF allowed us to establish their most probable origin and to characterize some of their properties.

The experiment was performed at GANIL. The target was metallic ^{232}Th of thickness 0.7 mg/cm^2. An advantage of having such a highly fissile target is that it allows the target recoils to be distinguished from the evaporation residues through the measurement of mass spectra. The experimental set-up around the target is given in Fig. 1. Heavy fragments were detected in a Si array at forward angles. To filter events at high excitation energy, data were captured in coincidence with light charged particles, LCP at back angles (BaF$_2$) 5. Similar BaF$_2$ crystals where also placed at forward angles. Intermediate mass fragments, IMF($4 \leq Z \leq 16$), where detected in 32 ion chambers-Si telescopes, TEGARA, covering a large range of angles. Details of the geometrical coverage is given in 6.

In measuring a velocity versus mass spectra at forward angles we expect 1 to find an island of HF at a mass $\sim 190u$ with velocity typically 1 cm/ns. It is abundantly clear from

Table 1. Experimental results

Beam energy	44 A.MeV	77 A.MeV
σ_{HF} ($V \geq 0.5$ cm/ns)	290 mb	250 mb
$\langle M \rangle$	$150 \pm 10u$	$140 \pm 10u$
$\langle V \rangle$	0.9 ± 0.1 cm/ns	0.84 ± 0.1 cm/ns
M_{LCP}	> 6	> 7.5
Temp. (α-HF)	5.1 MeV	5.0 MeV

Fig. 2a that the cross section for HF is rather small, with the fission fragments (mass $\sim 90u$), FF, having a considerably higher yield. This is true for 44 and 77 A.MeV and even more so at 27 A.MeV. It is only by demanding the coincidence Si-LCP($\theta_{LCP} > 140°$) or Si-IMF that the HF become apparent. As seen in the figure, the HF have mass and velocity which are rather smaller than the one expected from full Linear Momentum Transfer, LMT, or from the measured systematics for central collisions (180 MeV/c/nucl. of projectile 1, 2, 3, 4). By and large, data for 44 and 77 A.MeV are rather similar, showing only minor differences in the mass of the HF (see Table 1). The yield at 27 A.MeV is very small. To establish the existence of fission contaminant, particularly from asymmetric quasi-elastic events, data were taken in coincidence with a parallel plate (Fig. 1.) A LCP($\theta_{LCP} > 140°$)-HF-PPAC analysis shows that the HF events do not arise from fission contaminant. HF mean velocity and mass values over the measured distributions were extracted from the Si-LCP correlations and are summarized in Table 1. At 44 A.MeV the values are consistent with those of Utley et al. 7. The smaller value of the mean mass at 77 A.MeV would suggest a longer evaporation chain. The average LMT absorbed by the HF is typically 4–5 GeV/c.

The low velocity and the strong presence of fission fragments, FF, makes the extraction of the HF differential cross-section somewhat delicate and reduced due to the velocity threshold. Nevertheless, using an unfolding procedure to separate the fission from the HF 6 and integrating over the measured velocity range give the angular distributions, Fig. 3. At 27 A.MeV the cross-section is too small to extract from the singles data. The angular distributions are forward peaked as would be expected from an evaporation residue. The solid line represents a simulation using the evaporation code 8, EVAP, with a nucleus of mass $200u$, an excitation energy of 650 MeV, an angular momentum of $80\hbar$ and a recoil velocity of 1.0 cm/ns. The simulation exhibits a fall in differential cross-section inconsistent with the data. The integrated differential cross-section are given in Table 1 and show that within the given velocity threshold they are constant with incident energy. Comparing these values with Schwinn et al. 2 gives a good agreement at 44 A.MeV but a large variance at 77 A.MeV.

The LCP at back angles apart from allowing us to mark the HF, yield energy spectra and LCP multiplicities. The energy spectra for alphas were transformed in the rest frame of the HF and fitted using a Maxwell–Boltzmann function. The fits give apparent temperatures of ~ 5 MeV at both 44 and 77 A.MeV. The proton spectra have a shape which cannot be fitted with a Maxwell–Boltzmann function over the full energy range. In other words, if we restrict the energy range up to 30 MeV this is a Maxwell–Boltzmann function consistent with 6 MeV. Beyond this range we find an enhancement, reminiscent of pre-equilibrium forward angle data 9. This enhancement is more significant at 77 A.MeV. To obtain the mean LCP multiplicity from the LCP-HF data, a simulation was undertaken assuming velocity spectra for the HF and LCP distributions which when passed through the experimental filter reproduced the data. LCP angular distributions are assumed to be isotropic in the emitter frame with ratios for $p/d/t/^3$He/α adopted from the measured values. The experimental

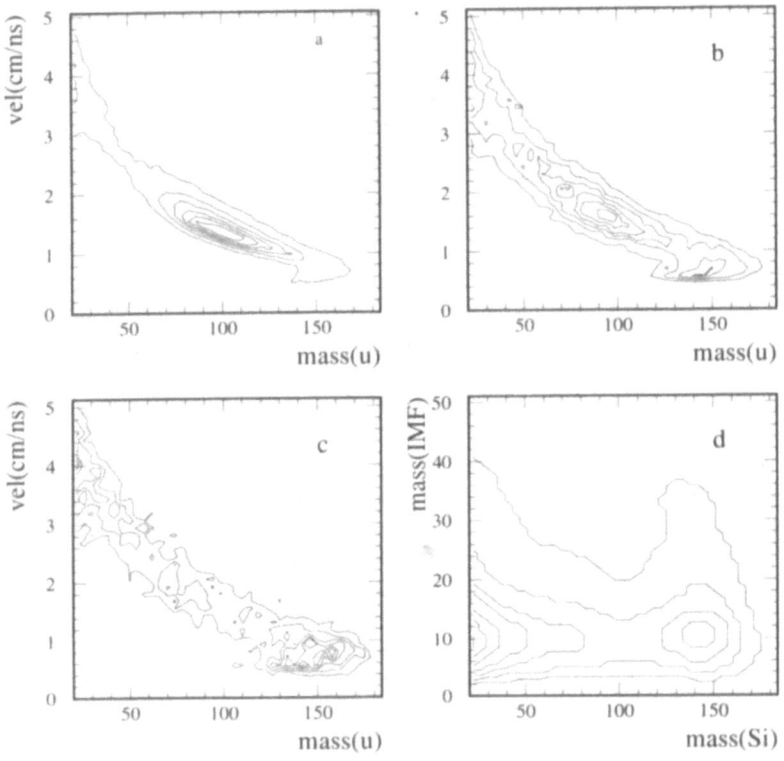

Figure 2. (a) Mass-Velocity in singles mode in Si. (b) as in a) in coincidence with LCP. (c) as a) for angles between 8° and 25° in coincidence with LCP and IMF. (d) Mass(Si)-Mass(TEGARA) as for c).

geometry was taken into account using the code GEANT 3.15. Extracted multiplicity values are given in the Table 1 and show a unit enhancement at 77 A.MeV. It is considered that the multiplicities are minimal values because no background substraction for the FF-LCP is included. Such consideration does not include variations in the angular distribution of the LCP. To give an order of magnitude, the measured multiplicities can be reproduced by assuming an evaporation residue of mass 230u recoiling at a velocity of 0.8 cm/ns, angular momentum 80$\hbar$ and excitation energies from 650 to 750 MeV. Similar values for the multiplicities and excitation energies are obtained for central collision data 2.

In Fig. 2c the mass-velocity plot for the Si-LCP-IMF coincidences is given along with the corresponding mass(Si) vs. mass(IMF), Fig. 2d. The striking feature is that the correlation is indeed quite strong with HF. The mass distribution of the IMFs has a somewhat extended range for HF that for fission. Kinematics reconstruction between IMFs and HF gives a strong LMT unbalance of $\sim$ 50%. For small θ_{IMF} the relative velocity between IMF and HF is higher than the Coulomb repulsion between two spheres, but beyond 48° the IMFs have a completely relaxed energy distribution. It is interesting that although our data do not extend to small enough positive angles, θ_{IMF}, the plot for both the LMT and relative velocity between IMF and HF superimpose those of the Ar + Ag system in a similar incident energy range 9. The angular correlation integrated over all Z_{IMF} is given in Fig. 4 and is forward peaked with an enhancement on the opposite side of the HF ($-8° \leq \theta_{HF} \leq -25°$). Performing

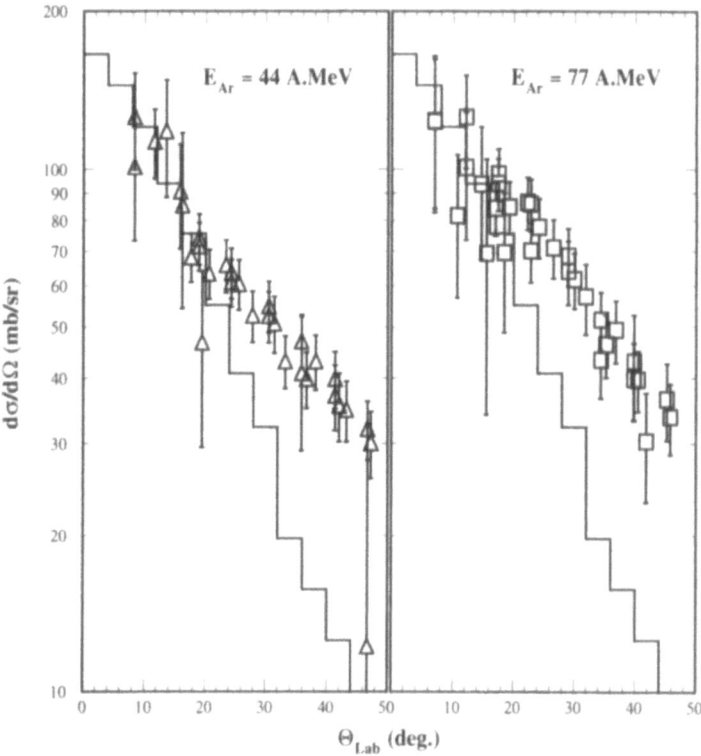

Figure 3. Angular distributions for heavy fragments. The solid line represents a simulation using the evaporation code EVAP.

HF-IMF correlations with EVAP and parameters noted above, does not reproduce the data (see hashed zone in Fig. 4). These results indicate that on average the HF are not evaporation residues but issue of a mechanism reminiscent of deep inelastic collisions, occurring over a large range of incident energy. This is in good agreement with the results of Lips et al. 10 for the ^{40}Ar(31 A.MeV) + ^{232}Th system and Rivet et al. 9.

Integrating the angular HF-IMF correlations using out of plane results of Lips et al. 10 give an IMF multiplicity of approximately two. To exploit the IMF multiplicity the correlation HF-IMF-IMF was analyzed where one of the IMFs is detected in the forward BaF$_2$(θ_{IMF} ≤ 10°). The resulting angular correlation is very similar to that for HF-IMF, however with a moderately stronger asymmetry, favoring the angles opposite to the HF angles. The 44 and 77 A.MeV data are again similar with the exception that the events at the higher energy have IMFs (θ_{IMF} ≤ 10°) which on average are close to beam velocity. Assuming that the forward IMF has mass equal to $10u$, the summed momenta of the three fragments yield a mean LMT of ~ 7.5 GeV/c at 44 and 77A.MeV.

To appreciate the dynamics that are involved in producing the heavy fragments we have considered the Boltzmann–Nordheim–Vlasov, BNV, equation. The calculation allows entrance channel effects ranging from complete to incomplete fusion or deep inelastic reactions to be described 11.

For the present system at 27 and 44 A.MeV, the most important properties (mass A, velocity V_{lab}, predicted excitation energy E^*, intrinsic angular momentum J) of the primary

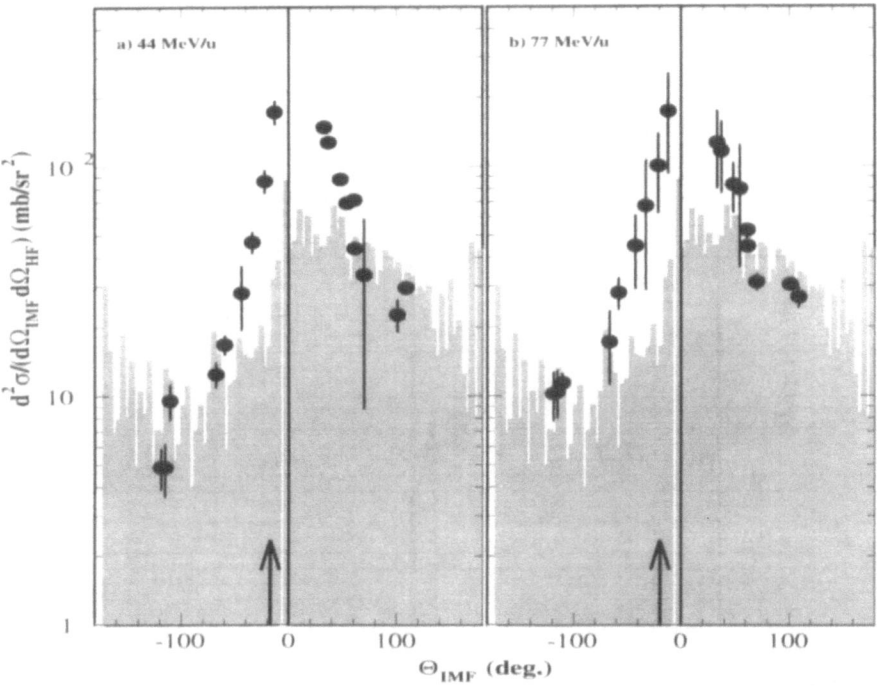

Figure 4. Angular correlations between HF and IMFs. The HF are detected in the angular range of −8° to −25° (arrow). Hashed zone represents a simulation using EVAP.

fragments formed through the dynamics are shown in Tables at 2 and 3 a time when pre-equilibrium effects have ceased, T_D, for different impact parameters, b. It is to be remarked that within this time about half of the available energy in the center of mass is removed through the emission of approximately 30 nucleons. This effect is uniform over b values up to ~ 7 fm 11. It should be noted here that in the range of $b = 6$–8 fm, the properties of the primary fragments given in Tables 2 and 3 are calculated at 180 fm/c for, at the equilibration time of 120 fm/c they are not yet formed. Therefore the given values include the evaporation between 120 and 180 fm/c.

From low to intermediate b values, for 44 and 77 A.MeV a transition from incomplete fusion to "deep inelastic" mechanism occurs. The "deep" is characterized by the formation of a HF and two IMFs which originate from the overlapping nuclear zone and the projectile remnant. Unlike at 44 A.MeV, at 77 A.MeV one of the IMFs has close to beam velocities ($b = 4$–8 fm). Further, the LMT of the three fragments add up to 7–8 GeV/c. These results

Table 2. BNV calculations for ^{40}Ar + ^{232}Th at 27 A.MeV

b (fm)	T_D (fm/c)	$A(u)$	E^* (MeV)	$J(\hbar)$	V_{lab} (cm/ns)
4	120	251	463	121	0.93
6	120	253	424	173	0.87
7	180	60	37	32	2.04
		181	0	54	0.57

Table 3. BNV calculations for ^{40}Ar + ^{232}Th at 44 A.MeV

b (fm)	T_D (fm/c)	$A(u)$	E^* (MeV)	$J(\hbar)$	V_{lab} (cm/ns)
2	120	241	705	72	1.18
4	120	245	684	127	1.14
6	180	5	12	1	3.60
		44	45	27	1.32
		177	178	107	0.90
8	180	5	28	1	7.20
		14	76	3	3.60
		216	26	65	0.48

are in good qualitative agreement with the data. In particular the recoil velocity and the mass after evaporation 8 are also consistent with the data.

The calculations yield HF with low mass and excitation energy thus the fission probability is attenuated. The opposite effect is calculated at 27 A.MeV where essentially two fragments are present in the exit channel producing a relatively heavy fragment with large J and thus overcome by fission (intermediate values of b). Again this offers a good description of what is observed experimentally

A difficulty arises in the predicted E^*. For intermediate b the calculated predicted E^* for the HF are relatively small to reproduce the measured LCP multiplicities. Part of the discrepancy is removed by considering evaporation starting at about 120 fm/c. We are studying the possibility that the LCP spectra contain a significant pre-equilibrium in the calculation component, as suggested by the proton data. Also, the nucleon-nucleon cross-section could be adjusted so as to limit the number of pre-equilibrium nucleons.

In conclusion, the data shows that in the ^{40}Ar + ^{232}Th system at 44 and 77 A.MeV, HF are observed and are the products from a nuclear complex at high excitation energy. The neutron 2, 7 and LCP multiplicities support this view. As for the apparent temperatures, they are average values in a long chain of emission, nonetheless they are consistent with high temperatures being reached. Further, the HF, HF-IMF and HF-IMF-IMF data at forward angles are not consistent with an evaporation residue being formed and interpreted as arising from a highly dissipative mechanism at intermediate impact parameters. The cross-sections for HF with the given experimental thresholds are constant with incident energy and even if they were to arise from central collision they are relatively small to explain the missing values 1. At 27 A.MeV the cross-section for HF is small. BNV calculations are presented and show an IMF emission from the overlap between the projectile and target. The calculations give a good overall qualitative description of the data.

REFERENCES

1. M. Conjeaud et al., Phys. Lett., **159B**, 244 (1985).
2. E. Schwinn et al., Nucl. Phys., **A568**, 169 (1994).
   D. X. Jiang et al., Nucl. Phys., **A503**, 560 (1989).
3. C. Volant et al., Phys. Lett.,**195B**, 72 (1987).
4. E. C. Pollacco et al., Z. Phys., **A346**, 63 (1993).
5. G. Lanzanó et al., Nucl. Inst. & Meth., **312**, 515(1992).
6. E. C Pollacco et al., Nucl. Phys., **A583** (1995)441.
7. D. Utley et al., Phys. Rev., **C49**, 1737 (1994).
8. D. Durand, private communication.

9. M. F. Rivet et al., Bormio 93.
10. V. Lips et al., Phys. Rev., **C 49**, 1214 (1994).
11. M. Colonna et al., Nucl. Phys., **A541**, 295(1992).

26

THERMAL e^+e^- PRODUCTION IN HIGH-ENERGY NUCLEAR COLLISIONS

John J. Neumann,[1*] David Seibert,[2†] and George Fai[1‡]

[1]Center for Nuclear Research
Department of Physics
Kent State University, Kent, OH
[2]Department of Physics
McGill University, Montréal, QC, H3A 2T8, Canada

ABSTRACT

We use a boost-invariant one-dimensional (cylindrically symmetric) fluid dynamics code to calculate e^+e^- production from the $\rho-\omega$ peak in the central rapidity region of S + Au and Pb + Pb collisions at SPS energy ($\sqrt{s}$ = 20 GeV/nucleon). We assume that the hot matter is in thermal equilibrium throughout the expansion, but consider deviations from chemical equilibrium in the high temperature (deconfined) phase. We use equations of state with a first-order phase transition between a massless pion gas and quark gluon plasma, with transition temperatures in the range $150 \leq T_c \leq 200$ MeV.

1. INTRODUCTION

Earlier predictions of e^+e^- transverse mass spectra from several resonances[1,2,3] show a useful sensitivity to the QCD transition temperature when the equation of state incorporates a strong first-order phase transition and transverse expansion is neglected. The correlation between the apparent temperatures of these spectra and the transition temperature was suggested as a thermometer for the transition temperature.[1] In an earlier work[4] we found that when transverse expansion was taken into account, the apparent temperature of photon spectra displayed only a very weak correlation with the transition temperature. We wish to study here whether a useful correlation can be found for e^+e^- pairs if we include transverse expansion.

In our calculation we assume a boost-invariant longitudinal expansion as discussed by Bjorken,[5] coupled to a cylindrically symmetric transverse expansion. We assume thermal

*Electronic mail (internet): neumann@scorpio.kent.edu.
†Electronic mail (internet): seibert@hep.physics.mcgill.ca.
‡Electronic mail (internet): fai@ksuvxd.kent.edu.

equilibrium throughout the evolution of the system, but consider deviations from chemical equilibrium in the high-temperature phase by allowing the quark and antiquark densities to be a (fixed) fraction of the equilibrium value. In the e^+e^- production rate we include the bulk and freeze-out contributions from ρ and ω decays. We calculate the transverse mass distribution for e^+e^- pairs in the range $0.8 \leq m_T \leq 1.8$ GeV. We investigate the sensitivity to different assumptions about the initial temperature, freeze-out temperature, and quark fraction, and compare production rates to the preliminary NA45 data.[6] We use standard high-energy conventions, $c = \hbar = k_B = 1$.

2. FLUID-DYNAMICAL EVOLUTION

Here we describe the initial conditions and assumptions about the evolution; the details of the fluid-dynamical calculation are described elsewhere.[4] We assume a central collision of two large nuclei at SPS energy. For such high collision energies we expect approximate longitudinal boost invariance,[5] so the behavior of the produced matter at different rapidities is the same in the longitudinally comoving frame for fixed proper time $\tau = \sqrt{t^2 - z^2}$, where z is the distance along the beam axis. At $\tau = 0$ the colliding nuclei reach the point of maximum overlap and are assumed to form a longitudinally expanding pancake. The hot matter has thermalized at $\tau = \tau_0$ (≈ 0.2 fm/c),[7,8] when a cylindrically-symmetrical transverse expansion begins, coupled to the longitudinal expansion.

From $\tau = 0$ until the transverse expansion starts at $\tau = \tau_0$, we assume a boost-invariant cylinder of radius $R_<$ (the radius of the smaller nucleus), filled uniformly with QGP at temperature $T = T_0$. This is approximately compatible with the initial entropy density for short times.[9] We determine T_0 by assuming entropy conservation for $\tau > \tau_0$, hence

$$s(T_0) = \frac{3.6 dN_\pi/dy}{\pi R_<^2 \tau_0}, \tag{2.1}$$

where s is the entropy density, with total (charged plus neutral) multiplicity density $dN_\pi/dy = 240$ and 1600 for central S + Au and Pb + Pb collisions, respectively. The NA45 experimenters estimate $dN_{ch}/d\eta = 125$ for their S + Au collision sample, so for comparison of our model to the NA45 invariant mass spectrum we use $dN_\pi/dy = (3/2)(125) = 188$ in our calculations.

The equations of state (EOS's) that we use here are of the form

$$T < T_c: \quad \begin{cases} e & = \frac{\pi^2}{10}T^4, \\ P & = \frac{e}{3}, \end{cases} \tag{2.2}$$

$$T = T_c: \quad \begin{cases} \frac{\pi^2}{10}T_c^4 \leq e \leq \frac{\pi^2}{30}g_q T_c^4 + B, \\ \frac{\pi^2}{30}T_c^4 = P = \frac{\pi^2}{90}g_q T_c^4 - B, \end{cases}$$

$$T > T_c: \quad \begin{cases} e & = \frac{\pi^2}{30}g_q T^4 + B, \\ P & = \frac{\pi^2}{90}g_q T^4 - B, \end{cases}$$

where g_q is the number of massless degrees of freedom in the deconfined phase. We treat only the case of zero baryon density, so the entropy density is

$$s = \frac{e + P}{T}, \tag{2.3}$$

independent of the phase of the matter. Below T_c, the EOS is that of a massless pion gas. Because recent calculations have predicted that the quarks may reach only a fraction of their equilibrium number by the beginning of transverse expansion,[7,8] we take $g_q = 16 + 21x$, where x is a parameter that we vary between 0 and 1 to simulate the effect of reducing the quark density in the QGP below the equilibrium value ($x = 1$ is equilibrium for two flavors of massless quarks). The vacuum energy density in the deconfined phase, B, is related to the transition temperature, T_c, by requiring equal pressures in the deconfined and hadronic phases at $T = T_c$:

$$B = \frac{\pi^2(g_q - 3)}{90} T_c^4. \tag{2.4}$$

The value of B we use in the non-equilibrium case is calculated by assuming $x = 1$ and supplying an equilibrium transition temperature T_c'. It is this T_c' that is actually used on the graphs and is useful for comparison purposes.

3. ELECTRON–POSITRON PAIRS FROM BULK MATTER

We calculate the central rapidity region m_T distribution from both bulk and freeze-out. The contribution from the bulk of the boost-invariant hot matter is

$$\left. \frac{d^2 N_{e^+e^-}^{(\rho+\omega)}}{m_T dm_T dy} \right|_{y=0} = \int d\eta \int d\tau\,\tau \int dr\, 2\pi r \int_0^\pi d\theta \, \sin\theta \int_0^{2\pi} d\phi \int_0^\infty dp\, p^2 \frac{dR}{d^3 p} \tag{3.1}$$

$$\times \delta\left(\frac{1}{2}m_T^2 - \frac{1}{2}m_T'^2\right) \delta\left(\eta + \tanh^{-1}\left[\frac{p\cos\theta}{\gamma\left(\sqrt{p^2 + m^2} + pv\sin\theta\cos\phi\right)}\right]\right).$$

Here $m_T = \sqrt{m^2 + p_T^2}$, R is the e^+e^- production rate per unit four-volume in the fluid frame, v is the transverse velocity of the fluid in the cell characterized by proper time τ, space-time rapidity η and radial position r (measured in the frame moving with transverse velocity zero and longitudinal velocity $\tanh\eta$ in the lab), and $\gamma = (1 - v^2)^{-1/2}$. Evaluating the integrals over η and p with the δ-functions, we obtain

$$\left. \frac{d^2 N_{e^+e^-}^{(\rho+\omega)}}{m_T dm_T dy} \right|_{y=0} = \int d\tau\,\tau \int dr\, 2\pi r \int_0^\pi d\theta\, \sin\theta \int_0^{2\pi} d\phi\, \frac{p_*^2}{J(p_*)} \left(\frac{dR}{d^3 p}\right)_{p=p_*}, \tag{3.2}$$

$$p_T^2 = \gamma^2\left[v^2(p_*^2 + m^2) + p_*^2\sin^2\theta\cos^2\phi + 2v\sin\theta\cos\phi\, p_*\sqrt{p_*^2 + m^2}\right] \tag{3.3}$$
$$+ p_*^2\sin^2\theta\sin^2\phi,$$

where p_* is the momentum, measured in the frame moving with the fluid, that corresponds to the given p_T of the lab frame, and the Jacobian is

$$J(p) = \frac{1}{2}\left|\frac{\partial}{\partial p}p_T^2\right|. \tag{3.4}$$

The e^+e^- production rate from ρ and ω mesons in thermally and chemically equilibrated hadron gas is

$$\frac{dR}{d^3 p} = \frac{m(g_\rho\Gamma_{\rho\to e^+e^-} + g_\omega\Gamma_{\omega\to e^+e^-})}{E(2\pi)^3}\left(e^{E/T} - 1\right)^{-1}, \tag{3.5}$$

where $g_\rho = 9$ and $g_\omega = 3$ are the degeneracies of the ρ and ω mesons, and the partial widths are $\Gamma_{\rho\to e^+e^-} = 6.77$ keV and $\Gamma_{\omega\to e^+e^-} = 0.60$ keV. E is the energy measured in the fluid frame, and we use the average mass of the ρ and ω, $m = 0.775$ GeV. No mesons live in the QGP, and hence the e^+e^- production rate for resonances is zero for that part of the fluid.

The integral of Eq. (3.2) is performed over the space-time evolution of the collision and results in nearly-exponential spectra for the bulk matter contribution. We then mimic the experimental procedure and extract an apparent temperature (the fit temperature, T_{fit}) from the obtained bulk spectra. For fitting purposes, we take (3.2) with $v = 0$ (i.e. ignoring transverse expansion), resulting in the formula[1]

$$\frac{d^2 N_{e^+e^-}^{(\rho+\omega)}}{m_T dm_T dy} \sim \left(\frac{m_T}{T}\right)^{1/2} \exp\left(-\frac{m_T}{T} + 0.4\frac{T}{m_T}\right). \tag{3.6}$$

4. ELECTRON–POSITRON PAIRS FROM FREEZE-OUT

Mesons which pass through the freeze-out surface $T = T_{fo}$ into cooler matter are no longer subject to thermal equilibrium. As these frozen-out mesons free-stream toward the detectors a fraction equal to the branching ratio will decay into e^+e^- pairs. The contribution from the decay of frozen-out ρ and ω mesons is

$$\frac{d^2 N_{e^+e^-}^{(\rho+\omega)}}{m_T dm_T dy}\bigg|_{y=0} = \left(\frac{g_\rho \Gamma_{\rho \to e^+e^-}}{\Gamma_\rho} + \frac{g_\omega \Gamma_{\omega \to e^+e^-}}{\Gamma_\omega}\right) \int d\tau\, \tau 2\pi r(\tau) \int_{-1}^{1} d(\cos\theta) \int_0^{2\pi} d\phi$$
$$\times \int_0^\infty dp\, p^2 \gamma (2\pi)^{-3} \left(e^{E/T} - 1\right)^{-1} \left(v_x - \frac{dr}{d\tau}\right) \Theta\left(v_x - \frac{dr}{d\tau}\right) \delta(\frac{1}{2}m_T^2 - \frac{1}{2}m_T'^2) \tag{4.1}$$

where

$$v_x = \frac{p\sin\theta\cos\phi + v\sqrt{p^2 + m^2}}{\sqrt{p^2 + m^2} + pv\sin\theta\cos\phi}, \tag{4.2}$$

and $\Gamma_\rho = 151$ MeV and $\Gamma_\omega = 8.43$ MeV are the total widths for the ρ and ω. $dr/d\tau$ is the radial velocity of the freeze-out surface and $r(\tau)$ is its location. The Θ function insures that only mesons going through the freeze-out surface in the outward direction are counted.

5. RESULTS

Our "standard" calculation is a S + Au collision with $\tau_0 = 0.2$ fm/c, $x = 1$, freeze-out temperature $T_{fo} = 100$ MeV, and any given transition temperature T_c; if a different value of one of the parameters is given, it should be assumed that the others are held at the standard values.

Fig. 1 shows three typical spectra: the bulk contribution from the standard calculation with $T_c = 170$ MeV, the same calculation including freeze-out, and the bulk contribution from the standard calculation without transverse expansion. The spectra from the bulk contributions are nearly exponential, and can be fitted with an apparent temperature as described above.

In Fig. 2, we vary T_c from 150 to 200 MeV and calculate the resulting values of the temperature, T_{fit}, needed in Eq. (3.6) to fit the resulting bulk matter spectrum for a central S + Au collision at SPS energy. Our standard value for the equilibrium time, $\tau_0 = 0.2$ fm/c, implies an initial temperature of $T_0 = 355$ MeV for the S + Au system.

In the absence of transverse expansion, T_{fit} is a monotonically increasing function of T_c, so one can infer T_c given T_{fit} from the measured e^+e^- spectrum,[1] though the small slope

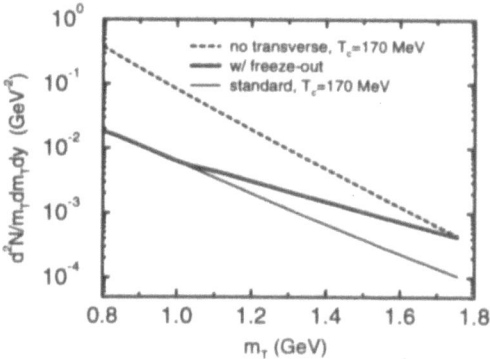

Figure 1. Spectra from S + Au with no transverse expansion, standard calculation (bulk only), and standard calculation (bulk and freeze-out).

means the uncertainty in T_c is about twice that of T_{fit}. This can be seen in the lowest curve, where we have run our simulation without transverse expansion.

Including transverse expansion, we find that T_{fit} is higher than in the case of longitudinal-only expansion. T_{fit} vs. T_c is still monotonic, but as in the case of fitting photon spectra[4] there is some positive curvature at small T_c. This curvature is a result of the relationship between T_0, T_c, and the amount of radial fluid motion: The further T_c is below T_0, the more time the system has to expand before entering the mixed phase, where a vanishing pressure gradient causes a lack of acceleration outward. Thus lowering T_c increases the amount of transverse expansion. The increase in m_T caused by the fluid motion compensates somewhat for the decrease of T_{fit} with decreasing T_c found without transverse expansion. The effect is least noticeable for $\tau_0 = 1$ fm ($T_0 = 208$ MeV), which we attribute to the fact that T_0 is close to T_c.

Contrary to experience without transverse expansion,[1] increasing T_{fo} to 150 MeV decreases T_{fit}, another indication that the transverse motion has a large effect on the spectrum. Apparently the cool hadronic matter near the outside of the cylinder expands outward at a velocity high enough that the contribution to m_T from the fluid motion more than makes up for the low temperature. Thus, removing the contribution from the late stages (by raising T_{fo}) decreases T_{fit}.

The curve for $x = 0.9$ shows that T_{fit} is about 3 MeV higher than for $x = 1$ (the standard calculation). The increase occurs because the temperature of the mixed phase, which makes a large contribution to the e^+e^- spectrum, is raised when x is lowered and B is held constant.

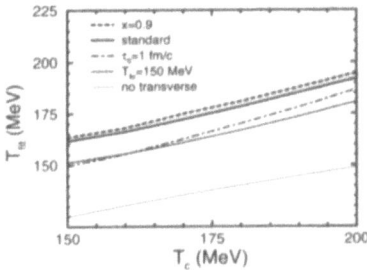

Figure 2. T_{fit} vs. T_c for the bulk contribution, S + Au.

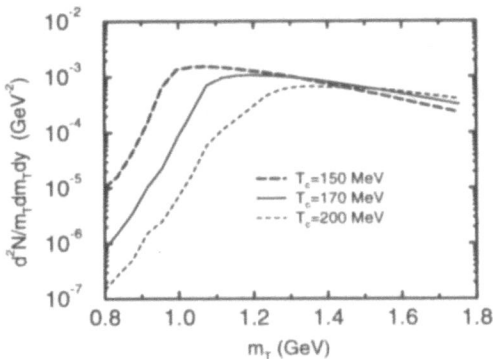

Figure 3. Freeze-out contributions with T_{fo} = 100 MeV, for different T_c (standard calculation).

Fig. 3 shows the freeze-out contribution for a standard calculation with various T_c.

The freeze-out spectra go to zero at $m_T \approx m$ ($p_T \approx 0$), and approach an exponential shape at higher m_T. At $m_T \approx m$, the freeze-out contribution is much smaller than the bulk contribution, and the opposite is true at larger m_T, where the spectra take on a more exponential shape. As Fig. 4 shows, the total (bulk + freeze-out) spectra can be characterized as having two regions in m_T, a bulk-dominated (0.8 < m_T < 1.0 GeV) and a freeze-out-dominated region (1.4 < m_T < 1.8 GeV). We have found that it is possible to get reasonable estimates of the bulk and freeze-out contributions over the entire range of m_T by fitting the bulk-dominated region of the total with a Boltzmann-like function. It is unclear whether such a separation of the contributions will be possible in actual experimental data.

We compared our overall production rates with the invariant mass spectrum from the NA45 experiment. By integrating over the ρ/ω peak in the data using a Breit–Wigner distribution, we obtained a value of $dN_{e^+e^-}^{(\rho+\omega)}/d\eta = 3.77 \times 10^{-4}$. We calculated the same quantity for our simulated data from the standard calculation with T_c = 160 MeV and dN_π/dy = 188, by integrating over the range of m_T shown on the graphs. We also took into account the experimental cuts in the rapidity, where only e^+e^- pairs whose members are both in the range 2.1 < η < 2.65 are counted. This gave us $dN_{e^+e^-}^{(\rho+\omega)}/d\eta = 3.70 \times 10^{-4}$. The same calculation for the case of no transverse expansion gives a value that is at least a factor of 5 to 10 too high,

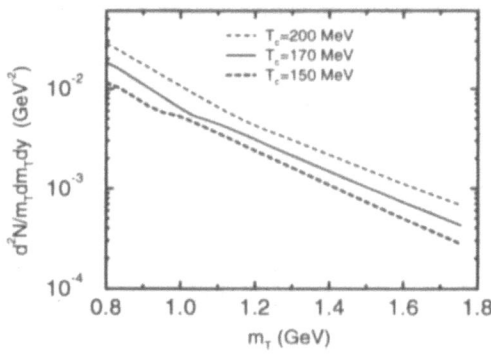

Figure 4. Total production for the standard calculation, for several T_c.

and increasing τ_0 or decreasing x below 1 also gave a value of $dN_{e^+e^-}^{(\rho+\omega)}/d\eta$ that was slightly large. However, the estimated error in the NA45 data is of order unity, and so would not rule out any of the calculations except that with no transverse expansion. Though not shown here, the e^+e^- yield is much less sensitive to τ_0 than the photon yield,[4] so using e^+e^- spectra may be appropriate when uncertainties in estimating τ_0 are a problem.

Our conclusion is that transverse expansion does not destroy the correlation suggested in Ref. 1, so that the shape of the $e^+e^-m_T$ spectrum, as parameterized by T_{fit}, should make a good thermometer to measure T_c. The preliminary NA45 data appear to rule out models in which transverse expansion plays no role, but otherwise place no significant constraints on the collision dynamics.

ACKNOWLEDGMENTS

We thank T. Ullrich for providing the preliminary NA45 data, and C. Gale for useful comments. We also thank the Institute for Nuclear Theory at the University of Washington for its hospitality and partial support. This work was supported in part by the U. S. Department of Energy under Grant No. DOE/DE-FG02-86ER-40251, in part by the Natural Sciences and Engineering Research Council of Canada, and in part by the FCAR fund of the Québec government.

REFERENCES

1. D. Seibert, Phys. Rev. Lett. **68**, 1476 (1992); D. Seibert, V. K. Mishra, and G. Fai, Phys. Rev. C **46**, 330 (1992).
2. M. Asakawa and C. M. Ko, Phys. Lett. B **322**, 33 (1994); C. M. Ko and M. Asakawa, Nucl. Phys. **A566**, 447c (1994).
3. C. M. Ko and D. Seibert, Phys. Rev. C **49**, 2198 (1994).
4. J. J. Neumann, D. Seibert, and G. Fai, Phys. Rev. C **51**, 1460 (1995).
5. J. D. Bjorken, Phys. Rev. D **27**, 140 (1983).
6. T. Ullrich, private communication.
7. E. Shuryak, Phys. Rev. Lett. **68**, 3270 (1992).
8. K. Geiger and J. I. Kapusta, Phys. Rev. D **47**, 4905 (1993).
9. D. Seibert, Phys. Rev. Lett. **67**, 12 (1991).
10. D. Seibert, Z. Phys. C **58**, 307 (1993).

27

FIRST RESULTS FROM EXPERIMENT NA49 AT THE CERN SPS WITH 158 eV/NUCLEON Pb ON Pb COLLISIONS

H. Rudolph and the NA49 collaboration

Lawrence Berkeley Laboratory
University of California
Berkeley, CA

ABSTRACT

CERN experiment NA49 had its first beam time in November/December 1994 with a ^{208}Pb beam of 158 GeV/nucleon. The experimental setup to study Pb + Pb collisions is described and first results on two particle correlations and transverse energy production are discussed.

1. INTRODUCTION

Experiments using heavy ion collisions to study the properties of highly compressed matter have gone to higher energies and heavier projectiles in recent years. Energies of 200 GeV per nucleon for oxygen and sulfur projectiles could be reached when the CERN SPS accelerator was modified to allow for heavy ion acceleration beginning in 1986.

Several experiments investigated the collisions of those projectiles in fixed target reactions. Tracking of the charged particles emerging from the collision in a magnetic field allows for example to reconstruct pseudorapidity and transverse momentum distributions of those particles and thus learn about the dynamics of the reaction. Quantities such as stopping and the energy density in the collision can then be deduced. At the same time reconstruction of longer-lived strange particles such as Kaons and Lambdas is possible from the tracks, giving important information about possible strangeness enhancement in heavy ion collisions which could be a signature for new phenomena in these reactions.

After a series of experiments at the SPS it became clear that the phase transition of nuclear matter into a so-called quark gluon plasma (QGP) could probably not be produced or at least observed with sulfur and oxygen projectiles at SPS energies. Still heavier systems or higher energies are needed to produce the necessary energy densities or system size to generate and observe the phase transition. This has led to new projects for heavy ion

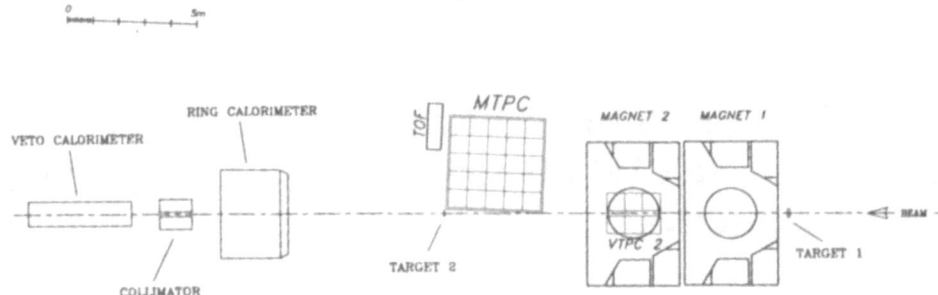

Figure 1. Setup of experiment NA49 in 1994. The beam enters from the right. The detectors from right to left are: the vertex tpc, the main tpc and the hadronic calorimeters. Target 2 was used for the calorimeter configuration and target 1 for the normal configuration. The 1995 setup will have an additional TPC in magnet 1 and a second MTPC.

collisions: acceleration of Au beams at the AGS (BNL), Pb beams at the CERN-SPS, the RHIC accelerator at the Brookhaven National Laboratory and the heavy ion program at the LHC. NA49 is one of the experiments studying collisions of Pb beam particles with a fixed target at the SPS.

2. THE EXPERIMENTAL SETUP

The 1994 configuration of experiment NA49 can be seen in Fig. 1. The lead beam enters the experimental area from the right hand side coming from the SPS accelerator. It is defined by a 0.2 mm quartz Cherenkov counter in combination with a veto scintillation counter. Its position is recorded by a silicon strip detector with 200 μm resolution. The magnets 1 and 2 have a field strength of 1.5 T and 1.1 T adding up to a total bending power of 7.8 Tm.

The main detectors are two large time protection chambers, one placed in the second vertex magnet (VTPC 2) the other downstream of the magnet on the right hand side of the beam (MTPC). These TPCs enable tracking of charged particles in three dimensions over the entire detector volume which is $200 \cdot 72 \cdot 260$ cm^3 for VTPC2 and $384 \cdot 129 \cdot 384$ cm^3 for the MTPC.

TPCs operate in a similar way to drift detectors. The charged particle traverses a large drift volume depositing charge in the chamber gas, which is then drifted under the influence of a uniform electric field towards the readout plane. There the charge is amplified in the vicinity of very thin field wires and a signal is capacitively induced on pads which are positioned close to the wire plane. The coordinates of the initial charge in three dimensions can then be deduced from the charge distribution over several pads and the measurement of the drift time. The signal amplitude on the pads is a measure of the amount of charge deposited by a charged particle. In the MTPC particle identification can be performed using the particle's momentum, mean energy deposition and the relativistic rise of the energy deposition in a gas with increasing momentum.

The pad size for the Vertex TPC is $3.13 \cdot 39$ mm^2. For the Main TPC the pad size is $3.13 \cdot 39$ mm^2 for the pads close to the beam and $4.95 \cdot 39$ mm^2 for those farther away from the beam. The total number of pads read in the 1994 run is 91008.

The "Ring" Calorimeter[1] downstream of the MTPC consists of an electromagnetic

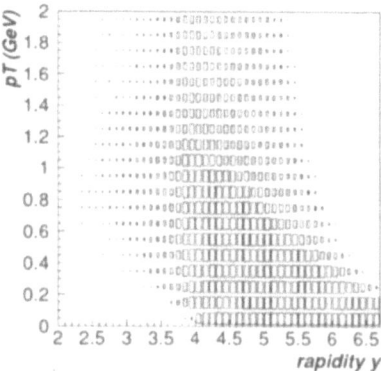

Figure 2. Transverse momentum p_T and rapidity y distribution in the MTPC for reconstructed pions of a simulated flat phase space pion distribution.

Lead/Scintillator calorimeter 16 radiation lengths (X_0) or 1 interaction length (λ_{int}) thick, followed by a hadronic Iron/Scintillator calorimeter of 6 λ_{int}. It is tube-shaped with an inner/outer radius of 0.28/1.50 meters, and it is divided into 240 cells, 24 in azimuth and 10 radially, with the radial size chosen to cover equal units in pseudorapidity.

Downstream of the Ring Calorimeter, an iron collimator defines the acceptance of the forward ("Veto") calorimeter.[1] It has a hole of $10 \cdot 10$ cm^2 at 11 meters from target 2 that allowed only particles with an emission angle of less than about 0.3^0 ($\approx 5^0$ in the c.m. frame) to reach the Veto calorimeter. This small solid angle covers the projectile spectator region.

A time of flight (TOF) detector positioned at the far right hand side behind the MTPC is used for particle identification in the momentum range where this cannot be achieved by ionization measurement in the MTPC.

3. NA49 RUN IN 1994

Experiment NA49 took its first data in November/December 1994. The run was subdivided into several periods in order to take data under different apparatus conditions. For the "calorimeter runs" the target was positioned behind the Main TPC (Target 2 in Fig. 1) to have a hadron acceptance of the calorimeter of $2.1 < \eta < 3.4$. For the TPC runs the target was positioned in front of the first vertex magnet (Target 1). Several magnet settings were used in this configuration. In the normal mode the magnets were set to B(VT1) = 1.5 T and B(VT2) = 1.1 T to accept positive hadrons into the MTPC and with opposite polarity for measuring negative hadrons. In the "HBT" mode the fields were set to the lower values of B(VT1) = 0.3 T and B(VT2) = 1 T in order to obtain better acceptance for low momentum pions in the TPCs.

4. TPC ANALYSIS

The coordinates as obtained from the measured charge signals on the pad plane and the drift time can be used to reconstruct the tracks of charged particles traversing the TPCs. Adjacent charge signals first are combined in a plane perpendicular to the beam to form clusters and then tracks are built from these clusters.

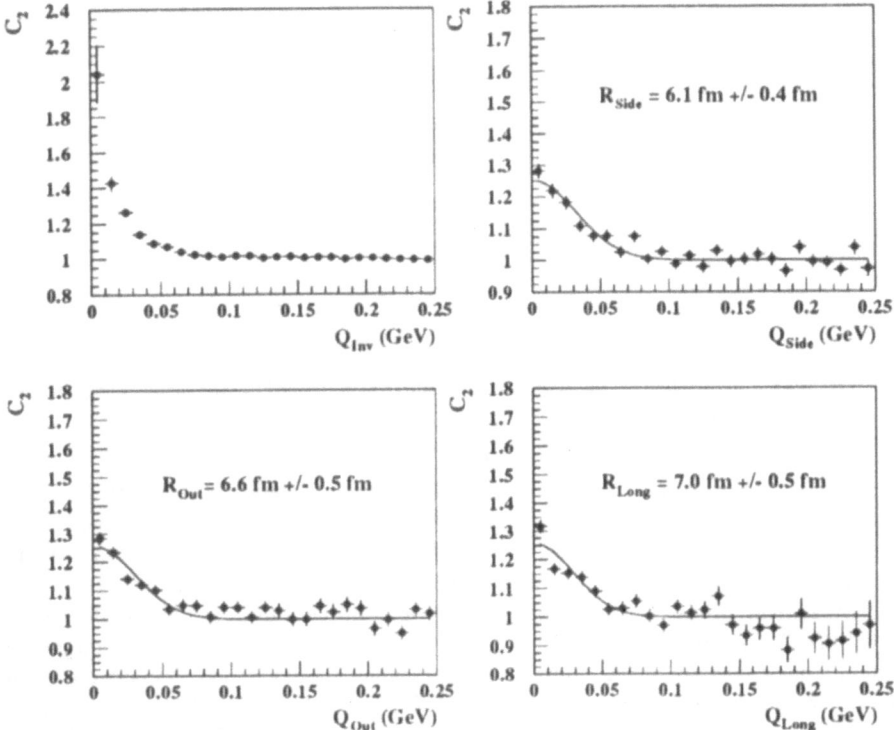

Figure 3. The correlation function in momentum difference Q_{Inv} and its projections on the components Q_{Long}, Q_{Out} and Q_{Side} with the correlation length values obtained from a fit of function (1) to the data. (Preliminary).

The measurement of the bending of a track in the field of the two vertex magnets can be used to calculate the particle momentum. Fig. 2 shows a p_T/y distribution obtained from simulated pions traversing the Main TPC. It shows the very wide p_T and y acceptance of this detector.

5. TWO PARTICLE CORRELATIONS

Using tracks reconstructed in the TPC and their measured momenta, a two particle Bose–Einstein correlation analysis can be performed on the data taken with the HBT configuration of the magnets. This type of analysis can provide information about the space–time geometry of the particle emitting source. In the case of an expanding source, as expected in heavy ion reactions, it can be used to study the source dynamics, based upon some model dependent assumptions.

In this analysis[2] a parametrization of the correlation length is used which was proposed by S. Pratt.[7] The correlation function is defined as:

$$C_2 = 1 + \lambda e^{-(Q_{Side}^2 R_{Side}^2 + Q_{Out}^2 R_{Out}^2 + Q_{Long}^2 R_{Long}^2)/2} \tag{5.1}$$

where Q_{Long} is the 4-momentum difference of the two particles in the beam direction, Q_{Out} in particle direction and Q_{Side} perpendicular to the other two. The measured correlation function

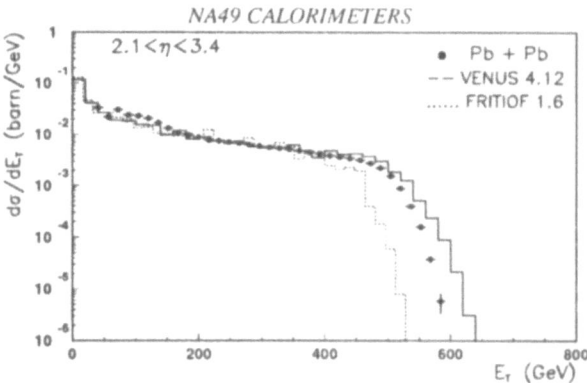

Figure 4. Transverse energy distribution produced in Pb + Pb collisions as measured by the Ring calorimeter.[3] The experimental results are here compared with two nuclear collision models. (Preliminary).

is corrected for losses due to limited two track resolution of the detector, the missing particle identification, the Coulomb repulsion (Gamov factor) and limited momentum resolution. The correlation functions[2] for 550 central NA49 events are shown in Fig. 3 with the resulting values for the correlation lengths.

The values for R_{Long}, R_{Side} and R_{Out} are about 40% higher than those obtained from an analysis on S + Au data, in agreement with the expectation of a bigger particle emitting source. The difference $R_{Out} - R_{Side}$ is consistent with zero, hinting at a very short lifetime of the emitting source. Further analysis of this data, in particular the rapidity and transverse momentum dependance of the correlation length, is in progress. This will give more insight into the reaction dynamics, as explained for example within a model by Sinyukov.[8]

6. CALORIMETER ANALYSIS

The calorimeters (Veto and Ring) used in NA49 had previously been used in other experiments (NA5, NA35). Details about the hardware and their calibration for the 1994 run can be found elsewhere.[4,5] The veto calorimeter is used in all NA49 runs to define the centrality of the reaction by measuring the deposited energy of projectile-like particles. The ring calorimeter is used to measure the transverse energy produced in a reaction.

The transverse energy differential cross section measured[3] in NA49 is given in Fig. 4. In the same figure the predictions of FRITIOF and VENUS are shown in the same acceptance. The data favors the VENUS model predictions. The mean E_T for near head–on collisions can be calculated according to Ref. 5 and is found to be $E_T^{b \approx 0} = 520$ GeV. The number of participants in central Pb + Pb collisions is 390±5 and therefore the mean E_T per participant is 1.33 GeV which is very similar to 1.31 GeV, the corresponding number for central S + Au collisions at the slightly higher projectile energy of 200 GeV/nucleon. We observe that the average produced E_T per participant is the same in all systems. Since the mean number of collisions each participant undergoes is higher by 45(85)% in Pb + Pb as compared to S + Au (S+S),[3] it appears that the production of E_T is roughly independent of the number of collisions a participant nucleon undergoes. This is in agreement with previous NA35 results where a similar observation was made concerning negative hadron production.

7. CONCLUSIONS

Experiment NA49 had a successful first run in 1994. All detector components including the newly built TPCs worked well. A first look at the data shows that physics analysis is possible in the high multiplicity environment of Pb-Pb collisions. Calorimetry and tracking in the TPC give first insights into observables such as transverse energy and source size.

ACKNOWLEDGMENTS

This work was supported by the Director, Office of Energy Research, Division of Nuclear Physics of the Office of High Energy and Nuclear Physics of the U.S. Department of Energy under Contract DE-AC03-76SF00098.

REFERENCES

1. C. de Marzo et al.; Nucl. Instr. and Meth. 217 (1983) 405
2. T. Alber et al.; Proceedings of the Quark Matter 1995 Conference, to be published in Nucl. Phys. B
3. S. Margetis et al.; Proceedings of the Quark Matter 1995 Conference, to be published in Nucl. Phys. B
4. I. Huang et al.; 'Reduction of Calorimeter data in NA49', LBL-36877, LBL, Berkeley, 1995
5. J. Bächler et al.; Z. Phys. C52(1991)239
6. J. D. Bjorken; Phys. Rev. D27(1983)140
7. S. Pratt; Phys. Rev. D33(1986)1314
8. Yu. M. Sinyukov; Nucl. Phys. A498 (1989) 151c

28

THE EFFECT OF PROJECTILE SHAPE ON CROSS SECTIONS AND MOMENTUM DISTRIBUTIONS OF FRAGMENTS FROM HEAVY-ION REACTIONS

Bernard Harvey,[1] Klara Shitikova,[2]* and Grandon Yen[3]

[1]Nuclear Science Division
Lawrence Berkeley Laboratory
Berkeley, CA
[2]Institute of Nuclear Physics
Moscow State University
Moscow, Russia
[3]Institute of Physics
Academica Sinica
Nanking, Taipei 11529, Taiwan

ABSTRACT

Reaction cross sections, inclusive fragment cross sections and momentum distributions were calculated using two different theoretical models for the proton end neutron density distributions in the projectile nucleus. The results are compared with the experimental values for reactions of ^{12}C and ^{11}Li on a carbon target.

1. INTRODUCTION

In reactions of ^{12}C or ^{11}Li on targets of ^{12}C, it has been shown that the total reaction cross section, inclusive fragment cross sections and fragment momentum distributions are sensitive to the assumed proton and neutron radial density distributions in the projectile 1.

In the present work, cross sections and momenta were calculated from Monte Carlo nucleon–nucleon collisions. Two models were used to obtain the theoretical density distributions in ^{11}Li, the hyperspherical functions model 2 and a harmonic oscillator shape to which was added an exponential tail to the neutron density distribution. In all calculations, the

---

*Invited talk, 11th Winter Workshop on Nuclear Dynamics. Key West, Florida (1995).

Advances in Nuclear Dynamics
Edited by Wolfgang Bauer and Alice Mignerey, Plenum Press, New York, 1996

^{12}C target was assumed to have a Fermi shape with half-density radius and diffusivity from Hartree–Fock-Strutinsky calculations 3.

In Section 2, the density distributions are described. In Section 3, the results of calculations with the two models are compared with experimental cross sections, and Section 4, calculated and experimental fragment momentum distributions are compared.

2. DENSITY DISTRIBUTIONS

a) In the hyperspherical functions method, the wave function of a nucleus is expressed in the form of an expansion in the K-harmonic polynomials:

$$\Psi(1, 2...A) = \rho^{-(3A-4)/2} \sum_{K,\gamma} X_{K,\gamma}(\rho) Y_{K,\gamma}(\theta) \tag{2.1}$$

where the hyperspherical harmonics are the eigenfunctions of the angular part of the Laplacian:

$$\Delta_{\Omega_n} Y_{K,\gamma}(\theta_i) = -K(K + n - 2) Y_{K,\gamma}(\theta_i). \tag{2.2}$$

The value of K is the analog of the angular momentum at $n = 3$; it is called the global moment.

In a basis of hyperspherical functions in which the symmetries have been properly taken into account, a better description of the asymptotic part of the wave function is possible. In 4, an attempt was made to provide a unified description of ^{6,7,8,9,11}Li rather than just a single isotope. No attempt was made to parametrize the effective interation used for each isotope: rather a simple parametrization for all the isotopes was used. Furthermore, in this treatment, there is no inert core and all the nucleons are properly antisymmetrized. Lastly, because Jacobian coordinates were used, no problems were encountered with the treatment of the center of mass.

b) Since the Monte Carlo calculations require a separate density distribution for nucleons with different binding energies, a harmonic oscillator model was used for ^{11}Li. The s- and p-shell parameters were chosen to be 1.619 and 2.0 fm respectively. It was also assumed that there was one proton and four neutrons in the $p_{3/2}$-shell. The outer two neutrons were assumed to have an exponential density distribution beyond a radius of 2.5 fm:

$$\rho(r) = 0.3795 \frac{e^{-2r/L}}{r^2} \tag{2.3}$$

with $L = 3.5$ fm. The normalization, 0.3795, came from the requirement that the volume integral contain just two neutrons.

The values of the harmonic oscillator parameters and of L were chosen to fit the experimental reaction cross section and the ^{9}Li, ^{8}He inclusive cross sections for the reaction ^{11}Li + C at 790 MeV/A.

Fig. 1 compares the total density distributions for ^{11}Li calculated by the two models. Out to about 6 fm, the results are quite similar but beyond that radius, the harmonic oscillator with exponential neutron tail gives a much higher density.

3. TOTAL AND INCLUSIVE CROSS SECTIONS AND NUCLEAR SHAPE

In Ref. 5, it was shown that Monte Carlo calculations using nucleon–nucleon scattering cross sections could reproduce the experimental ^{12}C + C total reaction cross section for beam energies from the Coulomb barrier to 2 GeV/A.

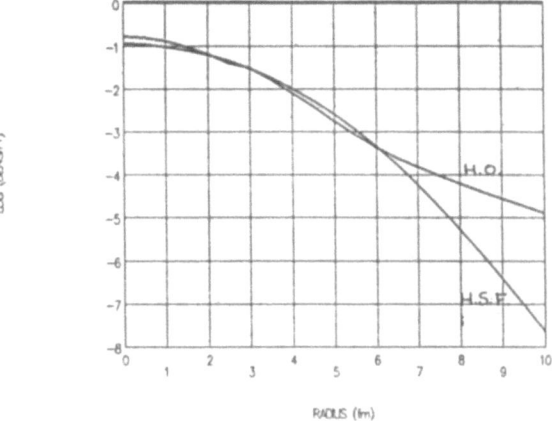

Figure 1. Comparison of total density of ^{11}Li from hyperspherical functions (HSF) and harmonic oscillator model (HO) with exponential "tail" in the neutron density.

Table 1. Comparison of Experimental and Calculated Total Reaction Cross Section and Inclusive Cross Sections for the Reaction ^{11}Li + C, 790 MeV/A. Calculations were made with hyperspherical function (HSF) and harmonic oscillator + exponential neutron "tail," (HO) density distributions. Cross sections are in mb

	σ_R	σ_{9Li}	σ_{8Li}	σ_{8He}
Experiment	1042	213	62	26
HSF	1057	155	11	47
HO	1035	214	13	29

Table 2. Comparison of rms Radii from Experiment with Radii Calculated with HSF and HO. All radii are in fm

	protons	neutrons	total
Ref[a]			3.16
HSF	3.01	3.22	3.13
HO	2.43	3.35	3.13

[a]Tanihata et al., Phys. Lett. **B206**, 592 (1988)

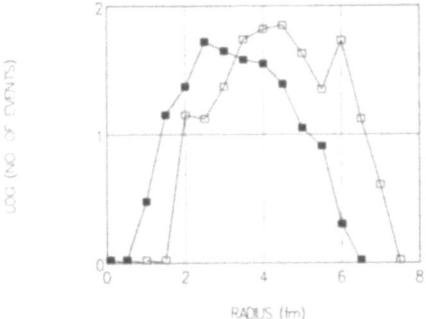

Figure 2. Radius in ^{11}Li of the p-shell proton knocked out to form ^{10}He. Open squares: HSF model. Filled squares: HO model.

The effect of changing the shape of the ^{12}C projectile has been investigated[1]. The reaction cross section and the ^{11}C inclusive cross section were both extremely sensitive to the assumed surface diffusivity, and so, of course, was the rms matter radius. The shape of the ^{11}C momentum distribution also changed with the assumed projectile surface diffusivity.

We have therefore made Monte Carlo calculations for the reaction ^{11}Li + C at 790 MeV/A to compare the results obtained with hyperspherical function (HSF) density distributions for ^{11}Li and the harmonic oscillatory (HO) shape with an added exponential neutron "tail" as described in Section 1. The reaction cross sections, the ^{9}Li inclusive cross sections and the rms radii are compared in Tables 1 and 2.

For both models, the total cross section is the same and agrees with experiment. For protons, though, the rms radius from the HSF model is greater than from the HO model. Therefore the cross section for ^{8}He is too high (^{8}He is made almost entirely by knock-out of a single proton to make ^{10}He which is unbound to decay to ^{8}He).

For neutrons, the HSF rms radius is too low and therefore the cross section for ^{9}Li is smaller than the HO and experimental values.

Figs. 2 and 3 show the distribution of radii in ^{11}Li of the proton and neutron that were knocked out to produce ^{10}He and ^{10}Li primary fragments. Nucleons knocked out from the s-shell are not included since a primary fragment with an s-shell hole would be too highly excited to decay just to ^{8}He or ^{9}Li. For protons, the HSF distribution extends to larger radii than for the HO model. This result is consistent with the large proton rms radius and ^{8}He cross section from the HSF model.

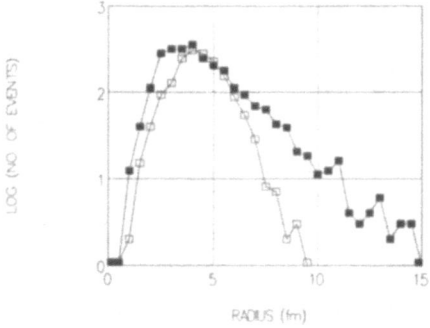

Figure 3. As Fig. 2, but for knocked out p-shell neutron

Table 3. Comparison of Transverse Momentum Widths for ^{9}Li Fragments from the Reaction ^{11}Li + C, 790 MeV/A, Using Hyperspherical Function (HSF) and Harmonic Oscillator + Exponential Neutron "Tail" (HO) Density Distributions. σ_B and σ_N are the widths of the broad and narrow momentum components. B/N is the ratio of their areas

	σ_B(MeV/c)	σ_N(MeV/c)	B/N
Experiment	80	21	1.5
HSF	75	48	1.9
HO	81	29	1.1

Fig. 3 shows that the HO neutron distribution extends to larger radii than the HSF distribution. This too is consistent with the larger ^{9}Li cross section and neutron rms radius from the HO model.

4. TRANSVERSE MOMENTUM DISTRIBUTION OF FRAGMENTS

The Monte Carlo method permits the calculation of fragment momentum distributions parallel with or perpendicular to the projectile direction. The momentum of a fragment is assumed to arise from three sources:

1) The fragment recoils with the Fermi momentum of a knocked-out nucleon. The maximum Fermi momentum $P_f(r)$(max) of a nucleon at radius $\vec{r}$ is assumed to be proportional to the cube root of the total density at $\vec{r}$. The actual Fermi momentum P_f of the nucleon can have any value from 0 to $P_f(r)$(max) with a spherical distribution of probability:

$$\text{Prob}\,(P_f) = \frac{P_f^2}{P_f(r)(max)^3} \tag{4.1}$$

The Fermi momentum is assumed to be isotropic.

2) The struck nucleon was bound to the projectile. As it leaves, it must therefore lose some kinetic energy (momentum) and the lost momentum is transferred to the residual fragment. The contribution from this "friction" adds about 20 MeV/c to the width of the momentum distribution of ^{11}C from ^{12}C projectiles but has a negligible effect for ^{9}Li from ^{11}Li, where the binding energy is very small.

3) If a primary fragment is formed in a excited state, it may decay by emission of nucleons or other fragments. The final fragment therefore recoils with the momentum of the emitted particles.

The momentum distribution of ^{9}Li fragments from ^{11}Li + C at 790 MeV/c has been studied 6; it has broad and narrow Gaussian components with widths of 80 and 21 MeV/c and a ratio of areas B/N of 1.5. Since the shape of the momentum distribution is sensitive to the projectile surface diffusivity, we have used the HSF and HO models to calculate the ^{9}Li momentum distribution. The results are summarized in Table 3 and shown in Fig. 4.

The width of the broad component is about the same for the two models, but the narrow component is too wide with the HSF density distribution. This result is consistent with the absence of a long neutron "tail" at large radii and low local neutron densities which give a small value for the local Fermi momentum.

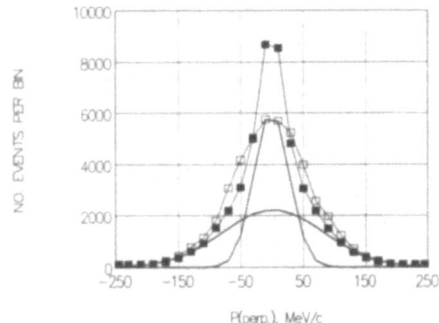

Figure 4. Comparison of momentum distribution of ^{9}Li from the reaction ^{11}Li + C, 790 MeV/A, from the hyperspherical functions and harmonic oscillator models. Filled squares: harmonic oscillator. Open squares: hyperspherical functions. The heavy lines are the broad and narrow momentum components from the harmonic oscillator model.

Thus both the cross sections and the momentum distributions lead to the conclusion that the HSF density distributions extend too far for protons and not far enough for neutrons. The inclusion of several more terms in the expansion (Eq. (1)) will give an extension of the asymptotic density to larger radii. This will give better agreement between the hyperspherical functions method and the experimental data.

5. CONCLUSION

Measurements of the reaction cross section and the inclusive cross sections for fragments in which the projectile has lost one neutron or one proton provide a powerful method for investigating the neutron, proton and total densities in the surface of the projectile.

ACKNOWLEDGMENTS

This work was supported in part by the Director, Office of Energy Research, Division of Nuclear Physics of the Office of High Energy and Nuclear Physics of the U. S. Department of Energy under Contract No. DE-AC03-76SF00098.

REFERENCES

1. B. G. Harvey, Phys. Rev. C **49**, 2890 (1994).
2. R. I. Djibouti, K. V. Shitikova, M. Energoatomizdat, p. 272 (1993).
3. F. Tondeur, J. P. Arcoragi and J. M. Pearson, J. Phys. (Paris), Coloq. **45**, C6–125 (1984).
4. V. V. Burov et al., Preprint JINR E2-93-41 (1993).
5. B. G. Harvey, Nucl. Phys. **A444**, 498 (1985).
6. T. Kobayashi et al., Phys. Rev. Lett. **60**, 2599 (1988).

29

THE GIANT DIPOLE RESONANCE BUILT ON HIGHLY EXCITED STATES

T. Suomijärvi,[1]* J. H. Le Faou,[2] Y. Blumenfeld,[2] P. Piattelli,[3] C. Agodi,[3]
N. Alamanos,[4] R. Alba,[3] F. Auger,[4] G. Bellia,[3†], Ph. Chomaz,[5‡],
R. Coniglione,[3] A. Del Zoppo,[3] P. Finocchiaro,[3] N. Frascaria,[2]
J. J. Gaardhøje,[6] J. P. Garron,[2] A. Gillibert,[4] M. Lamehi-Rachti,[2§],
R. Liguori-Neto,[4¶], C. Maiolino,[3] E. Migneco,[3] J. C. Roynette,[2]
D. Santonocito,[3] P. Sapienza,[3] and J. A. Scarpaci[2]

[1]National Superconducting Cyclotron Laboratory
Michigan State University
East Lansing, Michigan
[2]Institut de Physique Nucléaire
IN$_2$P$_3$-CNRS, 91406 Orsay, France
[3]INFN-Laboratorio Nazionale del Sud
Via S. Sofia 44, 95123 Catania, Italy
[4]SEPhN, DAPNIA, CEA Saclay
91191 Gif sur Yvette, France
[5]GANIL, BP 5027, 14021 Caen
France
[6]The Niels Bohr Institute, University of Copenhagen
DK-2100 Ø, Denmark

ABSTRACT

Gamma-rays emitted from hot nuclei of mass around 115 and excitation energies above 300 MeV, formed in the ^{36}Ar + ^{90}Zr at 27 MeV/u, ^{36}Ar + ^{94}Zr at 32 MeV/u and ^{36}Ar + ^{98}Mo at 37 MeV/u, have been measured with the MEDEA multidetector in coincidence with light charged particles and evaporation residues. The γ-ray yield from the decay of the Giant Dipole Resonance is independent of excitation energy and of bombarding energy.The best agreement with the data is

*On leave from Institut de Physique Nucléaire, IN$_2$P$_3$-CNRS, 91406 Orsay, France
†Dipartimento di Fisica dell'Università, Catania, Italy.
‡On leave from Division de Physique Théorique, Institut de Physique Nucléaire, Orsay, France.
§On leave from University of Teheran, Iran.
¶On leave from University of São Paolo, Brazil.

Advances in Nuclear Dynamics
Edited by Wolfgang Bauer and Alice Mignerey, Plenum Press, New York, 1996

obtained by assuming a cut-off of the resonance γ-emission above an excitation energy of 250 MeV.

1. INTRODUCTION

Giant resonances are collective excitations of the nucleus. They occur throughout the whole periodic table and their characteristics such as the mean energy, width and strength are smooth functions of the mass number of the nucleus. Giant resonances are related to macroscopic properties of the nucleus such as the compression moduli, symmetry energy and spin-isospin sound velocity and their width gives information on the energy dissipation in nuclear matter. Giant resonances can be built on all nuclear states offering an interesting possibility to study the bulk properties of excited nuclear matter through the measurement of their characteristics. Moreover, the measurement of the strength of giant resonances in very hot nuclei gives information on the possibility to maintain well ordered collective motion at high nuclear temperatures.

Up to now experimental efforts concerning excited nuclei have been focused onto measuring the Giant Dipole Resonance due to its rather easy observation through gamma decay. The most complete systematics have been obtained for medium mass nuclei in the vicinity of $A \approx 115$.[1] It is shown that at least up to 300 MeV excitation energy, the GDR remains remarkably stable exhausting all the sum rule strength at around 15 MeV corresponding to its ground state energy. With increasing excitation energy the GDR width first increases from its ground state width 4.8 MeV to about 12 MeV mostly due to angular momentum effects and then saturates for energies above 130 MeV.[2] Above 300 MeV excitation energy a saturation of the γ yield has been measured[3,4] and has given rise to divergent interpretations.[5,6,7]

In order to study the collective motion at very high temperatures, we recently pursued investigations of the GDR in hot nuclei formed in three different reactions: ^{36}Ar + ^{90}Zr at 27 MeV/u, ^{36}Ar + ^{94}Zr at 32 MeV/u and ^{36}Ar + ^{98}Mo at 37 MeV/u. Experiments were performed at the GANIL facility (Caen, France) by using the MEDEA detector.[8] Most of the results that will be presented here concern the reaction at 27 MeV/u, some preliminary data at 37 MeV/u will also be shown. The data at 27 MeV/u has already been published in Ref. 9, 10.

2. EXPERIMENTAL METHOD

Hot nuclei were formed in incomplete fusion reactions of ^{36}Ar + ^{90}Zr at 27 MeV/u, ^{36}Ar + ^{94}Zr at 32 MeV/u and ^{36}Ar + ^{98}Mo at 37 MeV/u. The experimental setup was designed to observe the γ-rays arising from GDR decay with high efficiency and to measure simultaneously the excitation energy and temperature of the decaying nuclei through the measurement of evaporation residue velocities, light charged particle (LCP) spectra and LCP multiplicities.

Gamma-rays and light charged particles were detected with the MEDEA detector[8] which consists of a ball built with 180 barium fluoride (BaF$_2$) crystals that covers the angular range between 30° and 170° and a forward phoswich wall covering the angles between 10° and 30°. To allow for the simultaneous measurement of γ-rays and light charges particles, the entire system operates under vacuum inside a large scattering chamber. Gamma-rays and light charged particles were identified in the BaF$_2$ detectors using a pulse shape analysis technique and the time of flight information.

In order to measure fusion-like residues, two parallel plate avalanche counters (PPAC) covering between 6° and 22° on either side of the beam were added into the MEDEA detector. The time of flight and energy loss information from these counters allowed to clearly distinguish fusion-like residues from projectile-like and target-like fragments. Moreover, the time of flight measurement yielded information on the velocity of the fusion-like residues.

The trigger was given by one PPAC firing in coincidence with at least one BaF_2 detector. This trigger requirement eliminates all cosmic ray contamination of the γ spectra.

3. EXPERIMENTAL RESULTS

A widely used method for characterizing the composite system formed in incomplete fusion reactions is the determination of the Linear Momentum Transfer (LMT). This can be done by applying a massive transfer model which allows to determine the LMT fraction as well as the mass and excitation energy of the composite system from the measured residue velocity. For the reactions at 27 MeV/u and 37 MeV/u, the mean linear momentum transfers are measured to be 74% and 65%, respectively, which according to the massive transfer model, correspond to excitation energies of 560 MeV and 660 MeV and residue masses of 115 and 120.

Figure 1 shows the γ-spectra measured for momentum transfer bins centered around the mean momentum transfer of each reaction. At low energies statistical γ-rays emitted by the compound nucleus at the end of its decay chain give rise to a steep exponential decay. The high energy photons are interpreted as due to the nucleon-nucleon bremsstrahlung during the initial stages of the collision process. This high energy γ yield can be represented by an exponential function, fit to the spectrum for $E_\gamma > 35$ MeV. The slope parameters are 9.5 ± 1 MeV and 11 ± 1 MeV for 27 and 37 MeV/u respectively, in good agreement with previous systematics.[11] At about 15 MeV, a pronounced bump can be seen in both spectra corresponding to γ-rays emitted from the GDR. The remarkable feature is that after subtraction of the bremsstrahlung component the two spectra are identical, showing that the γ-yield from hot nuclei no longer depends on the bombarding energy in the high energy region considered here.

The most relevant parameter to study the evolution of the GDR γ-yield is the excitation energy rather than the bombarding energy. In order to investigate this evolution as a function of excitation energy, the data at 27 MeV/u were divided into 3 bins of linear momentum transfer and γ-spectra were extracted for each bin. A combination of the linear momentum transfer measurement and a moving source fit of the coincident proton spectra furnishes excitation energies of 350, 500 and 550 MeV for the 3 bins.[9] After the bremsstrahlung component was subtracted from the spectra, the γ-multiplicity was integrated over two regions: from 12 to 20 MeV and 20 to 35 MeV. The first region encompasses the bulk of the GDR γ-rays but the second is also dominated by the GDR decay. Therefore, to constrain different theoretical interpretations it is important to also investigate the high energy region of the spectra. As shown in Fig. 2, both integrated yields are constant within the error bars as a function of excitation energy, confirming the saturation of the GDR γ-multiplicity at these high excitation energies. This saturation is clearly in contradiction with standard statistical model predictions. The solid lines on Fig. 2 show the results of a standard CASCADE calculation assuming $E_{GDR} = 76.5A^{-1/3}$ MeV, $\Gamma_{GDR} = 12$ MeV, and $S_{GDR} = 100\%$ sum rule strength,[9] which reproduced earlier experimental results up to 230 MeV excitation energy.[1] The calculated multiplicities strongly overshoot the present data and increase with excitation energy.

The elucidation of the mechanism responsible for the γ-ray saturation should fur-

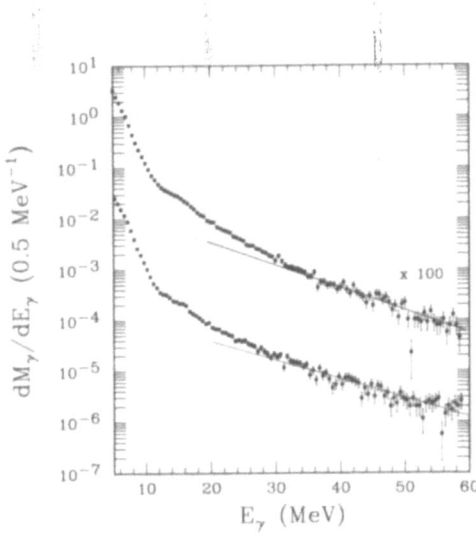

Figure 1. Normalized γ-spectra measured for the reactions ^{36}Ar + ^{90}Zr at 27 MeV/u (top) and ^{36}Ar + ^{98}Mo at 37 MeV/u (bottom). The solid lines are a fit to the high energy component of the spectra ($E_\gamma > 35$ MeV).

ther our understanding of the limits of well-ordered collective motion in nuclei.[5,6,7] It has been suggested that the saturation could be due to a rapid increase of the GDR width with temperature.[4,6,7] Indeed, the broadening of the GDR leads to a depletion of the γ-yield between 12 and 20 MeV, as shown by CASCADE calculations including a width increasing with excitation energy as $\Gamma_{GDR} = 4.8 + 0.0026E^{*1.6}$ MeV (long dashed line in Fig. 2 top). However, at higher γ energies, the predicted multiplicity continues to increase, in contradiction to the experimental results (long dashed line in Fig. 2 bottom). Calculations performed by including the predictions of theoretical models[6,7] in CASCADE confirm the above conclusions.[9,10] Therefore a rapid increase of the width of the GDR cannot alone be responsible for the observed saturation of the γ-yield.

The simplest way to reproduce the integrated γ-yields is to introduce a sharp suppression of the γ-emission above 250 MeV excitation energy. The yields calculated under this hypothesis are presented as short dashed lines in Fig. 2. A good agreement with the experimental results is obtained for both γ-energy regions. Figure 3 shows a comparison between the spectrum calculated under the cut-off hypothesis and the experimental spectra for the 500 MeV excitation energy bin measured at 27 MeV/u and the spectrum measured at 37 MeV/u, from which the bremsstrahlung components have been subtracted. Indeed, both experimental spectra are very well reproduced above 12 MeV γ-energy. In Ref. 5 it was proposed that a certain time is needed to equilibrate dipole oscillations with the compound nucleus during which a considerable cooling down of the compound nucleus occurs by particle emission. An alternative explanation would be a transition from collective to chaotic motion for the GDR.[6]

For γ-energies between 8 and 12 MeV, however, a marked underestimation of the experimental spectrum is apparent. This could be an indication that the GDR is replaced by some low-lying strength at very high excitation energies. A similar shift of the resonance strength has been predicted by the Random Phase Approximation calculations of Ref. 12, 13.

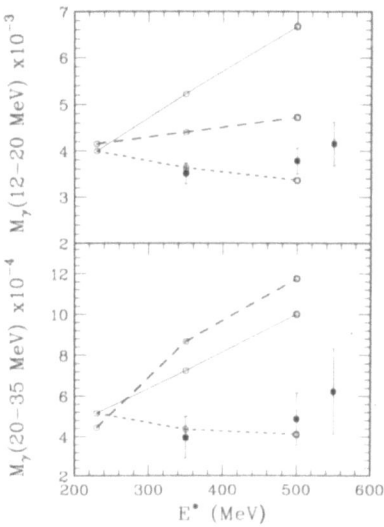

Figure 2. Calculated and experimental γ-yields between 12 and 20 MeV (top) and 20 and 35 MeV (bottom) as a function of excitation energy. Solid line: standard CASCADE calculation, long dashed line: CASCADE calculation with increasing GDR width, short dashed line: CASCADE calculation with cut-off (see text).

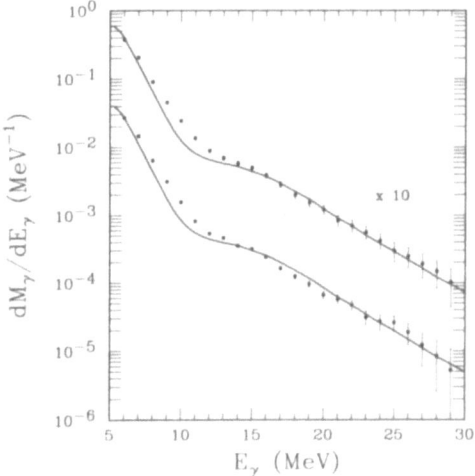

Figure 3. Comparison of experimental spectra measured at 27 (top) and 37 MeV/u (bottom) after subtraction of the bremsstrahlung component compared to CASCADE calculations including a cut-off of the GDR γ-emission above 250 MeV excitation energy.

4. CONCLUSIONS

In conclusion, a saturation of the γ-yield from the GDR decay is observed above 300 MeV excitation energy, confirmed by measurements at two bombarding energies, 27 and 37 MeV/u. This saturation cannot be explained by assuming a strong increase of the GDR width at high excitation energies, but a cut-off of the γ-emission above 250 MeV must be invoked. The experimental γ spectra presented here were triggered by the presence of a heavy evaporation residue. Data were also taken by triggering on high light charged particle multiplicities, which should give access to an even higher temperature region where multifragmentation and vaporization are expected to set in.

REFERENCES

1. J. J. Gaardhøje, Annu. Rev. Nucl. Part. Sci. 42 (1992) 483.
2. A. Bracco et al., Nucl. Phys. A519 (1990)47c.
3. J. J. Gaardhøje et al., Phys. Rev. Lett. 59 (1987) 1409.
4. K. Yoshida et al., Phys. Lett. 245 B (1990) 7.
5. P. F. Bortignon et al., Phys. Rev. Lett. 67 (1991) 3360.
6. Ph. Chomaz, Nucl. Phys. A569 (1994) 203c.
7. A. Smerzi et al., Phys. Lett. B320 (1994) 216.
8. E. Migneco et al., Nucl. Instr. and Meth. A314 (1992) 31.
9. T. Suomijärvi et al., Nucl. Phys. A569 (1994) 225c.
10. J. H. Le Faou et al., Phys. Rev. Lett. 72 (1994) 3321.
11. H. Nifenecker and J. A. Pinston, Annu. Rev. Nucl. Part. Sci. 40 (1990) 113.
12. H. Sagawa and G. F. Bertsch, Phys. Lett. 146B (1984) 138.
13. P. F. Bortignon et al., Nucl. Phys. A 460 (1986) 149.

30

NEUTRINOS FROM PROTONEUTRON STARS: A PROBE OF HOT AND DENSE MATTER

Sanjay Reddy and Madappa Prakash

Physics Department
State University of New York at Stony Brook
Stony Brook, NY

ABSTRACT

Neutrino processes in dense matter play a key role in the dynamics, deleptonization and the early cooling of hot protoneutron stars formed in the gravitational collapse of massive stars. Here we calculate neutrino mean free paths from neutrino–hyperon interactions in dense matter containing hyperons. Significant contributions to the neutrino opacity arise from scattering involving the Σ^- hyperon, and absorption processes involving the neutral and Σ^- hyperons. The estimates given here emphasize the need for (a) opacities which incorporate many-body effects in a multi-component mixture, and (b) new calculations of thermal and leptonic evolution of protoneutron stars with neutrino transport and equations of state with strangeness-rich matter.

1. INTRODUCTION

The general nature of the neutrino signature expected from a newly formed neutron star (hereafter referred to as a protoneutron star) has been theoretically predicted[1] and confirmed by the observations[2] from supernova SN1987A. Although neutrinos interact weakly with matter, the high baryon densities and neutrino energies achieved after the gravitational collapse of a massive star (≥ 8 solar masses) cause the neutrinos to become trapped on the dynamical timescales of collapse.[3,4] Trapped neutrinos at the star's core have Fermi energies $E_\nu \sim$ 200–300 MeV and are primarily of the ν_e type. They escape after diffusing through the star exchanging energy with the ambient matter, which has an entropy per baryon of order unity in units of Boltzmann's constant. Eventually they emerge from the star with an average energy $\sim$ 10–20 MeV and in nearly equal abundances of all three flavors, both particle and anti-particle.

Although the composition and the equation of state of the hot protoneutron star matter are not yet known with certainty, QCD based effective Lagrangians have opened up intriguing

Advances in Nuclear Dynamics
Edited by Wolfgang Bauer and Alice Mignerey, Plenum Press, New York, 1996

possibilities. Among these is the possible existence of matter with strangeness to baryon ratio of order unity. Strangeness may be precipitated either in the form of fermions, notably the Λ and Σ^- hyperons, or, in the form of a Bose condensate, such as a K^--meson condensate (see Ref. 5 for detailed discussion and extensive references). In the absence of trapped neutrinos, strange particles are expected to appear around 2–4 times the nuclear matter density of $n_0 = 0.16$ fm^{-3}. Neutrino-trapping causes the strange particles to appear at somewhat higher densities than in neutrino-free matter. The compositions shown in Figs. 1 and 2 highlight the influence of hyperons in the neutrino trapped regime. The results shown in these figures were calculated[5] using a field-theoretical model in which baryons interact via the exchange of σ, ω and ρ mesons. With the appearance of hyperons in matter, the electron–neutrino fraction increases with density in contrast to the case in which matter contains nucleons and leptons only. A similar behavior is observed in kaon condensed matter[6] and also in matter where a phase transition to quark matter occurs.[7] This behavior is associated with the presence of non-leptonic negatively charged particles in matter,[5] such as the Σ^- hyperon, or K^-meson, or d and s quarks.

Keil and Janka[8] have recently investigated the influence of the equation of state on the cooling and evolution of the protoneutron star. They find that the neutrino luminosity depends sensitively on the composition and on the stiffness of the equation of state at high density. In particular, the influence of hyperons, which introduces a softening of the high density equation of state, was examined. In many cases, the protoneutron star collapsed to a black hole causing an abrupt cessation of the neutrino signal. Although a clear distinction between the different equations of state could be achieved on the basis of the calculated neutrino signals, Keil and Janka conclude that "none of the models could be considered as a good fit of the neutron star formed in SN 1987A".

We note, however, that in these studies, the sensitivity of the neutrino signals due only to the structural changes caused by the equation of state was assessed. In the presence of hyperons, the transport of neutrinos is also affected due to the changes in the composition, and additionally, to the interactions of neutrinos with strange particles. These effects were ignored in Ref. 8. Previous work[9,10] involving neutrino interactions with hyperons was concerned with charged-current reactions only. Studies of the opacities and the transport processes of neutrinos out of the core containing nucleons, leptons, and in some cases, pion condensates may be found in Refs. 11 through 26.

It is our purpose here to study neutrino mean free paths in matter containing hyperons. Specifically, we will estimate scattering and absorption mean free paths of neutrinos when matter is under degenerate conditions. We find that significant contributions to the neutrino opacity arise from scattering involving the Σ^- hyperon, and absorption processes involving the neutral and Σ^- hyperons. Compared with ordinary matter, the presence of strangeness in matter is expected to lead to an excess of ν_e neutrinos relative to other types. Calculations of the complete thermal and leptonic evolution of a newly formed neutron star incorporating neutrino transport and equations of state with strangeness-rich matter will be taken up separately. With new generation neutrino detectors capable of recording thousands of neutrino events, it may be possible to distinguish between different scenarios observationally.

2. NEUTRINO INTERACTIONS WITH STRANGE BARYONS

Neutrino interactions with matter proceed via charged and neutral current reactions. The neutral current processes contribute to elastic scattering, and the charged current reactions result in neutrino absorption. The interaction Lagrangian for these reactions is given by the

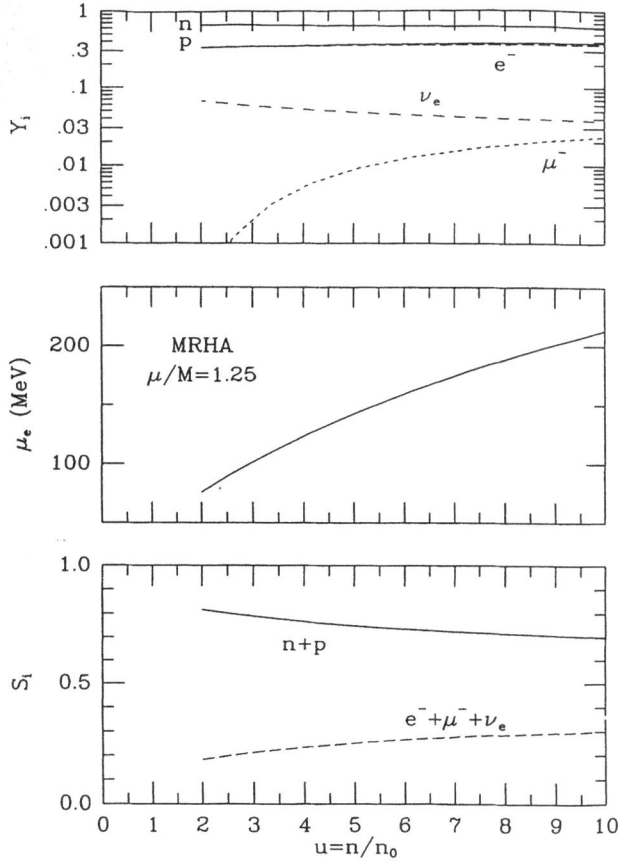

Figure 1. The composition, electron chemical potential and entropy per particle in nucleons only matter with a lepton fraction $Y_{Le} = Y_e + Y_{\nu_e} = 0.4$, where $Y_i = n_i/n_b$. Results are from Ref. 5.

Weinberg–Salam theory:

$$\mathcal{L}^{nc}_{int} = (G_F/2\sqrt{2})l_\mu j^\mu_z \qquad \text{for} \qquad \nu + B \rightarrow \nu + B$$
$$\mathcal{L}^{cc}_{int} = (G_F/\sqrt{2})l^e_\mu j^\mu_w \qquad \text{for} \qquad \nu + B_1 \rightarrow \ell^- + B_2, \tag{2.1}$$

where $G_F \simeq 1.436 \times 10^{-49}$ erg cm^{-3} is the weak coupling constant, ν is a neutrino, B_1 and B_2 are baryons, and ℓ^- is a lepton. The leptonic and hadronic currents appearing above are:

$$l_\mu = \overline{\psi}_\nu \gamma_\mu (1 - \gamma_5) \psi_\nu$$
$$l^e_\mu = \overline{\psi}_e \gamma_\mu (1 - \gamma_5) \psi_\nu$$
$$j^\mu_w = \overline{\psi}_i \gamma^\mu (g_{Vi} - g_{Ai}\gamma_5) \psi_i$$
$$j^\mu_z = \overline{\psi}_i \gamma^\mu (C_{Vi} - C_{Ai}\gamma_5) \psi_i, \tag{2.2}$$

where $i = n, p, \Lambda, \cdots$. The neutral current process couples neutrinos of all types (e, μ and τ) to the weak neutral hadronic current j^μ_z. The charged current processes of interest here are electron and muon neutrinos coupled to the charged hadronic current j^μ_w. The vector and axial

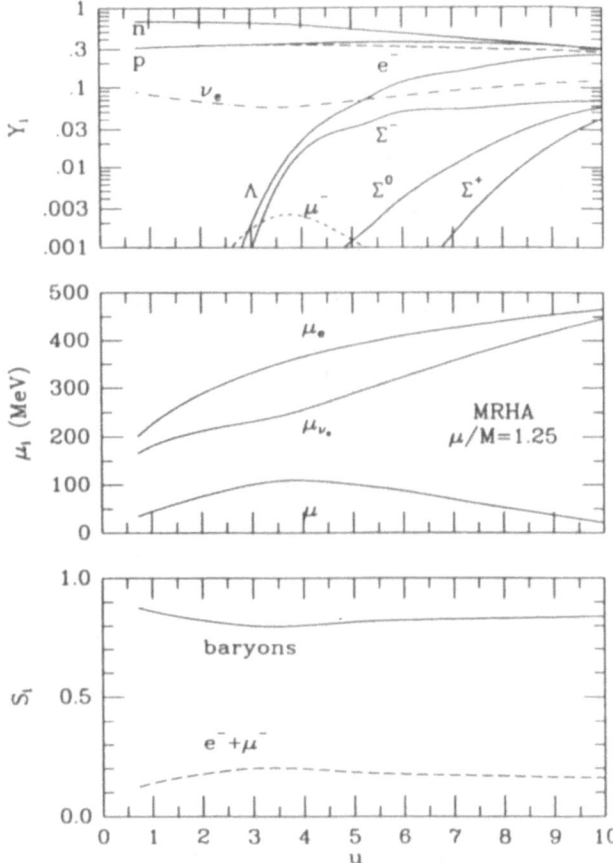

Figure 2. The composition, lepton chemical potentials and entropy per particle in strangeness-rich matter with a lepton fraction $Y_{Le} = Y_e + Y_{\nu_e} = 0.4$, where $Y_i = n_i/n_b$. Results are from Ref. 5.

Table 1. Neutral Current Vector and Axial Couplings

Reaction	C_V	C_A	$\mathcal{R}_{nc}^{(1)}$	$\mathcal{R}_{nc}^{(2)}$
$\nu_i + n \rightarrow \nu_i + n$	-1	$-D - F$	1	1
$\nu_i + p \rightarrow \nu_i + p$	$(1 - 4\sin^2\theta_W)$	$D + F$	0.7504	0.8597
$\nu_i + \Lambda \rightarrow \nu_i + \Lambda$	0	0	0	0
$\nu_i + \Sigma^- \rightarrow \nu_i + \Sigma^-$	$(-2 + 4\sin^2\theta_W)$	$-2F$	0.7392	0.6788
$\nu_i + \Sigma^+ \rightarrow \nu_i + \Sigma^+$	$(2 - 4\sin^2\theta_W)$	$2F$	0.7392	0.6788
$\nu_i + \Sigma^0 \rightarrow \nu_i + \Sigma^0$	0	0	0	0
$\nu_i + \Xi^- \rightarrow \nu_i + \Xi^-$	$(-1 + 4\sin^2\theta_W)$	D	0.2845	0.3238
$\nu_i + \Xi^0 \rightarrow \nu_i + \Xi^0$	1	$-D + F$	0.2861	0.1852

NOTE. — The quantity $\mathcal{R}_{nc}^{(1)} = [C_V^2 + 2C_A^2]_{\nu+B\rightarrow\nu+B}/[C_V^2 + 2C_A^2]_{\nu+n\rightarrow\nu+n}$ and $\mathcal{R}_{nc}^{(2)} = [C_V^2 + 4C_A^2]_{\nu+B\rightarrow\nu+B}/[C_V^2 + 4C_A^2]_{\nu+n\rightarrow\nu+n}$.

Table 2. Charged Current Vector and Axial Couplings

Reaction	g_V	g_A	$\mathcal{R}_{cc}$
$\nu_i + n \to e + p$	1	$F + D$	1
$\nu_i + \Lambda \to e + p$	$-\sqrt{3/2}$	$-\sqrt{3/2}(F + D/3)$	0.0394
$\nu_i + \Sigma^- \to e + n$	-1	$-(F - D)$	0.0125
$\nu_i + \Sigma^- \to e + \Lambda$	0	$\sqrt{2/3}$	0.2055
$\nu_i + \Sigma^- \to e + \Sigma^0$	$\sqrt{2}$	$\sqrt{2}F$	0.6052
$\nu_i + \Xi^- \to e + \Lambda$	$\sqrt{3/2}$	$\sqrt{3/2}(F - D/3)$	0.0175
$\nu_i + \Xi^- \to e + \Sigma^0$	$\sqrt{1/2}$	$(F + D)/\sqrt{2}$	0.0282
$\nu_i + \Xi^- \to e + \Xi^0$	1	$F + D$	0.0564
$\nu_i + \Xi^0 \to e + \Sigma^+$	1	$F - D$	0.2218

NOTE. — The quantity $\mathcal{R}_{cc} = [C^2(g_V^2 + 3g_A^2)]_{\nu+B_1 \to \ell+B_2}/$
$[C^2(g_V^2 + 3g_A^2)]_{\nu+n \to \ell+p}$.

vector coupling constants are listed in Table 1. Numerical values of the parameters that best fit the experiments are: D=0.756 , F=0.477, $\sin^2\theta_W$=0.23 and $\sin\theta_c = 0.231$.

In what follows, we consider the lowest order (tree level) processes for both elastic and absorption reactions. The squared matrix element for neutral current reactions is given by

$$\overline{|\mathcal{M}_{12\to34}|^2} = 16G_F^2[(C_V + C_A)^2(p_1 \cdot p_2)(p_3 \cdot p_4)$$
$$+ (C_V - C_A)^2(p_1 \cdot p_4)(p_2 \cdot p_3)$$
$$- (C_V^2 - C_A^2)(p_2 \cdot p_4)(p_1 \cdot p_3)], \tag{2.3}$$

where the overline on $\mathcal{M}$ denotes a sum over final spins and an average over the initial spins, and p_i denotes the four momenta of particles $i = 1, 4$. The squared matrix element for the charged current reactions is given by a similar relation, but with the replacement $C_V \to g_V$, $C_A \to g_A$, and $G_F \to G_F C$, where $C = \cos\theta_C$ for a change of strangeness $\Delta S = 0$ and $C = \sin\theta_C$ for $\Delta S = 1$, consistent with the Cabibbo theory.

3. NEUTRINO MEAN FREE PATHS IN DEGENERATE MATTER

We turn now to consider the mean free path of neutrinos in stellar matter comprised of degenerate baryons (neutrons, protons and hyperons) and leptons under conditions of charge neutrality and chemical equilibrium. For the estimates below, we employ the non-relativistic approximation for the baryons, so that the squared matrix element takes a simple form. We treat the neutrinos only in the degenerate and non-degenerate limits. Results for arbitrary neutrino degeneracy and with the full matrix element in Eq. (2.3) will be reported elsewhere.

For elastic collisions $1 + 2 \to 3 + 4$, where 1(3) denotes the initial (final) neutrino and 2(4) denotes the initial (final) baryon B, the scattering relaxation time may be calculated by linearizing the Boltzmann equation. For small departure from equilibrium, the inverse relaxation times from the various components are additive; thus

$$\frac{1}{\tau_s} = \sum_2 g_2 \int \prod_{i=2}^{4} \frac{d^3p_i}{(2\pi)^3} W_{fi}[n_2(1 - n_3)(1 - n_4) - (1 - n_2)n_3n_4]. \tag{3.1}$$

Above, the sum is over all species of baryons, g_2 is their degeneracy, n_i are the equilibrium

Fermi–Dirac distributions, and $\mathcal{W}_{fi}$ is the scattering rate

$$\mathcal{W}_{fi} = \left(\prod_{i=1}^{4} 2E_i\right)^{-1} (2\pi)^4 \delta^4(p_1 + p_2 - p_3 - p_4)\overline{|\mathcal{M}_{12\to34}|^2}. \tag{3.2}$$

The relaxation time τ_s characterizes the rate of change of the distribution function n_1 due to interactions with species 2, and may be used to define a scattering mean free path $\lambda_s = c\tau_s$.

For elastic scattering on heavy fermions, the momentum transfer is small. Thus, the scattering rate may be expressed as a function of the neutrino energy E_ν and the neutrino scattering angle θ. In degenerate matter, where the participant particles lie on their respective Fermi surfaces, the phase space integration can be separated into angle and energy integrals.[20,22] Thus for *degenerate neutrinos* and when $k_B T \ll E_\nu v_{Fi}/c$, where v_{Fi} is the velocity at the Fermi surface of species i, the inverse relaxation time is given by (see Ref. 19, 11 for the result in a single component system)

$$\frac{1}{\tau_s} = \sum_i \frac{G_F^2}{12\pi^3}(C_{Vi}^2 + 2C_{Ai}^2)m_{Bi}^2(k_B T)^2 E_\nu \left[\pi^2 + \frac{(E_\nu - \mu_\nu)^2}{(kT)^2}\right], \tag{3.3}$$

where the sum is over the baryonic components present in the system.

For *non-degenerate neutrinos* and when $k_B T \ll E_\nu v_{Fi}/c$, the result of Ref. 19 may be generalized to give

$$\frac{1}{\tau_s} = \sum_i \frac{G_F^2}{15\pi^3}(C_{Vi}^2 + 4C_{Ai}^2)p_{Fi}^2 E_\nu^3. \tag{3.4}$$

The relaxation time for absorption through charged current reactions can be calculated in a similar fashion. When neutrinos are degenerate, absorption on neutrons[18] and similarly on hyperons,[10] is kinematically allowed. In this case, the inverse absorption length is given by[10]

$$\frac{1}{\tau_a} = \sum_j \frac{G_F^2 C^2}{4\pi^3}(g_{Vj}^2 + 3g_{Aj}^2)m_{B_1}m_{B_2}(k_B T)^2\mu_e \left[\pi^2 + \frac{(E_\nu - \mu_\nu)^2}{(kT)^2}\right]\Xi$$

with

$$\Xi = \theta(p_{B_2} + p_e - p_{B_1} - p_\nu)$$
$$+ \frac{p_{B_2} + p_e - p_{B_1} + p_\nu}{2E_\nu}\theta(p_\nu - |p_{B_2} + p_e - p_{B_1}|) \tag{3.5}$$

where $\theta(x) = 1$ for $x \geq 1$ and zero otherwise.

When neutrinos are *non-degenerate* and when absorption is kinematically allowed, the relaxation time is given by

$$\frac{1}{\tau_a} = \sum_j \frac{G_F^2 C^2}{4\pi^3}(g_{Vj}^2 + 3g_{Aj}^2)m_{B_1}m_{B_2}(kT)^2\mu_e \left[\pi^2 + \frac{E_\nu^2}{(kT)^2}\right]\frac{1}{1 + e^{-E_\nu/kT}}. \tag{3.6}$$

We note that for nucleons-only matter, neutrino absorption on single nucleons can proceed only if the proton concentration exceeds some critical value in the range (11–15)% (see Ref. 27). For matter with lower proton concentrations, neutrino absorption occurs on two nucleons. However, as shown in Ref. 10, neutrinos may be absorbed on single hyperons as long as the concentration of Λ hyperons exceeds a critical value that is less than 3% and is typically about 1%.

4. RESULTS AND DISCUSSIONS

The calculations above give the mean free path in a mixture in which all baryons are under degenerate conditions. Several effects of strong interactions must be included before the results in Eq. (3.3) through Eq. (3.6) may be utilized. The renormalization of the density of states at the Fermi surfaces results in the baryon masses m_B being replaced by $m_B^* = p_F/v_F$. As pointed out by Iwamoto and Pethick,[19] baryon–baryon interactions introduce further Fermi liquid corrections. Specifically, the axial vector interaction is significantly suppressed, which increases the mean free path of neutrinos in dense matter. So far, such an analysis has been restricted to pure neutron matter only. In a multicomponent system, this formalism must be extended to include correlations between the different species present. The effect of density correlations in the long wavelength limit can be related to the compressibility of the system.[11] In addition, the medium dependence of g_A must also be considered. For example, it has been argued[28] that $|g_A|$ is quenched for nucleons in a medium. Whether g_A for hyperons is similarly affected is not known and is worth studying. Finally, depending on the momentum transfers involved, final-state interactions may modify the weak-interaction matrix element. Of the many corrections mentioned above, those due to the effective mass are the easiest to incorporate, since it would be contained even in a mean field description of the equation of state. Pending a more complete analysis, we consider below the modifications introduced by the multicomponent nature of the system incorporating only the effective mass corrections.

When neutrinos are degenerate, the relative abundances of the individual components do not play a significant role in determining the total mean free path. However, the extent to which the neutrino mean free path is altered by the presence of hyperons depends on the number of hyperonic species present. This may be illustrated by using the results for $\mathcal{R}_{nc}^{(1)} = [C_V^2 + 2C_A^2]_{\nu+B\to\nu+B}/[C_V^2 + 2C_A^2]_{\nu+n\to\nu+n}$ and $\mathcal{R}_{cc} = [C^2(g_V^2 + 3g_A^2)]_{\nu+B_1\to\ell+B_2}/[C^2(g_V^2 + 3g_A^2)]_{\nu+n\to\ell+p}$ listed in Tables 1 and 2. For example, in matter containing Λ, Σ^- and Σ^0 hyperons, the scattering mean free path is reduced by about 30–50% from its value in nucleons only matter. This reduction is achieved mainly by scattering on the Σ^- hyperon, since the Λ and Σ^0 hyperons do not contribute to scattering in lowest order. Similarly, up to 50% reduction may be expected from absorption reactions. Again, reactions involving the Σ^- hyperon, particularly the one leading to the Σ^0 hyperon, give the largest contribution.

The relative importance of the scattering and absorption reactions by degenerate neutrinos may be inferred by noting that

$$\frac{\lambda_a}{\lambda_s} = \frac{1}{3}\frac{E_\nu}{\mu_e}\left(\frac{\sum_i(C_{Vi}^2 + 2C_{Ai}^2)m_{B_i}^{*2}}{\sum_j C^2(g_{Vj}^2 + 3g_{Aj}^2)m_{B_1}^*m_{B_2}^*}\right). \tag{4.1}$$

For a fixed lepton fraction and for neutrinos of energy $E_\nu \sim \mu_\nu$, the factor $\mu_\nu/\mu_e \leq 1$ (see Figs. 1 and 2). Inasmuch as the baryon effective masses are all similar in magnitude, the factor in the parenthesis above may be approximated by the ratio of the coupling constants. From the results in Tables 1 and 2, it is easy to verify that the factor containing the coupling constants is of order unity (but generally less than unity) and is not very sensitive to the number of hyperonic species present. We thus arrive at the result that the absorption reactions dominate over the scattering reactions by a factor of about three to five, even in the presence of hyperons. This conclusion is not affected by the inclusion of the baryon effective masses in the calculation of λ_a/λ_s.

For non-degenerate neutrinos, and when $k_B T \ll E_\nu \nu_{Fi}/c$,

$$\frac{\lambda_a}{\lambda_s} = \frac{4}{15} \left(\frac{\sum_i (C_{Vi}^2 + 4C_{Ai}^2) p_{Fi}^2 E_\nu^3 (1 + e^{-E_\nu/kT})}{\sum_j C^2 (g_{Vj}^2 + 3g_{Aj}^2) m_{B_1}^* m_{B_2}^* (kT)^2 \mu_e \left[\pi^2 + \frac{E_\nu^2}{(kT)^2} \right]} \right). \tag{4.2}$$

In this case, the abundances of the various particles play a significant role in determining whether or not the absorption reactions dominate over the scattering reactions. Some insight may be gained by examining the concentrations in Figs. 1 and 2, and also the ratio $\mathcal{R}_{nc}^{(2)} = [C_V^2 + 4C_A^2]_{\nu+B \to \nu+B}/[C_V^2 + 4C_A^2]_{\nu+n \to \nu+n}$ listed in Table 2.

5. OUTLOOK

Neutrino signals in terrestrial detectors offer a means to determine the composition and the equation of state of dense matter. Many calculations of the composition of dense matter have indicated that strangeness-rich matter should be present in the core of neutron stars. Possible candidates for strangeness include hyperons, a K^- condensate, and quark matter containing s-quarks. Neutrino opacities for strange quark matter have been calculated previously.[18] In this work, we have identified the relevant neutrino–hyperon scattering and absorption reactions that are important new sources of opacity. Much work remains to be done, however. Many-body effects which may introduce additional correlations in a mixture are worth studying. Full simulations, of the type carried out in Refs. 1, 8 including possible compositions and the appropriate neutrino opacities, will be taken up separately.

ACKNOWLEDGMENTS

This work was supported by the U.S. Department of Energy under contract number DOE/DE-FG02-88ER-40388. We thank Jim Lattimer for helpful discussions.

REFERENCES

1. A. Burrows and J. M. Lattimer, *Astrophys. Jl.*, **307** (1986) 178; A. Burrows, *Ann. Rev. Nucl. Sci.*, **40** (1990) 181; and references therein.
2. K. Hirata, et. al., *Phys. Rev. Lett.*, **58** (1987) 1490; R. M. Bionta, et. al., *Phys. Rev. Lett.*, **58** (1987) 1494.
3. K. Sato, *Prog. Theor. Phys.* **53** (1975) 595.
4. T. J. Mazurek, *Astrophys. Space Sci.* **35** (1975) 117.
5. M. Prakash, et. al., *Phys. Rep* (1995) to be published; and references therein.
6. V. Thorsson, M. Prakash and J. M. Lattimer, *Nucl. Phys.* **A572** (1994) 693.
7. M. Prakash, J. Cooke and J. M. Lattimer, SUNY preprint, SUNY-NTG-95-1.
8. W. Keil and H. T. Janka, *Astronomy and Astrophysics*, 1994, to be published.
9. O. V. Maxwell, *Astrophys. Jl.* **316** (1987) 691.
10. M. Prakash, et. al., *Astrophys. Jl.* **390** (1992) L80.
11. R. F. Sawyer, *Phys. Rev.* **D11** (1975) 2740; *Phys. Rev.* **C40** (1989) 865.
12. D. L. Tubbs and D. N. Schramm, *Astrophys. Jl.* **201** (1975) 467.
13. D. Q. Lamb and C. J. Pethick, *Astrophys. Jl.* **209** (1976) L77.
14. D. Q. Lamb, *Phys. Rev. Lett* **41** (1978) 1623.
15. D. L. Tubbs, *Astrophys. Jl. Suppl.* **37** (1978) 287.
16. S. Bludman and K. Van Riper, *Astrophys. Jl.* **224** (1978) 631.
17. R. F. Sawyer and A. Soni, *Astrophys. Jl.* **230** (1979) 859.

18. N. Iwamoto, *Ann. Phys.* **141** (1982) 1.
19. N. Iwamoto and C. J. Pethick, *Phys. Rev.* **D25** (1982) 313.
20. B. T. Goodwin and C. J. Pethick, *Astrophys. Jl.* **253** (1982) 816.
21. A. Burrows and T. J. Mazurek, *Astrophys. Jl.* **259** (1982) 330.
22. B. T. Goodwin, *Astrophys. Jl.* **261** (1982) 321.
23. S. W. Bruenn, *Astrophys. Jl. Suppl.* **58** (1985) 771.
24. L. Van Den Horn and J. Cooperstein, *Astrophys. Jl.* **300** (1986) 142.
25. J. Cooperstein, *Phys. Rep.* **163** (1988) 95.
26. A. Burrows, *Astrophys. Jl.* **334** (1988) 891.
27. J. M. Lattimer, et. al., *Phys. Rev. Lett.* **66** (1991) 2701.
28. D. H. Wilkinson, *Phys. Rev.* **C7** (1973) 930; M. Rho, *Nucl. Phys.* **A231** (1974) 493; G. E. Brown and M. Rho, *Phys. Rev. Lett.* **66** (1991) 2720.
29. G. E. Brown and H. A. Bethe, *Astrophys. Jl.* **423** (1994) 659.

INDEX